UMTS

UMTS

Javier Sanchez
Mamadou Thioune

Part of this book adapted from the 2nd edition of "UMTS" published in France by Hermès Science/Lavoisier in 2004
First Published in Great Britain and the United States in 2007 by ISTE Ltd

ISTE Ltd
6 Fitzroy Square
London W1T 5DX
UK

ISTE USA
4308 Patrice Road
Newport Beach, CA 92663
USA

www.iste.co.uk

Library of Congress Cataloging-in-Publication Data

Sanchez, Javier.
 UMTS/Javier Sanchez, Mamadou Thioune.
  p. cm.
 ISBN-13: 978-1-905209-71-2
 ISBN-10: 1-905209-71-1
 1. Universal Mobile Telecommunications System. I. Thioune, Mamadou. II. Title. III. Title: Universal mobile telecommunications system.
 TK5103.4883.S36 2006
 621.3845'6--dc22

                                                                          2006035535

British Library Cataloguing-in-Publication Data
A CIP record for this book is available from the British Library
ISBN 10: 1-905209-71-1
ISBN 13: 978-1-905209-71-2

Printed and bound in Great Britain by Antony Rowe Ltd, Chippenham, Wiltshire.

Table of Contents

Preface

During the first decade of this millennium, more than €100 billion will be invested in third generation (3G) Universal Mobile Telecommunications System (UMTS) in Europe. This fact represents an amazing challenge from both a technical and commercial perspective. In the evolution path of GSM/GPRS standards, the UMTS proposes enhanced and new services including high-speed Internet access, video-telephony, and multimedia applications such as streaming.

Based on the latest updates of 3GPP specifications, this book investigates the differences of a GSM/GPRS network compared with a UMTS network as well as the technical aspects that ensure their interoperability. Students, professors and engineers will find in this book a clear and concise picture of key ideas behind the complexity of UMTS networks. This can also be used as a *starter* before exploring in more depth the labyrinth of 3GPP specifications which remain, however, the main technical reference.

Written by experts in their respective fields, this book gives detailed description of the elements in the UMTS network architecture: the User Equipment (UE), the UMTS Radio Access Network (UTRAN) and the core network. Completely new protocols based on the needs of the new Wideband Code Division Multiple Access (WCDMA) air interface are given particular attention by considering both Frequency- and Time-Division Duplex modes. Later on, the book further introduces the key features of existing topics in *Releases* 5, 6 and 7 such as High Speed Downlink/Uplink Packet Access (HSDPA/HSUPA), IP Multimedia Subsystem (IMS), Long Term Evolution (LTE), WLAN interconnection and Multicast/ Broadcast Multimedia Services (MBMS).

We would like to offer our heartfelt thanks to all our work colleagues for their helpful comments.

Some of the figures and tables reproduced in this book are the result of technical specifications defined by the 3GPP partnership (http://www.3gpp.org/3G_Specs/ 3G_Specs.htm). The specifications are by nature not fixed and are susceptible to modifications during their transposition in regional standardization organizations which make up the membership of the 3GPP partnership. Because of this, and as a result of the translation and/or adaptation of these points by the authors, these organizations cannot be considered responsible for the figures and tables reproduced in this book.

Chapter 1

Evolution of Cellular Mobile Systems

The purpose of this chapter is to describe the milestones in the evolution of cellular mobile systems. Particular attention is paid to the third generation (3G) systems to which the UMTS belong.

The performance of mobile cellular systems is often discussed with respect to the radio access technology they support, thus neglecting other important aspects. However, a cellular mobile communication system is much more than a simple radio access method, as illustrated in Figure 1.1. The mobile terminal is the vector enabling a user to access the mobile services he subscribed to throughout the radio channel. The core network is in charge of handling mobile-terminated and mobile-originated calls within the mobile network and enables communication with external networks, both fixed and mobile. Billing and roaming functions are also located in the core network. The transfer of users' data from the terminal to the core network is the role of the radio access network. Implementing appropriate functions gives to the core network and to the terminal the impression of communicating in a wired link. One or several radio access technologies are implemented in both the radio access network and the mobile terminal to enable wireless radio communication.

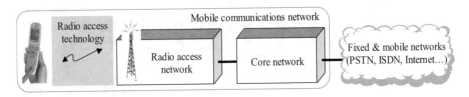

Figure 1.1. *Basic components of a mobile communications network*

1.1. Multiple-access techniques used in mobile telephony

Surveying the different multiple-access techniques is equivalent to describing the key milestones in the evolution of modern mobile communication systems. In the past, not all users of the radio spectrum recognized the need for the efficient use of the spectrum. The spectrum auctions for UMTS licenses have emphasized the fact that the radio spectrum is a valuable resource. Thus, the major challenge of multiple-access techniques is to provide efficient allocation of such a spectrum to the largest number of subscribers, while offering higher data rates, increased service quality and coverage.

1.1.1. *Frequency division duplex (FDD) and time division duplex (TDD)*

Conventional mobile communication systems use duplexing techniques to separate uplink and downlink transmissions between the terminal and the base station. Frequency division duplex (FDD) and time division duplex (TDD) are among the transmission modes which are the most commonly employed. The main difference between the two modes, as shown in Figure 1.2, is that FDD uses two separate carrier bands for continuous duplex transmission, whereas in TDD duplex transmission is carried in alternate time slots in the same frequency channel. In order to minimize mutual interference in FDD systems, a guard frequency is required between the uplink and downlink allocated frequencies (usually 5% of the carrier frequency). On the other hand, a guard period in TDD systems is required in order to reduce mutual interference between the links. Its length is decided from the longest round-trip delay in a cellular system (in the order of 20-50μs).

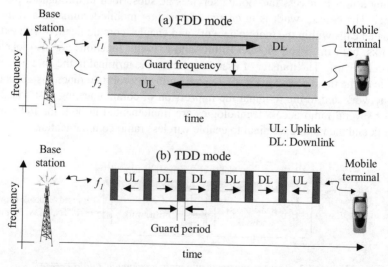

Figure 1.2. *Duplexing modes used in modern mobile communications systems*

1.1.2. *Frequency division multiple access (FDMA)*

FDMA is the access technology used for first generation analog mobile systems such as the American standard AMPS (*Advanced Mobile Phone Service*). Within an FDMA system, each subscriber is assigned a specific frequency channel as illustrated in Figure 1.3a. No one else in the same cell or in a neighboring cell can use the frequency channel while it is allocated to a user – when an FDMA terminal establishes a call, it reserves the frequency channel for the entire duration of the call. This fact makes FDMA systems the least efficient cellular systems since each physical channel can only be used by one user at a time. Far from having disappeared, the FDMA principle is part most of modern digital mobile communication systems where it is used as a complement to other radio multiplexing schemes.

1.1.3. *Time division multiple access (TDMA)*

GSM, TDMA/136 and PCS are second generation mobile standards based on TDMA. The key idea behind TDMA relies on the fact that a user is assigned a particular time slot in a frequency carrier and can only send or receive information in those particular times (see Figure 1.3b). When all available time slots in a given frequency are used, the next user must be assigned a time slot on another frequency. Information flow is not continuous for any user, but rather is sent and received in "bursts". The important factor to be considered while designing is that these time slices are so small that the human ear does not perceive the time being divided. In GSM up to 8 users may in theory share the same 200 kHz frequency band *almost* simultaneously, whereas in IS-136 different users can be allocated to 3 time slots within a 30 kHz frequency channel. The capacity of TDMA is about 3 to 6 times as much as that of FDMA [RAP 96].

1.1.4. *Code division multiple access (CDMA)*

In a CDMA system, unique digital codes, rather than separate radio frequencies, are used to differentiate users (see Figure 1.3c). The codes are shared by the terminal and the base station. All users access the entire spectrum allocation all of the time, that is, every user uses the entire block of allocated spectrum space to carry his/her message. CDMA technology is used in 2G IS-95 (cdmaOne) mobile communication systems and is also part of UMTS and cdma2000 3G standards.

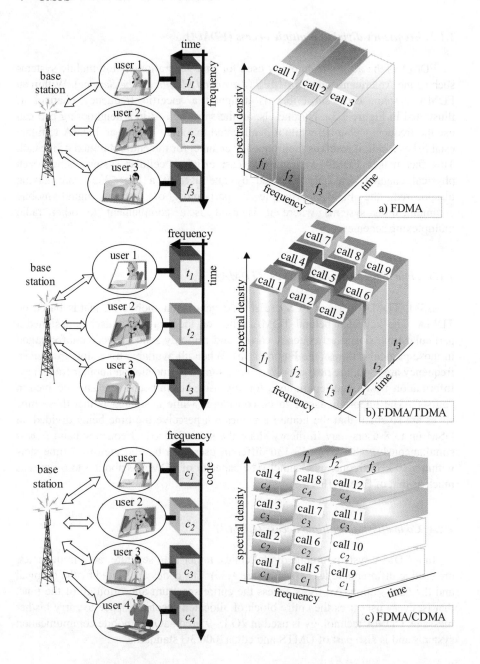

Figure 1.3. *FDMA, TDMA and CDMA multiplex access principle*

A very popular example used to stress the differences between FDMA, TDMA and CDMA is as follows.

Imagine a large room (frequency spectrum) intended to accommodate many pairs of people. Due to dividing walls, individual offices can be created within the room. They are then allocated to each pair of people so that their conversation can be isolated from noise generated by the other parties. Each office is like a single frequency/channel (principle of FDMA). No one else could use the office until the conversation was complete, whether or not the different pairs were actually talking. A better usage of each office can be achieved by accommodating multiple pairs of people within the same room. For this to work, each party shall respect a rule that is to keep silent while one pair is talking (principle of TDMA). The important factor to be considered is that these silence periods shall be small enough that the human ear cannot perceive the time slicing.

In the analogy with CDMA technology, all the offices are eliminated to create "open-spaces" instead, so that conversation can be carried out at any time. The rule is now for all pairs to hold their conversations in a different language – the brains of contiguous pairs of people being able to naturally filter interference from the other pairs. The languages are analog to the codes assigned by the CDMA system. In theory, it should be possible to accommodate in the large room as much pairs as permitted by the cubic-volume of each person and provided that the number of available languages is enough. Unfortunately, even if the parties speak in different languages, a sudden raise of the voice volume of one couple may disturb the conversations of all the neighboring pairs. This problem may be overcome by implementing an automatic mechanism to control the voice volume of each party. In a CDMA system, this is actually a power control scheme whose performance is of paramount importance for the system to operate properly. However, despite such mechanism, adding the contribution of all the voices, no matter how low their level is, may produce an overall background noise in the room that makes it too difficult to hold a clear conversation. In such a case, some couples may be required to get out from the room or just to remain silent.

1.1.5. *Space division multiple access (SDMA)*

SDMA technique is based on deriving and exploiting information on the spatial position of mobile terminals. The radiation pattern of the base station is adapted in both uplink and downlink directions to each different user in order to obtain, as illustrated in Figure 1.4, the highest gain in the direction of the mobile user. At the same time, radiation which is zero shall be positioned in the directions of interfering mobile terminals, the ultimate goal being the overall enhancement of capacity and coverage within the mobile system [RHE 96]. SDMA approach can be integrated with different multiple access techniques (FDMA, TDMA, CDMA) and therefore it can be optionally used in all modern mobile communications systems.

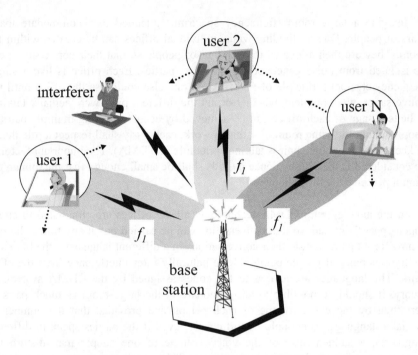

Figure 1.4. *Space division multiple access (SDMA) principle*

1.1.6. *Orthogonal frequency division multiplexing (OFDM)*

OFDM is a special case of multi-carrier modulation. The main idea is to split a data stream into N parallel streams of reduced data rate and transmit each of them on a separate sub-carrier. High spectral efficiency is achieved in OFDM since a large number of sub-carriers where overlapping spectra is used. OFDM can be combined with FDMA, TDMA and CDMA methods in order to obtain the access schemes referred to as MC-FDMA, MC-TDMA and MC-CDMA, respectively (see Figure 1.5).

OFDM has been adopted in the terrestrial digital video broadcasting (DVB-T) standard and in the digital audio broadcasting (DAB) standard followed by the wireless local area network standards IEEE 802.11a/g, IEEE 802.16, BRAN, HIPERLAN/2 and HIPERMAN. Although not used in current 2G/3G mobile radio systems, the successful deployment of the OFDM technique has encouraged several studies intended to design new broadband air interfaces for 4G mobile systems (see Chapter 13).

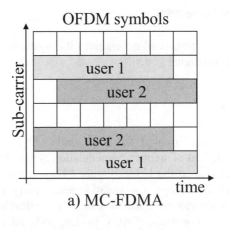

a) MC-FDMA

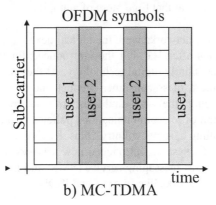

b) MC-TDMA

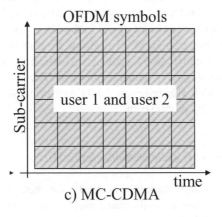

c) MC-CDMA

Figure 1.5. *Examples of multiple access for two users based on the OFDM principle*

1.2. Evolution from 1G to 2.5G

Rather than a revolution, third-generation (3G) systems are an evolution from second-generation (2G) digital systems.

1.2.1. *From 1G to 2G*

1G phones were analog, used for voice calls only, and their signals were transmitted by the method of frequency modulation (FM). AMPS was the first 1G system to start operating in the USA (in July 1978). It was based on the FDMA technique and FDD. In Europe, the situation was "every man for himself" and almost each country developed a standard in its own: Radiocom 2000 in France, NMT 900 in Nordic countries, TACS in England, NETZ in Germany, etc. International roaming was at that time more of a utopia.

2G mobile telephone networks were the logical next stage in the development of cellular mobile systems after 1G, and they introduced for the first time a mobile phone system that used purely digital technology. At the end of the last century, 2G mobile phones become a mass consumer product due to the amazing progress in semiconductors reducing the size and cost of electronic components. Also, aggressive deregulation of telecommunications policies enabled the development of several operators within a same country, thus leading to attractive subscription offers. Being digital, 2G systems introduced new services besides traditional voice transmission, such as short messaging service (SMS) and fax. They also enabled the access to digital fixed networks like Internet and ISDN.

One of the successful 2G digital systems is GSM, a European mobile phone standard based on the TDMA technique. Around 70% of mobile phone subscribers in the world have adopted GSM nowadays. In the USA, a different form of TDMA is used in the system known as TDMA/136 (formerly IS-136 or D-AMPS) and there is another US system called IS-95 (cdmaOne), based on the CDMA approach. Finally, the *Personal Digital Communications* (PDC) standard is the Japanese contribution to 2G, which also relies on the TDMA principle. Table 1.1 shows key radio characteristics of 2G mobile cellular systems.

1.2.2. *Enhancements to 2G radio technologies: 2.5G*

By the late 1990s the market was ready for new mobile communication technologies to evolve from 2G and created pressure for enhanced data delivery and telephony services, global roaming, Internet access, email, and even video. Unfortunately, standards for 3G systems were in the process of being developed. A more immediate solution to meet these demands was needed, thus leading to the so-called 2.5G.

Standard	TDMA/136 (D-AMPS)	IS-95 (cdmaOne)	GSM	PDC
Origin	USA	USA	Europe	Japan
Commercial launch	1992	1995	1992	1993
Main operation band (Mhz)	824-849 (UL) 869-894 (DL) 1,850-1,910 (UL) 1,930-1,990 (DL)	824-849 (UL) 869-894 (DL) 1,850-1,910 (UL) 1,930-1,990 (DL)	824-849 (UL) 869-894 (DL) 880-915 (UL) 925-960 (DL) 1,710-1,785 (UL) 1,805-1,880 (DL) 1,850-1,910 (UL) 1,930-1,990 (DL)	810-826 (UL) 940-956 (DL) 1,429-1,453 (UL) 1,477-1,501 (DL)
Access method	FDMA/TDMA	FDMA/CDMA	FDMA/TDMA	FDMA/TDMA
Duplexing	FDD	FDD	FDD	FDD
Channel bandwidth	30 kHz	1,250 kHz	200 kHz	25 kHz
Modulation	$\pi/4$ DQPSK	QPSK/O-QPSK	GMSK	$\pi/4$ DQPSK

Table 1.1. *Comparison of radio specifications for 2G cellular mobile systems*

As shown in Figure 1.6, three technologies have most often been proposed to upgrade GSM in the context of 2.5G: *High-Speed Circuit Switched Data* (HSCSD), *General Packet Radio Service* (GPRS) and *Enhanced Data Rates for Global Evolution* (EDGE). HSCSD enables transfer rates of up to 57.6 Kbps by allocating more than one time slot per user. GPRS enables the efficient use of the air interface by accommodating flexible user rates for packet oriented transfer using time slot assignment on demand (rather than via permanent occupation as in GSM and HSCSD). The result is an improved data rate of up to (theoretical) 171.2 Kbps (8 × 21.4 Kbps), versus the 9.6 Kbps rate of standard GSM networks. EDGE offers advanced modulation (8-QPSK in addition to GMSK) to achieve higher data rates (in the theoretical order of 384 Kbps). By applying EDGE to GPRS and EDGE to HSCSD, the hybrid techniques EGPRS and ECSD are obtained, respectively. While the role of ECSD in today's cellular market is marginal, EGPRS has been adopted by several GSM operators to make cost-effective the investments on GPRS and as a complement to UMTS network coverage.

IS-95B brings about improvements for handover algorithms in multi-carrier environments to the first version of the cdmaOne standard (IS-95A). Although IS-95B is based on CDMA, its logic for improvement is very similar to that of GPRS: rather than time slots given to the user as in GPRS, in IS-95B channels can still be aggregated to allow higher data rates of up to 115 Kbps bundling up to eight 14.4 or 9.6 Kbps data channels.

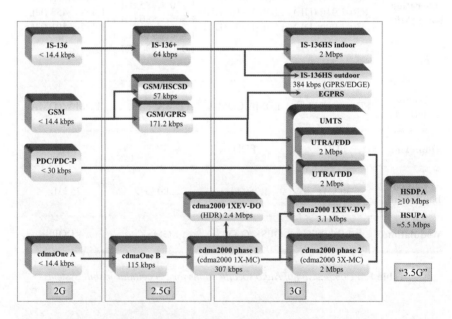

Figure 1.6. *Migration of 2G standards towards 3G. Some data rates are purely theoretical and they may be different depending on the hypothesis considered for their calculation*

TDMA/136 is also being enhanced to provide better voice capabilities, capacity, coverage, quality and data rates. In its evolution to 3G, an intermediate phase is envisaged where the bitrate of 30 kHz radio carrier will be increased by means of high-level modulation and by the use of 6 time slots rather than 3 within a 40 ms frame. This enhanced version of TDMA/136 is designated by 136+. By applying GPRS technology, packet transmission data rates of up to 64 Kbps can be obtained.

In 2001, Japan was the first country to introduce a 3G system commercially known as FOMA (*Freedom Of Mobile Multimedia Access*). It was based on an early version of UMTS standard specifications. Unlike the GSM systems, which developed various ways to deal with demand for improved services, Japan had no 2.5G enhancement stage to bridge the gap between 2G and 3G, and so the move into the new standard was seen as a fast solution to their capacity problems in PDC networks. Nevertheless, the standard implemented a packet mode variant to PDC (P-PDC), which gives packet data rates of up to 28.8 Kbps.

After the USA that leaded the 1G of mobile cellular systems; after Europe that played the first role in 2G, Asia, and more precisely, Japan, China and Korea, are willing to be the key players in the newborn 3G.

1.3. 3G systems in IMT-2000 framework

It is wrong to believe that UMTS is the only 3G system around the world, although this was the original purpose of the ITU (*International Telecommunications Union*). This "unique" 3G system should be called FPLMTS (*Future Public Land Mobile Telecommunications System*). The name being unpronounceable, it was changed to IMT-2000 (IMT stands for *international mobile telecommunications*[1]). A problem arose when, in 1998, no less that ten terrestrial radio access technologies were submitted to the ITU by its members – the regional standardization organisms. In the end, the term IMT-2000 generated not a single 3G standard but a family of standards, most of them associated to their numerous 2G predecessors. Note that IMT-2000 consists of both terrestrial component and satellite component radio interfaces. Only the analysis of its terrestrial radio interfaces is in the scope of this book.

---

1 The number 2000 was supposed to represent: the year 2000, when the ITU expected the system would become available; the data rates offering services around 2000 Kbps and the spectrum in the 2000 MHz region that the ITU hoped to make it available worldwide.

The key features of 3G systems in the IMT-2000 framework are:

– high data rates. Minimum 144 Kbps in high mobility environments (more than 120 km/h); 384 Kbps, in common mobility environments (less than 120 km/h) and 2 Mbps, achievable in stationary and low mobility environments (less than 10 km/h);

– support for circuit-switched services (e.g. PSTN- and ISDN-based networks) as well as packet-switched services (e.g. IP-based networks);

– capability for multimedia applications, involving services with different quality of services (bitrate, bit error rate, delay, etc.);

– high voice quality; similar to that provided by wired-networks;

– small terminal for worldwide use and with worldwide roaming capability;

– compatibility of services within IMT-2000 and with the fixed networks;

– interoperability with their 2G/2.5G predecessors.

Restraining the definition of a 3G system to radio data rates and mobility environments made the term *3G* become rather vague. This can be understood by keeping in mind that at that time (around 1992) usage of Internet was limited to academic and technical circles. The definition was originally quite specifically defined, as any standard that provided mobile users with the performance of ISDN or better. It is one of the goals of this book to show all the technical innovations behind a 3G network such as UMTS in addition to the amazing feats performed by its radio access technology (UTRA).

1.3.1. *IMT-2000 radio interfaces*

The radio interfaces for the terrestrial component of IMT-2000 are shown in Table 1.2, whereas Table 1.3 lists their major technical parameters. Parameters of DECT technology were deliberately omitted since this is not a major player in the 3G networks arena today, but this may change in the future.

NOTE.– besides the names given by IMT-2000, there is no global consensus on how to designate the 3G radio access technologies out of the ITU framework. For instance, UTRA/FDD and UTRA/TDD radio interfaces are often called "WCDMA" (*Wideband-CDMA*).

3G radio access technology	IMT-2000 designation
UTRA/FDD *Universal terrestrial radio access frequency division duplex*	IMT-2000 CDMA Direct Spread
UTRA/TDD *UTRA time division duplex* TD-SCDMA (low chip rate UTRA/TDD) *Time division synchronous CDMA*	IMT-2000 CDMA TDD
Cdma2000	IMT-2000 CDMA Multi-carrier
UWC-136 *Universal wireless communications*	IMT-2000 CDMA Single-carrier
DECT *Digital enhanced cordless telecommunications*	IMT-2000 FDMA/TDMA

Table 1.2. *Radio access technologies defined in the IMT-2000 framework. The terms Cdma2000, UWC-136 and DECT designate also a complete 3G mobile network. The term "multi-carrier" given to cdma2000 does not mean that the OFDM technique is implemented*

1.3.1.1. *IMT-2000 radio access technologies used by UMTS networks*

A UMTS network may use either UTRA/FDD, UTRA/TDD, TD-SCDMA or DECT radio access technologies.

IMT-2000 CDMA Direct Spread: UTRA/FDD

This radio interface is called *universal terrestrial radio access* (UTRA) FDD or *wideband CDMA* (WCDMA). UTRA/FDD employs CDMA as radio access technology. Based on direct sequence CDMA approach, the chip rate is 3.84 Mcps and operates in paired frequency bands with a 5 MHz bandwidth carrier in UL and the same in DL (see Figure 1.7a). It should be stressed that current UMTS systems deployed in Europe and in Japan are based exclusively on UTRA/FDD technology.

IMT-2000 CDMA TDD: UTRA/TDD

This radio interface is called UTRA/TDD and comprises two variants, called "1.28 Mcps TDD" (TD-SCDMA) and "3.84 Mcps TDD". Chinese standardization authorities originally proposed TD-SCDMA as an alternative to European UTRA/TDD. A harmonization process was then carried out and TD-SCDMA is now part of UTRA/TDD specifications and it is referred to as the "UTRA/TDD low-chip rate" option [TS 25.843, *R*4].

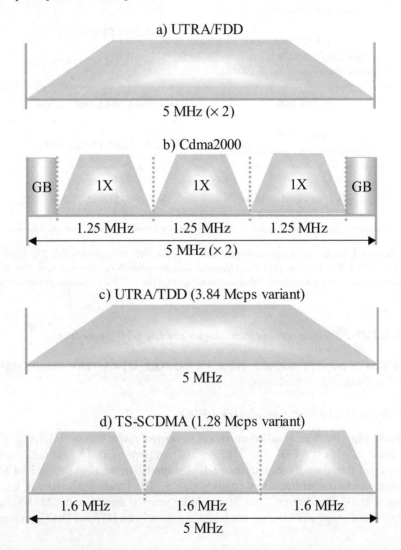

Figure 1.7. *Spectrum usage for IMT-2000 technologies based on CDMA (GB: guard band)*

As in UTRA/FDD, the variant "3.84 Mcps TDD" of UTRA/TDD technology employs direct sequence CDMA with a chip rate of 3.84 MHz on a 5 MHz bandwidth carrier (see Figure 1.7c). In contrast to UTRA/FDD, time division duplexing and TDMA component on top of a CDMA component are used. Consequently, a radio channel in UTRA/TDD is denoted by a time slot and a code. Time slots can be allocated to carry either downlink or uplink user data within an unpaired band, and as such, the spectrum requirement is just half the bandwidth of UTRA/FDD (see Chapter 12).

TD-SCDMA implements all the functions of European UTRA/TDD but is based on a chip rate of 1.28 Mcps on a 1.6 MHz bandwidth carrier, i.e. a third of the UTRA/TDD 3.84 Mcps chip-rate and carrier. This enables three carriers to be used within a given spectrum of 5 MHz (see Figure 1.7d). This feature enables operators with an extra degree of flexibility since the system may be operated with frequency reuse of one, two or three. TD-SCDMA can also be used where a contiguous 5 MHz band is unavailable.

The UTRA/TDD specifications have been developed with the strong objective of harmonization with UTRA/FDD in order to achieve maximum commonality. Both UTRA/FDD and UTRA/TDD are harmonized in terms of key physical layer parameters and they share a common set of protocols in the higher layers.

IMT-2000 FDMA/TDMA: DECT

DECT (*Digital Enhanced Cordless Telecommunications*) is in the evolution path towards 3G of the 2G European cordless system DECT (originally referred to as *Digital European Cordless Telephone*). The standard specifies a TDMA radio interface with TDD duplexing and the modulation scheme is either Gaussian frequency shift keying (GFSK) or differential phase shift keying (DPSK). Besides 2-level modulation, it is allowed to use 4-level and/or 8-level modulation. The resulting radio frequency bitrates are 1.152 Mbps, 2.304 Mbps and 3.456 Mbps. The standard supports symmetric and asymmetric connections, connection oriented and connectionless data transport as well as variable bitrates of up to 2.88 Mbps per carrier [Rec. ITU-R M. 1457-1].

1.3.1.2. *IMT-2000 CDMA multi carrier: cdma2000*

The IMT-2000 *CDMA multi-carrier* (MC) technology is called cdma2000. It has been promoted by CDG (*CDMA Development Group*) consortium (www.cdg.org). As in UTRA/FDD technology, multiple access is based on CDMA and the duplexing mode used is FDD. In contrast to UTRA/FDD, it includes a multi-carrier CDMA component designated for 1.25 MHz carrier bandwidth with a chip rate of 1.2288 Mcps. A harmonization process is taking place to enable service interoperability between cdma2000 and UMTS, thus enabling inter-system roaming capability.

Standardization of cdma2000 comprises several variants. In the variant known as cdma2000 1X (1X is referring to use of a single 1.25 MHz carrier), basic enhancements are done on IS-95 for high rate packet data and for doubling voice capacity. Data transmission of up to 307 Kbps can be achieved in downlink. Cdma2000 1X is considered by its promoters as the world's first IMT-2000 network commercially deployed in October 2000 in South Korea.

Another cdma2000 option proposes a technology based on three times the carrier rate of cdma2000 1X and is known as cdma2000 3X. The idea is that three 1.25 MHz carriers are used within a 5 MHz bandwidth (the extra 1.25 MHz are used for guard bands as shown in Figure 1.7b). The three channels within this band can be aggregated in order to obtain high data transfer rates. The standard makes it possible to use larger bandwidths than 5 MHz to support the highest data rates more efficiently. With a 5 MHz band, a chip rate of 3.6864 Mcps is defined and transfer rates of up to 2.048 Mbps can be achieved.

In order to enhance the data rate capability of cdma2000 1X still further, other variants were proposed named cdma2000 1XEV-DO (*Data Only*) and cdma2000 1XEV-DV (*Data and Voice*). The former, also referred to as HDR (*High Data Rate*), uses a separate 1.25 MHz carrier for data. It also implements a downlink shared channel where several users are served at different times within the same carrier. In combination with higher level adaptive modulation (8-PSK and 16 QAM), turbo encoding, fast scheduling and hybrid ARQ (*Automatic Repeat Request*), cdma2000 1XEV-DO promises peak data rates of up to 2.4 Mbps. On the other hand, cdma2000 1XEV-DV integrates voice and data transmission on the same carrier, thus avoiding the waste of capacity attributed to 1XEV-DO. It is also backward compatible with 1X and 1XEV-DO and enables, on paper, data rates of 3.1 Mbps for packet data transmission.

	UTRA/FDD	UTRA/TDD	Cdma2000	UWC-136
Multiple access	CDMA	TDMA and CDMA	CDMA	TDMA
Examples of regions where used	Europe, Japan, America	China (TD-SCDMA)	South Korea, America, Japan, China	America
Duplexing	FDD	TDD	FDD	– FDD – TDD optimal for 136HS *indoor*
Chip-rate (Mcps)	3.84	3.84 (1.28 TD-SCDMA)	$N \times 1.2288$ $N = 1, 3, 6, 9, 12$	
Carrier bandwidth	5 MHz	5 MHz (1.6 MHz TD-SCDMA)	$N \times 1.25$ MHz $N = 1, 3, 6, 9, 12$	– 136^+: 30 kHz – 136HS *outdoor*: 200 kHz – 136HS *indoor*: 1.6 MHz
Base station synchronization	Asynchronous (synchronous optional)	synchronous	synchronous	asynchronous
Frame length (ms)	10	10 (sub-frame of 5 ms TD-SCDMA)	5, 10, 20, 26, 66, 40, 80	– 136+: 40 – 136HS *outdoor*: 4.615 – 136HS *indoor*: 4.615
Data modulation	BPSK (UL) QPSK (DL)	QPSK (8-PSK in TD-SCDMA is optional)	BSPK, QPSK, 8-PSK, 16-QAM	– 136+: $\pi/4$ DQPSK, 8-PSK – 136HS *outdoor*: GMSK, 8-PSK – 136HS *indoor*: BOQAM, QOQAM
Website with specs	ftp://www.3gpp.org/ Specs	ftp://www.3gpp.org/ Specs	www.tiaonline.org/ standards	www.tiaonline.org/ standards

Table 1.3. *Key radio parameters of IMT-2000 terrestrial radio interfaces*

1.3.1.3. *IMT-2000 TDMA single carrier: UWC-136*

UWC-136 (*Universal Wireless Communications-136*) is the name of the consortium promoting the evolution of TDMA/136 towards 3G. Such an evolution is planned in three phases. The first phase corresponds to standard 136+ introduced in previous sections. In the second phase, EDGE is applied and 200 kHz carrier bandwidths are introduced (rather than 30 kHz) to obtain high-speed data rates. Modulation methods GMSK and 8-PSK are employed with a physical channel symbol rate of 270.833 ksps. This approach is referred to as 136HS outdoor since it enables operations in high mobility environments.

Higher speed data rates (2 Mbps) in low mobility scenarios are obtained in the third phase designated as 136HS indoor. This radio interface applies a 1.6 MHz carrier bandwidth. A variable channel modulation can be applied to provide an optimal adaptation of throughput versus channel robustness. In order to do this, two mandatory modulations methods shall be supported: B-O-QAM (*Binary Offset Quadrature Amplitude Modulation*), and Q-O-QAM (*Quadruple Offset Quadrature Amplitude Modulation*). The physical channel symbol rate is 2.6 Msps. Operation based on TDD duplexing may be optionally implemented.

1.3.1.4. *HSDPA/HSUPA: high-speed downlink/uplink packet access*

With the goal of enhancing the data rates offered by IMT-2000 systems in the radio interface, HSDPA and HSUPA technologies are proposed. Packet data rates of up to 10 Mbps are achieved in downlink with HSDPA and up to 5.5 Mbps in uplink with HSUPA. This is made possible by applying a higher modulation scheme (16-QAM) with adapting coding rates, turbo encoding and hybrid ARQ (H-ARQ) techniques. HSDPA is a feature defined in 3GPP *Release 5* standard whereas HSUPA is part of *Release 6*: they may be considered as part of the "3.5G" of mobile telecommunication systems (see Chapters 13 and 14).

1.3.2. *Core network approaches in 3G systems*

Although the IMT-2000 radio interfaces aforementioned show clearly a revolutionary step from 2 to 3G, the approach is evolutionary in the core network side. Two core network approaches are envisaged in the first deployments of 3G systems as illustrated in Figure 1.8:

– evolved GSM/MAP core network. The UMTS core network is based on the GSM/GPRS core network maintaining basically the same entities as well as functions and protocols across the network. It applies specifically to *Mobile Application Part* (MAP) protocol grouping call processing and mobility management functions;

– evolved ANSI-41 core network. In cdma2000 and UWC-136, call processing and mobility management functions are controlled by IS-41/AINSI-41 core network presented as an evolved version of the one used in their 2G predecessors cdmaOne and TDMA/41, respectively. As in UMTS systems, UWC-136 further integrates an evolved GPRS backbone for packet-services provision.

Under the coordination of ITU, the specifications of evolved GSM/MAP core network include the necessary capabilities for operation with an evolved ANSI-41-based core network. Since core network procedures are independent of the radio technology implemented in the RAN, a UMTS system operator may choose the IMT-2000 radio access technology that is best adapted to its needs in terms of capacity and coverage. Finally, let us note that all 3G systems propose options to evolve towards a core network offering circuit and packet multimedia services based on IP approach: the *IP multimedia sub-system* (IMS), studied in Chapter 13.

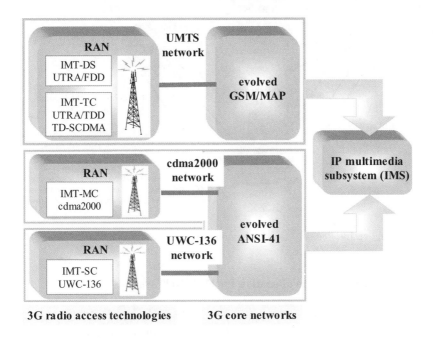

Figure 1.8. *3G core networks and interconnection with IMT-2000 radio technologies*

1.4. Standardization process in 3G systems

The specifications of UMTS systems including UTRA/FDD and UTRA/TDD radio interfaces are developed within the *Third Generation Partnership Project*

(3GPP, www.3gpp.org) whereas those of cdma2000 are developed in a similar partnership project named 3GPP2 (www.3gpp2.org). Both 3GPP and 3GPP2 are not, legally speaking, standardization organisms. Regional standardization bodies make up the 3GPP and the 3GPP2 and they are the owners of all the technical specifications produced by the working groups therein (see Figure 1.9). Since the creation of 3GPP at the end of 1998, the technical specifications of HSCSD, GPRS and EDGE are being also developed within the same 3GPP framework as UMTS. Let us note that UWC-135 technologies are developed by the American standardization organism TIA TR45.3 with input from UWC, while DECT technology is specified by a set of ETSI standards.

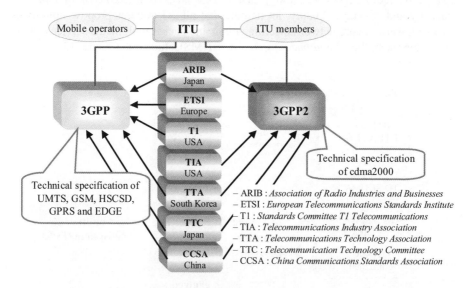

Figure 1.9. *Regional standardization bodies composing 3GPP and 3GPP2*

1.5. Worldwide spectrum allocation for IMT-2000 systems

The ITU is responsible for frequency band allocation of IMT-2000 systems. This has been achieved during the *World Administrative Radio Conference* held in 1992 (WARC-92) and in 2000 (WARC-2000)

1.5.1. *WARC-92*

WARC-92 identified the bands 1,885-2,025 MHz and 2,110-2,200 MHz for possible use of IMT-2000 radio interfaces. This represents 230 MHz to be shared by terrestrial and satellite components of IMT-2000 systems. In Europe, the available IMT-2000 bands are used by UTRA/FDD and UTRA/TDD radio interfaces exclusively as depicted in Figure 1.10. Conversely, in China and Japan, these UMTS radio interfaces will need to coexist with those of other 3G standards, such as cdma2000.

It can be observed in Figure 1.10 that 2G cordless standard DECT already occupies 1,880-1,900 MHz band and thus, the available bands for terrestrial UMTS deployment in Europe are 1,900-1,980 MHz. This gives 155 MHz in total to be used as follows:

– 120 MHz for paired UTRA/FDD radio interface, where 1,920-1,980 MHz is used in uplink, whereas band 2,110-2,170 MHz is used in downlink. Channel spacing is 5 MHz and raster is 200 kHz in each link direction;

– 35 MHz for unpaired UTRA/TDD radio interface using 1,900-1,920 MHz and 2,010-2,025 MHz frequency bands with 5 MHz channel spacing and 200 kHz raster;

– 60 MHz for IMT-2000 satellite component operating in 1,980-2,010 MHz and 2,170-2,200 MHz frequency bands.

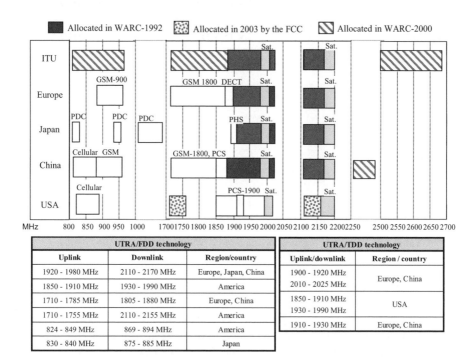

UTRA/FDD technology			UTRA/TDD technology	
Uplink	Downlink	Region/country	Uplink/downlink	Region / country
1920 - 1980 MHz	2110 - 2170 MHz	Europe, Japan, China	1900 - 1920 MHz	Europe, China
1850 - 1910 MHz	1930 - 1990 MHz	America	2010 - 2025 MHz	
1710 - 1785 MHz	1805 - 1880 MHz	Europe, China	1850 - 1910 MHz	USA
1710 - 1755 MHz	2110 - 2155 MHz	America	1930 - 1990 MHz	
824 - 849 MHz	869 - 894 MHz	America	1910 - 1930 MHz	Europe, China
830 - 840 MHz	875 - 885 MHz	Japan		

Figure 1.10. *Worldwide spectrum allocation for IMT-2000 systems*

The WARC-92 spectrum identified for IMT-2000 is being used for the first wave of commercial UMTS deployments. Table 1.4 shows examples of countries where UMTS licenses have been awarded. In Japan, 3 licenses with 2×20 MHz spectrum each were awarded to two UMTS operators and one cdma2000 operator.

The original idea of the ITU was that all countries in the world allocate the same frequencies to IMT-2000 systems, thus enabling easy global roaming. However, the only country that follows ITU's recommendation exactly is China, since in Europe and Japan that spectrum is partly used by GSM and cordless PHS (*Personal Handyphone System*) standards, respectively (see Figure 1.10). In the USA, such a 3G band is already being used by 2G PCS and fixed wireless networks, and as such, the operators will need to gradually deploy 3G networks in place of their existing 2G infrastructures.

Country	Number of licenses awarded	Paired UTRA/FDD spectrum per license	Unpaired UTRA/FDD spectrum per license
Finland	4	2×15 MHz	1×5 MHz
France	3	2×15 MHz	1×5 MHz
Germany	5	2×10 MHz	1×5 MHz
	1	2×10 MHz	…
United Kingdom	1	2×15 MHz	1×5 MHz
	1	2×15 MHz	…
	3	2×10 MHz	1×5 MHz

Table 1.4. *Number of licenses awarded to UMTS operators in certain countries*

1.5.2. *WARC-2000*

Even assuming that existing 2G networks will eventually be upgraded to 3G, the ITU considered that at least 160 MHz more will be needed in each region before 2010. Moreover, it has become clear that further spectrum is needed, as third generation services become more and more sophisticated, in order to incorporate future multimedia applications and as the number of users increases. This fact was the basis for the WARC-2000 decision to identify spectrum in the bands:

– 806-960 MHz. This band is currently being used by 2G systems in Europe, Japan, China and the USA. This means that existing 2G operators not owning 3G spectrum licenses will be able to deploy 3G systems in the 2G bands they already possess;

– 1,710-1,885 MHz. Some parts of this band are already used for mobile services in Europe and Asia. In the USA, this band was largely used by the Department of Defense. However, during 2003 a compromise was reached with the FCC (*Federal Communications Commission*) and two 45 MHz frequency bands were allocated for 3G service applications: 1,710-1,755 MHz and 2,110-2,155 MHz;

– 2,500-2,690 MHz. Different countries use this band for different applications, such as broadcasting and fixed networks. Some parts are also used by satellites communicating with Earth.

In addition to the above bands, China decided to allocate an additional band for IMT-2000 system deployment: 2,300-2,400 MHz.

Chapter 2

Network Evolution from GSM to UMTS

2.1. Introduction

The choice of the term "UMTS" that means *Universal Mobile Telecommunications System* was not fortuitous. It was a matter of marking the difference with its 2G predecessors, still considered as mobile *telephony* systems. The *Global System for Mobile communications* (GSM) strongly contributed to making mobile phones become an everyday life product. Its first goal that was of providing high-quality voice services was achieved. GPRS and EDGE technologies are part of the second phase named GSM Phase 2+ and they are setting up the basis for providing multimedia services and high-speed access to Internet – services enhanced within UMTS standard. Table 2.1 summarizes the key milestones in the recent history of UMTS.

This chapter provides a formal definition of UMTS and outlines the key steps in the evolution of GSM networks towards UMTS by considering the GSM Phase 2+ extensions as well as the design principles behind its network architecture.

2.2. UMTS definition and history

Sometimes the UMTS is considered as synonymous with IMT-2000. This is wrong since IMT-2000 encompasses a family of 3G networks of which UMTS is a member (see Chapter 1). In this book a UMTS network will be defined as follows:

UMTS is a mobile cellular communications system belonging to the family of third generation systems defined in the IMT-2000 framework. UMTS network architecture comprises a radio access network referred to as UTRAN (*Universal Terrestrial Radio Access Network*) and a core network based on the one specified for GSM Phase 2+. The UTRAN may implement one or two radio access technologies relying on (wideband) CDMA access mode: UTRA/FDD (*Universal Terrestrial Radio Access/Frequency Duplex Division*) and UTRA/TDD (*Universal Terrestrial Radio Access/Time Duplex Division*). The technical specifications of the UMTS networks are developed by the 3GPP.

February 1992	WARC-92 allocates 230 MHz spectrum around 2 GHz for IMT-2000 use
1987-1995	RACE I and RACE II projects are carried out in Europe aiming at the technical definition of UMTS
1995-1999	The EU funded ACTS/FRAMES projects striving for the definition of two radio access technologies denoted FMA1 and FMA2 which can be seen as the predecessors of UTRA/TDD and UTRA/FDD, respectively
December 1996	Start of the development of specifications for UMTS under ETS1 SMG2
June 1998	Up to 10 terrestrial radio access technologies are submitted to the ITU-R to be part of IMT-2000
October 1998	ITU-R selects the current IMT-2000 radio interfaces: UTRA/FDD, UTRA/TDD, TD-SCDMA, cdma2000, UWC-136 and DECT. A number 3G satellite components are also identified
December 1998	The first meetings of the 3GPP Technical Specification Groups take place aiming at the completion of UMTS technical specifications started by ETSI
March 1999	Finland is the first European country to give out UMTS licenses
2001 and 2002	Test of the first UMTS networks in Europe: Isle of Man and Monaco
October 2001	Commercial launch of 3G services in Japan with FOMA brand which is based on an early version of UMTS specifications employing UTRA/FDD
March 2003	Commercial launch of UMTS services in Europe, in the UK and Italy

Table 2.1. *Key milestones in UMTS network development*

2.3. Overall description of a UMTS network architecture

UMTS network design is based on a flexible and modular architecture. In theory, such architecture is neither associated to a unique radio access technology nor to a predefined set of services. The idea behind this philosophy is twofold: to guarantee interoperability with its 2G predecessors and with other 3G systems and to enable UMTS operators to make their network evolve as a function of their needs for the development of new services, as well as additional capacity and radio coverage.

The architecture of a UMTS network comprises a number of physical entities grouped in *domains* in accordance with their function within the network. Three domains compose the architecture of a UMTS network (see Figure 2.1): the *User Equipment* (UE) domain, the *Universal Terrestrial Radio Access Network* (UTRAN) domain and the *Core Network* (CN) domain. The UTRAN and CN domains are part of the network infrastructure domain.

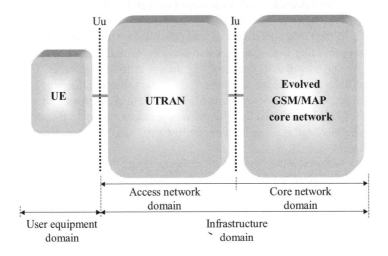

Figure 2.1. *UMTS network architecture*

User equipment is the name given to a mobile terminal within a UMTS network (similar to a mobile station in GSM/GPRS terminology). This is the vector enabling a subscriber to access UMTS services through the radio interface Uu. The elements and functional split of the UE are studied in detail in Chapter 3.

The *UTRAN* provides the user with the physical resources to access the core network. Within the Uu interface, it performs radio resource management and admission control functionality and sets up the "pipes" (bearers) enabling the UE to communicate with the CN. Chapter 6 analyses in details the functional entities and procedures implemented in the UTRAN.

The *core network* is in charge of the management of telecommunication services for each UMTS subscriber. This includes: the establishment, termination and modification of circuit and packet UE-terminated and UE-originated calls; the mechanisms for UE authentication; interconnection with external mobile and fixed networks; user's charging; etc. Chapter 4 addresses UMTS CN aspects.

In Figure 2.1, Uu and Iu interfaces enable the communication between the three domains above and serve as reference points so that the functions of each domain are bound. The interfaces are standardized, thus enabling a UMTS operator to select different equipment suppliers for each domain.

2.4. Network architecture evolution from GSM to UMTS

Phase 1 of the GSM specifications was published by the ETSI in 1990. The focus was put on providing high-quality voice services. Circuit switched data transfer at 9.6 kbps was also available, thus enabling access to ISDN networks. The provision of fax and SMS teleservices was also offered within this architecture. The choice of supplementary services was very limited. The specifications for GSM Phase 2 were released in 1995. It enhanced the offer of supplementary services by introducing cell broadcast service, call waiting/holding, multi-party calls, closed-user group, etc.

As data services have become more demanding, it turned out that data rates offered by GSM were insufficient. This fact motivated the ETSI to issue Phase 2+ specifications in 1996 that were amended in subsequent releases (R96, R97, R98, and R99). Phase 2+ brought out very important enhancements to GSM Phase 1 and Phase 2 in terms of network architecture and service offered as described in next sections.

2.4.1. GSM network architecture of Phases 1 and 2

The architecture of GSM networks of Phases 1 and 2 is shown in Figure 2.2. It can be divided into three functional entities: *Base Station Sub-System* (BSS), *Network and Switching Sub-System* (NSS) and the *Operation Support Sub-system* (OSS).

One or more BSSs form the radio access network in the GSM architecture. Each BSS consists of a *Base Station Controller* (BSC) remotely commanding one or several *Base Transceiver Stations* (BTS). They communicate through the Abis interface. The BTS comprises radio transmission and reception devices including antennae and physical layer signal processing procedures associated with Um radio interface. The BSC is in charge of the allocation and release of radio TDMA channels, it performs paging and maintains handover across and inside BSSs. The *Transcoder and Rate Adapter Unit* (TRAU) is also considered as part of BSS though, in practice, it is placed in the NSS. Its role is to convert the 64 kbps speech signal used in the NSS and in external networks to the speech rates supported in the GSM air interfaces, e.g. full (13 kbps) or half (6.5 kbps) speech coding rates.

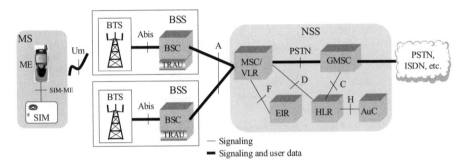

Figure 2.2. *GSM network architecture of Phases 1 and 2*

The NSS can be seen as the ancestor of UMTS core network, and more precisely, of its circuit-switched domain (see Chapter 4). It interacts with a BSS through A interfaces and includes the main switching function of GSM as well as the databases needed for user authentication, call control and mobility management.

The OSS is responsible for the network operation and maintenance including traffic monitoring, billing and status report of the network entities.

The *Mobile Station* (MS) is the wireless user's terminal that consists of the *Mobile Equipment* (ME) and the *Subscriber Identity Module* (SIM). All the radio functions and user's applications are in the ME (the handset), whereas the SIM card contains subscriber-related and radio information, both permanent and temporary.

2.4.2. *GSM network architecture of Phase 2+*

GSM Phase 2+ specifications introduced new teleservices (see Chapter 3) such as *Voice Group Call Service* (VGCS), *Voice Broadcast Service* (VBS) and the supplementary service *Enhanced Multi-Level Precedence and Preemption* (eMLPP). Improved voice quality was achieved with the standardization of *Enhanced Full*

Rate (EFR) and *Adaptive MultiRate* (AMR) codecs. HSCSD, GPRS and EDGE technologies brought out high-data rate services and new transmission principles.

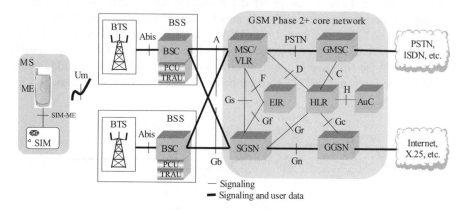

Figure 2.3. *GSM Phase 2+ network architecture*

A series of developments were initiated in Phase 2+ to enhance the functionality of GSM networks enabling the faster deployment of new services and services differentiation. This is the case of the *Virtual Home Environment* (VHE) concept, which means that users are consistently presented with the same personalized presentation, and features of subscribed services wherever they may roam (see Chapter 3). The development of new services in the VHE framework is achieved by the introduction of secure standardized execution environments in the SIM, the ME and the core network, respectively: the *SIM Application Toolkit* (SAT), the *Mobile Station Application Execution Environment* (MexE) and the *Customized Application for Mobile network Enhanced Logic* (CAMEL).

GPRS technology is one of the most important network features introduced in Phase 2+. It is designed to support bursty data traffic and it makes it possible to efficiently use the network and radio resources. User data packets can be directly routed from the GPRS terminal to packet switched networks based on the Internet protocol (e.g. Internet or private intranets) and X.25 networks. As shown in Figure 2.3, GPRS introduces new equipments as well as software upgrades of the existing ones. The most important new network elements are the GPRS supporting nodes (SGSN and GGSN) connected via an IP-based GPRS backbone network. Enhancements are necessary in the BSS requiring, among others, a *Packet Control Unit* (PCU) that allocates resources for the GPRS packet transmission over the air interface. In GPRS, charging is no longer based on the length of transmission time but rather on the transported data volume. A distinction is also made between different Qualities of Service (QoS) to suit different application requirements and usage situations.

The impact of EDGE on GSM/GPRS networks is limited to the BSS. The base station is affected by the new transceiver unit capable of handling EDGE modulation as well as protocol upgrades in lower layers, in both the BTS and the BSC. A more important impact on the BSS is in the definition of the GSM/EDGE radio access network (GERAN) currently standardized by the 3GPP (*Release 5*). It constitutes a RAN featuring EDGE modulation and coding modes that can be interconnected to a UMTS core network. This fact makes GERAN a UTRAN-*like* RAN that supports 3G services.

2.4.3. Architecture of UMTS networks: evolutionary revolution of GSM

In the evolution path of UMTS networks, a similar approach by phases as in GSM networks is adopted. The first UMTS networks deployed in Europe and in Japan belong to Phase 1. From the standard point of view, they are based on the first technical release of 3GPP specifications, i.e. *Release 99*. UMTS Phase 1 introduces the UTRAN and defines interfaces for inter-working with an evolved GSM Phase 2+ core network. This graceful upgrading of technology enables operators to make cost-effective the core networks investments made on GSM Phases 1, 2 and 2+. Operators also get the chance to test and develop possible 3G services and business models.

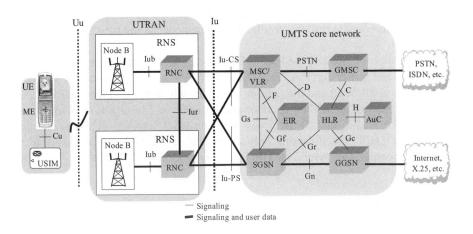

Figure 2.4. *UMTS Phase 1 network architecture*

By comparing Figures 2.3 and 2.4 it turns out that the main difference between UMTS Phase 1 and GSM Phase 2+ networks relies on the introduction of a new radio access network, the UTRAN. And we can talk here about a *revolution*, since everything changed compared with GSM BSSs: the radio access technology based on CDMA principles (instead of TDMA); the network transport technology based on ATM and all the new equipments and communication protocols implementing such innovations. Conversely, with respect to the UMTS core network, we can talk about an *evolution*, since it groups basically the same NSS and GPRS backbone functional entities defined in GSM Phase 2+.

As far as UMTS Phase 2 is concerned, it is not formally defined but it can be said that it will be based on *Release 4*, *Release 5*, *Release 6* and *Release 7* of 3GPP technical specifications. They aim at gradually introducing full IP-protocol usage in the transport layers of both the UTRAN and the core network (see chapter 13).

2.5. Bearer services offered by UMTS networks

A basic telecommunication service proposed by any UMTS network is the transport of user data trough the network. This is realized using one or multiple "pipes" referred to as bearer services that follow a hierarchical structure [TS 23.107]. The characteristics of each "pipe" are specified by a set of parameters that define the expected QoS such as the bit rate, the transfer delay and the bit-error rate (see Chapter 3). When a user requests a service, the UE application negotiates, via the core network, the UMTS bearer characteristics that are the most appropriate for carrying information. The requested QoS may be accepted, refused or re-negotiated according to resource availability in the network and subscription level.

As shown in Figure 2.5, the End-to-End service used by the UE will be realized through several bearer services: *TE/MT local bearer* service that conveys user data between the application and the radio-modem part within the UE; the *UMTS bearer* service, offered by the UMTS operator and an *external bearer service* such as the Internet. The End-to-End service may use non-UMTS transit networks. In this respect, the 3GPP specifications only cover the procedures and mechanisms to guarantee a certain QoS within the UMTS bearer service.

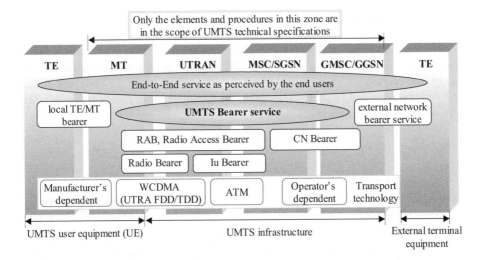

Figure 2.5. *UMTS bearer service architecture*

The *UMTS bearer service* consists of two parts: the *radio access bearer* service (RAB), that provides the transport of user data through Uu and Iu interfaces, i.e. from the mobile terminal to the core network and the *core network bearer service*, that utilizes the UMTS backbone network (circuit or packet-switched) to carry the user data within the core network all the way to the gateway or switch that interconnects with the external network.

The RAB service is realized by the *radio bearer* service and the *Iu bearer* service. The radio bearer service covers all aspects of the Uu radio interface transport and uses either UTRA/FDD or UTRA/TDD technologies. The Iu bearer service provides the transport between the UTRAN and the CN through Iu interface.

2.6. UMTS protocol architecture based on "stratum" concept

Uu and Iu interfaces are inspired by the OSI (*Open System Interconnection*) model to define the protocols that in the UMTS architecture enable reliable user data and signaling message exchange between domains. In UMTS terminology, a set of protocols having the same functional role in the network are referred to as "stratum".

2.6.1. *Access stratum*

The *access stratum* (AS) groups together the protocols and functions enabling the transport of user data and network control signaling generated by upper layers, i.e. by the *non-access stratum*. AS protocols are present in the lower layers of Uu and Iu interfaces, and as such, they enable the transport of information between the UE and the CN, as depicted in Figure 2.6.

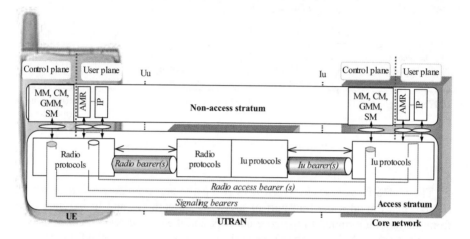

Figure 2.6. *Scope of access stratum and non-access stratum in UMTS architecture*

In Uu interface, AS supports transfer of detailed radio related information to coordinate the use of radio resources between the UE and the UTRAN. AS protocols in Uu interface are referred to as "radio protocols" and enable the set-up and the control of radio bearers.

In Iu interface, AS applies mechanisms to convey data and signaling information from the UTRAN to the CN serving access node (MSC or SGSN). AS protocols in Iu interface are referred to as "Iu protocols" and enable the set-up and the control of Iu bearers. The radio protocols are studied in Chapter 7, whereas Iu protocols are analyzed in Chapter 6.

The set of AS protocols in Uu and Iu interfaces, permit the realization and the control of the radio access bearers. From a functional point of view, those protocols are distributed in two planes:

– User plane which contains the protocols involved in the materialization of the radio access bearers. The data stream(s) (voice, video, data, etc.) generated by upper (application) layers are adapted to the physical media by applying AS mechanisms such as data formatting, transcoding and error correction and recovery.

– *Control plane*[1] which includes the protocols that generate control signaling messages aiming at the establishment, modification and release of radio access bearers within the user plane.

2.6.2. *Non-access stratum*

The *non-access stratum* (NAS) includes the protocols outside the access stratum, i.e. the protocols not related to the transport of user data or network signaling. They concern all data and signaling messages exchanged between the UE and the CN independently on the radio access technology implemented within the UMTS network. The UTRAN does not analyze these NAS messages and just plays the role of a relaying function. The user plane in NAS comprises basically user application protocols generating data streams to be transported by AS, that may or not adhere to GSM/UMTS standards. On the other hand, the NAS control plane encompasses all the protocols associated to *Connection Management* (CM) and *Mobility Management* (MM) functions for circuit-switched services, and *Session Management* (SM) and *GPRS Mobility Management* (GMM) functions for packet-switched services. Chapter 8 focuses on non-access stratum control functions.

Table 2.2 gives examples of access stratum and non-access stratum functions in a UMTS network.

---

1 The control plane is sometimes referred to as a signaling plane.

Function	Non-access stratum	Access stratum	Studied in Chapter
Call control	Yes	No	8
Bearer service	Yes (activation)	Yes (realization)	6, 7, 8
Supplementary services	Yes	No	3, 8
Mobility management	Yes (CN location and routing areas)	Yes (cell and UTRAN routing areas)	6, 7, 8
Attachment/detachment	Yes	No	8
Handover/ cell reselection	Yes (RAB relocation)	Yes	5, 6, 7, 8, 11
Ciphering/integrity	Yes (activation)	Yes (execution)	7, 8
Authentication	Yes	No	9
Voice and video coding	Yes	No	Appendix 1 (AMR)
Radio channel coding	No	Yes	7, 9
Location services	Yes, handling	Yes, position estimation	
Charging	Yes	No	

Table 2.2. *Examples of functions performed by access-stratum and non-access stratum*

Chapter 3

Services in UMTS

3.1. Introduction

It is with the services offered by UMTS that the end-users can perceive the differences from 2G/2.5G predecessors: mobile service subscribers are less sensitive to the amazing technical feats of each new technology than to the benefits this can provide to them. We have to keep in mind that besides video-telephony, other UMTS services like voice telephony, browsing (WAP, HTTP), messaging (SMS, MMS, email, push-to-talk), streaming (music, video), data synchronization (calendars, contacts, files) and location-based services can also be offered with GPRS and EGPRS technologies. What is clear is that UMTS additionally proposes higher data rates and offers more flexibility in the radio interface to handle simultaneous connections (combined packet and circuit) with different levels of QoS (bit rate, error rate, delay, etc.).

GSM and GPRS have not yet succeeded in breaking the common service provision scheme where the mobile operator offers access and switching functions as well as *all* the associated mobile services. This approach restrains the role that third party service providers could play in accelerating the deployment of new services and in promoting service differentiation. Generally speaking, UMTS standards do not specify all the services the network shall provide but rather, they define a set of tools enabling (in theory) anybody to develop any kind of new service that can be put on top of the UMTS bearer services discussed in Chapter 2. This chapter lists and describes the services standardized in UMTS Phase 1 (based on *Release 99* specifications) and investigates those that can be created using standardized *service capabilities*.

3.2. UMTS mobile terminals

The UMTS *User Equipment* (UE) is not a simple mobile phone but rather, a mobile multimedia terminal able to provide simultaneously voice, video and data services. The UMTS standards do not impose the physical aspect of the UE (size, weight, type of display/keyboard, etc.). The look-and-feel of the UE depends on each mobile equipment manufacturer. The difference between UEs may come from their capability to provide a certain number of services – there is no mandatory set of services imposed by the UMTS standards either. Therefore, in the mobile market we can find UEs offering: speech-only, video-telephony only, Internet-access only, streaming-only, or a combination of all above.

The UEs may also differ from each other according to their radio access capability. They can be dual-mode and allow seamless service provision across 2G and 3G networks. Examples are GSM + UTRA/FDD, GSM + GPRS + UTRA/FDD, GSM + EGPRS + UTRA/FDD, GSM + EGPRS + UTRA/FDD + HSDPA. This possibility shall take into consideration the bearer limitation of each technology (e.g. an active video-telephony call cannot be maintained when roaming from UMTS to GSM).

3.2.1. *UE functional description*

From a functional point of view, the UE is composed of two parts, the *Mobile Equipment* (ME) and the *Universal Subscriber Identity Module* (USIM). The reference point bounding ME/USIM functions is referred to as "Cu".

3.2.1.1. *Mobile equipment*

The UE without a USIM/SIM card inserted is named *Mobile Equipment* and further comprises a *Mobile Termination* (MT) and a *Terminal Equipment* (TE).

The MT constitutes the UMTS radio-modem functionality and performs reliable data and signaling message transfer throughout the radio interface. Channel coding/decoding, spread spectrum, RF modulation and packet retransmission are examples of the functions carried out by the MT. In general, it can be said that the MT groups together all the *access stratum* and *non-access stratum* functions and protocols. The TE is the part within the UE where user data is generated in uplink and processed in the downlink. Application protocols such as WAP/IP and the video/audio codecs are located in the TE. Typically, the MT and the TE functions are physically placed together within the same ME (see Figure 3.1a), but they can also be separated. In this case, the MT may be a PCMCIA inserted into a portable computer (see Figure 3.1b), and communication between the MT and the TE is carried out via standard PCMCIA interface.

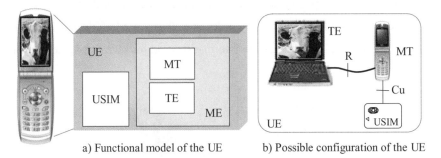

a) Functional model of the UE   b) Possible configuration of the UE

Figure 3.1. *UE functional model and possible configurations*

3.2.1.2. *USIM and UICC*

Although often referred to as USIM-*card*, the USIM is actually an application that resides on the UICC (*Universal Integrated Circuit Card*) smartcard. Besides the support for one application (optionally more than one), the UICC also provides a platform for other IC card applications: support of one or more user profile on the USIM, update USIM specific information over the air, security functions, user authentication, optional inclusion of payment methods, optional secure downloading of new applications. The UICC can also support a SIM application so that it may be used in GSM/GPRS handsets enabling backward compatibility, as shown in Figure 3.2. By analogy with a personal computer, the UICC can be considered as the hard disk and each USIM as a directory located therein containing a number of files and applications. The 3GPP specifications ensure interoperability between a USIM and an ME independently of the respective manufacturer by describing the characteristics of the USIM/ME interface [TS 31.102]. Examples of the information located in USIM are:

– the *Personal Identification Number* (PIN) which makes it possible to associate the UICC to a given subscriber regardless of the ME she/he uses;

– the preferred languages of the user in order of priority;

– one or several IMSI(s) and MSISDN(s);

– the user's temporary identities allocated by the core network when attached to the circuit domain (TMSI) and the packet domain (P-TMSI);

– circuit-switched and packet-switched temporary location information: LAI (*Location Area Identification*) and RAI (*Routing Area Identification*);

– the keys for ciphering and integrity procedures (see Chapters 7 and 8);

– the codes to enable emergency calls;

– information comprising short messages and associated parameters;

– the preferred PLMNs in priority order as well as those forbidden (FPLMN);

– frequency and scrambling code information used for accelerating cell selection.

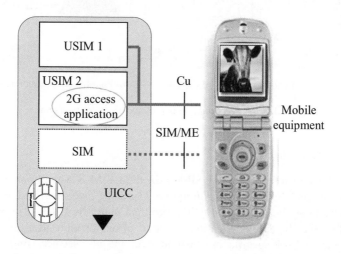

Figure 3.2. *Example of a UICC configuration containing various USIMs and one SIM application. The latter can also be embedded within the USIM*

3.2.1.3. *UMTS user and UE identities*

As in GSM, in UMTS the *International Mobile Subscriber Identity* (IMSI) makes it possible to unambiguously identify a particular UMTS subscriber. This is a fixed number not visible to the user and allocated by the network operator when the user subscribes to UMTS services. The IMSI number is located in the USIM and comprises: the *Mobile Country Code* (MCC); the *Mobile Network Code* (MNC), i.e. the code of the operator's network that holds the user's subscription information (Home PLMN) and the *Mobile Subscriber Identification Number* (MSIN).

The MSISDN is the dialable number that UMTS callers use to reach a mobile subscriber to establish a voice or video-call. More than one "phone number" or MSISDN may be associated with the same IMSI. As in GSM, the format of this number is based on the E.164 numbering plan [TS 23.003]. Let us note that a UMTS subscriber may also be reached via a dynamic or static IP address allocated by the network for packet-switched service provision (see Chapter 4).

When the UE exchanges data or signaling information with the core network, it identifies itself with the IMSI. However, for security reasons, the standards recommend to send the IMSI over the radio interface as rarely as possible (e.g. only the first time the UE establishes a signaling connection with the core network after having been switched-on). The use of temporary identities is then preferred and there are two types: the *Temporary Mobile Subscriber Identity* (TMSI) used when data and signaling exchanges concern circuit-switched services (voice, video-telephony, etc.), and the *Packet Temporary Mobile Subscriber Identity* (P-TMSI) when the exchanges concern packet-switched services (streaming, Internet surfing, etc.). There is a direct relationship between IMSI, MSISDN, TMSI and P-TMSI and it is up to the core network to manage such identity mapping for each UMTS subscriber.

As in GSM, the UMTS Mobile Equipments are also provided with a unique number: the *International Mobile Equipment Identity* (IMEI). This number is assigned during the terminal manufacturing phase and it is effectively the serial number. By requesting the UE to transmit the IMEI number, the UMTS network has means of prohibiting its access to UMTS services even if the IMSI number corresponds to a valid subscription. This may happen when the mobile equipment turns out to be stolen or not homologated. Figure 3.3 illustrates the relationship between the different terminal and subscription identities in a UMTS network.

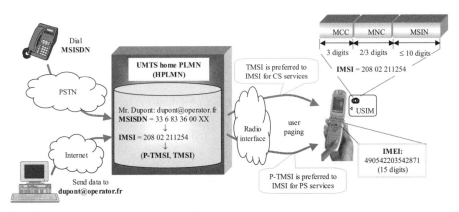

Figure 3.3. *Relationship between user and terminal identities in a UMTS network (example)*

3.2.2. *UE maximum output power*

The UEs are classified as a function of their maximum output power (see Table 3.1) [TS 25.101, TS 25.102]. The UTRAN knows the class or the classes supported by the UE during the establishment of a radio connection (RRC connection). Only classes 3 and 4 are part of 3GPP specifications in *Release 99* and both are typically supported by any UE. This gives the UTRAN a degree of flexibility such that the UE output power can be controlled in a larger range. For instance, voice and video-telephony services may require a low power level (class 4) whereas high rate packet data services may need the UE to transmit at its maximum power level (class 3). The knowledge of the power classes supported by the UEs is also an input parameter used for dimensioning the radio network.

Power class	Maximum output power		Tolerance	
	UTRA/FDD	**UTRA/TDD**	**FDD**	**TDD**
1	+ 33 dBm (2 W)	+ 30 dBm (1 W)	+ 1/– 3 dB	+ 1/– 3 dB
2	+ 27 dBm (0.5 W)	+ 24 dBm (0.25 W)	+ 1/– 3 dB	+ 1/– 3 dB
3	+ 24 dBm (0.25 W)	+ 21 dBm (0.126 W)	+ 1/– 3 dB	± 2 dB
4	+ 21 dBm (0.126 W)	+ 10 Bm (0.01 W)	± 2 dB	± 4 dB

Table 3.1. *UE power classes as a function of the UTRA mode*

3.2.3. *Dual-mode GSM/UMTS terminals*

Most European operators require the UE to be dual-mode, i.e. to support GSM *Radio Access Technology* (RAT) in addition to UTRA/FDD. This is because UMTS is initially deployed to cover urban areas, and operators must feel the need to provide nationwide coverage from the beginning. A dual-mode UE equipped with both UTRA and GSM technology would enable those operators who already have a GSM network to capitalize on their investments when introducing UMTS services.

In [TR 21.910], a classification is proposed for multi-mode UEs, which is larger than the simple supporting of UTRA and GSM RATs. Table 3.2 summarizes the key ideas of this classification. It should be stressed, however, that Type 3 and Type 4 UEs are *concepts* which are not defined in 3GPP specifications. UMTS mobile market trend is in favor of Type 2 UEs with UTRA/FDD-GSM radio capability, although Type 1 UEs were commercialized in Europe in early UMTS deployments.

UE Type	Dual-mode UTRA/GSM behavior
Type 1	Does not perform measurements on UTRA cells while camped on a GSM cell and vice versa. With respect to the network, the UE is seen as a single mode terminal and changes from one radio network to another is done manually, with the user's intervention
Type 2	Performs measurements on UTRA cells while camped on a GSM cell and vice versa. The results are reported to the network via the active mode. Switching from one radio network to another (inter-RAT handover/cell reselection) is done automatically, without the user's intervention
Type 3	Similar to Type 2 but it additionally offers the possibility to receive simultaneously user-data or signaling information in more than one mode. However, it does not transmit simultaneously in more than one mode
Type 4	Compared to Type 3, this UE makes it possible to simultaneously receive and transmit user-data or signaling information in UTRA and GSM technologies

Table 3.2. *Classification of UEs according to their dual-mode radio capabilities*

3.2.4. *UE radio access capability*

UMTS terminals are also categorized according to their radio capability. Indeed, when the UE wants to start sending user data or network control information, a radio signaling connection shall be established (RRC connection). During this phase, the UE declares its radio capability comprising [TS 25.306]: UMTS frequency operation bands, support for UTRA/FDD and/or UTRA/TDD, dual-mode capability, kind of ciphering and integrity algorithms supported, power class, support for UE positioning, maximum data rate, etc.

With respect to the maximum data rate, this is not explicitly declared by the UE to the network but is done via a number of radio parameters related to physical and upper radio layers performance. From these parameters, the network determines the data rate class of the UE and, at the same time, its service capabilities, independently for the uplink and downlink directions (see Figure 3.4). For instance, "64 kbps class" in both directions is the minimum required to support video-telephony. If the UE is further required to support a packet-switched data service (e.g. web browsing) in parallel with video-telephony, a "128 kbps class" in uplink and "384 kbps class" in downlink may be needed. Note that network signaling information can be transferred simultaneously with user data. Finally, "768 kbps class" and "2,048 kbps class" require costly UE hardware resources, and they will not be commercialized in first UMTS network deployments. Most of the current UMTS terminals are "384 kbps class" in both downlink and uplink directions.

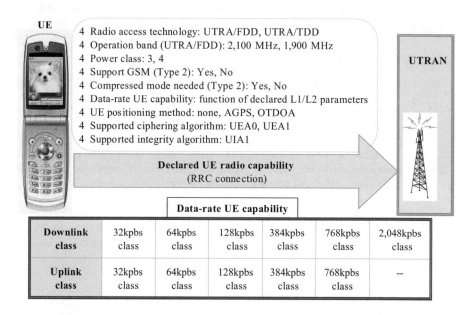

Figure 3.4. *UE radio capability declared when establishing an RRC connection*

3.3. Services offered by UMTS networks

Creating a new service may need a modification of standards: a long and laborious process. In UMTS, the approach is appreciably different: only certain services are standardized in order to maintain backward compatibility with GSM/GPRS predecessors and new ones are developed using standardized tools (*service capabilities*).

3.3.1. *Standard UMTS telecommunication services*

Figure 3.5 illustrates the UMTS telecommunication services [TS 22.105]. This classification is common to major mobile communication networks and comprises:

– bearer services;

– teleservices;

– supplementary services.

3.3.2. *UMTS bearer services*

UMTS bearer services are the "pipes" allowing reliable data transfer from the source to the destination (see Chapter 2). These "pipes" are not perceived by the user as a service, but rather the application or the teleservice that is on top of them. Multiple bearers may be active in a multimedia call – each of them having different *Quality of Service* (QoS).

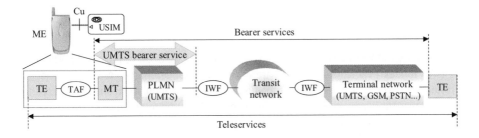

Figure 3.5. *Bearer services and teleservices in a UMTS network*

3.3.2.1. *QoS attributes of UMTS bearer services*

The UMTS bearer services are characterized by a limited number of attributes defining their QoS [TS 23.107]:

- traffics class: conversational, streaming, interactive, background;

- maximum and guaranteed bit rate: < 2,048 kbps;

- delivery order: in sequence SDU delivery or not: Yes, No;

- maximum SDU size: ≤ 1,500 bytes;

- SDU format information (bits): list of possible exact sizes of SDUs;

- SDU error ratio and residual bit error ratio: 10^{-6} to 10^{-2};

- delivery or discard of erroneous SDU: Yes, No;

- transfer delay: 100 ms to non-specified;

- allocation/retention priority of the UMTS bearer: 1, 2, 3;

- traffic handling priority compared to other bearers: 1, 2, 3.

3.3.2.2. *Circuit-switched and packet-switched bearer services*

Circuit-switched data (CSD) services are such that a dedicated physical circuit path must exist between source and destination for the duration of the call. Conversely, packet-switched data (PSD) services use either dedicated or common physical resources and only when there is actual data to be received or transmitted. UMTS inherited CSD bearers from GSM and PSD bearers from GPRS and additionally introduced new ones given the higher-data rates UMTS can support. Note that the UE independently requests CSD and PSD services to the network since the UMTS core network counts on separate entities to handle each of them (see Chapter 4). It is up to each operator to specify whether a given service supported by the UE shall be provided using the CSD or the PSD network infrastructure. Finally, multiple UMTS radio bearer services may be handled simultaneously by the UE PSD and/or CSD, according to the UE radio capability. Figure 3.6 shows possible configurations for the UMTS bearer.

UMTS CSD bearer services comprise transparent data and non-transparent data transfer modes [TS 22.002]. Transparent data transfer is characterized by constant bit rate and delay between the source and destination, which are features required to provide real-time bearer service for video and multimedia applications. For instance, in order to provide video-telephony, a transparent bearer using on top 3G-324M multimedia codec is used in UMTS. The supported air interface user rates for a multimedia call vary from 28.8 kbps to 64 kbps. In contrast, non-transparent data transfer offers a higher reliable data transfer with low bit-error rate due to the implementation of *Radio Link Protocol* (RLP) – this makes it possible to retransmit data when errors are detected at reception [TS 24.022]. RLP is implemented in the UE and in the network inter-working function (IWF) and its use may induce data rate variations and increase the transfer delay. Non-transparent bearers are typically used for connecting the UE with an external computer using an asynchronous transfer mode.

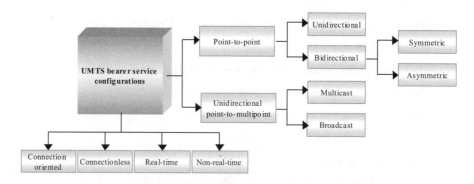

Figure 3.6. *Possible configurations for the UMTS bearer service*

End-to-End CSD transfer over a digital channel is referred to as *Unrestricted Digital Information* (UDI). However, if an analog transit network is interposed between the source and the destination, an IWF named "3.1 kHz audio" is needed to enable digital-to-analog signal adaptation. As an example, video-telephony service requires a UDI transparent UMTS bearer.

PSD bearer services in UMTS are inherited from GPRS [TS 22.060]. They may be connection-oriented or connectionless. In the first case, the source and destination exchange signaling information before starting user data transfer (e.g. multimedia messaging service). On the other hand, connectionless services do not need the establishment of a connection beforehand and source and destination addresses are specified in each message (e.g. short message service). The UMTS PSD bearer service may also be point-to-point or point-to-multipoint, unidirectional or bidirectional. In the latter, the bearer service can be symmetric (same bit rate in the transmission and the reception paths) or asymmetric (different bit rate in transmission and reception paths). Finally, it is possible in UMTS to offer services using multicast and broadcast schemes where a message is transmitted from a single source entity to all subscribers currently located within a geographical area (see Chapter 13). In contrast to the broadcast approach, where the subscribers addressed are not under the control of the source, multicast technique enables the source to select the sink parties before the connection is established, or by subsequent operations to add or remove parties from the connection [TS 22.105]. Multicast services are not part of Phase 1 UMTS services (*Release 99*) – they are being developed in *Release 6*.

3.3.2.3. *Real-time and non-real-time bearer services*

The control of delay variation (jitter) of the data transferred over the bearer is a required feature to support real-time services (e.g. voice service). This is not the case for non-real-time services where time relation (variation) between information entities of the stream is not an issue (e.g. email). Table 3.3 summarizes the range of QoS parameters tolerated to provide real-time and non-real-time UMTS bearer services [TS 22.105]. The CSD infrastructure in a UMTS network (CS domain) is the most appropriate to provide real-time services. Since the PSD infrastructure in UMTS (PS domain) is based on IP protocol and as such, it may generate delays caused by buffering and processing delays in routers, buffering delays at network edges and transmission delays. It can be clearly seen that non-real-time services are more suited to be provided via the PSD UMTS infrastructure.

			Real-time (constant delay)	Non-real-time (variable delay)
Operating environment	Speed	Data rate	BER/max transfer delay	BER/max transfer delay
Rural (outdoor)	500 km/h	144 kbps	BER: 10^{-3}-10^{-7} Delay: 20-300 ms	BER: 10^{-5}-10^{-8} Delay: $\geq$ 150 ms
Urban/suburban (outdoor)	< 120 km/h	384 kbps	BER: 10^{-3}-10^{-7} Delay: 20-300 ms	BER: 10^{-5}-10^{-8} Delay: $\geq$ 150 ms
Indoor/low range outdoor	< 10 km/h	2,048 kbps	BER: 10^{-3}-10^{-7} Delay: 20-300 ms	BER: 10^{-5}-10^{-8} Delay: $\geq$ 150 ms

Table 3.3. *QoS network requirements for real-time and non-real-time UMTS bearers*

3.3.2.4. *QoS mechanisms for negotiating and setting up UMTS bearer services*

The UE, the UTRAN and the core network are provided with QoS management functions intended to control the establishment and the modification of UMTS bearer services. In the example shown in Figure 3.7, a certain application in the UE detects the need to establish a service either on the CS or PS domain and makes the request by specifying the QoS profile (i.e. delay, service priority, bit-error rate, average data rate, etc.). The corresponding network domain (CS or PS) performs admission control based on the QoS profile requested and on the availability of resources. As in GSM/GPRS, the actual algorithms used for admission control in UMTS are not specified. The UMTS bearer service manager function in the core network translates UMTS QoS attributes into RAB, Iu and CN bearer attributes, and requests the respective managers to provide the corresponding bearer services. The CN may also consult the subscription control database. Each element involved in the network may consult the admission control entity before providing the bearer service. If the overall network resources are available and if the user's subscription is in order, the core network positively acknowledges the UE request, which implies that network agrees to provide the QoS requested in the QoS profile.

An important feature in UMTS standards relies on the possibility for the UE or the core network to dynamically modify the QoS profile for each UMTS radio bearer. This makes it possible to renegotiate QoS attributes at any time as a function of the network resources or the needs of the user's application within the UE.

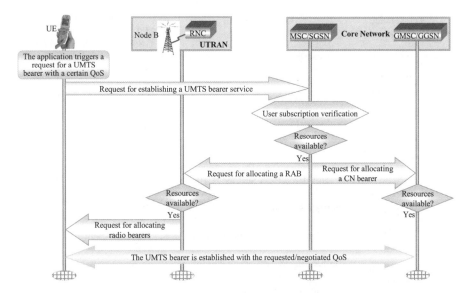

Figure 3.7. *Simplified QoS mechanism for setting up UMTS bearer services*

3.3.3. *Teleservices*

Teleservices can be seen as standardized applications on top of the UMTS bearer services that specify the functions of the terminal equipment. Table 3.4 lists the teleservices specified for UMTS.

3.3.3.1. *Speech transmission*

Speech is likely to be (at least at the very start) the most used UMTS teleservice. UMTS standards prescribe the adoption of AMR (*Adaptive Multi-Rate*) [TS 26.071] as default speech codec in UMTS in order to provide speech service across the UTRAN and GSM radio access networks [TS 22.100] (see Appendix 1 for more details on AMR speech codec).

Teleservice category	Designation in [TS 22.003]	Support in		
		GSM	**GPRS**	**UMTS**
Speech transmission	Telephony	Yes	No	Yes
	Emergency calls	Yes	No	Yes
Short message service	Short message MT/PP	Yes	Yes	Yes
	Short message MO/PP	Yes	Yes	Yes
	Cell Broadcast Service (CBS)	Yes	Yes	Yes
Facsimile transmission	Alternate speech and facsimile group 3	Yes	No	No
	Automatic facsimile group 3	Yes	No	No
Voice group service	Voice Group Call Service (VGCS)	Yes	No	No
	Voice Broadcast Service (VBS)	Yes	No	No

Table 3.4. *Teleservices specified in GSM, GPRS and UMTS*

3.3.3.2. *Emergency calls*

UMTS offers, as in GSM, the possibility to pass phone emergency calls from a UE containing or not a valid USIM. In Europe, the code for emergency calls is commonly "112" whereas in the USA such a code is "911".

3.3.3.3. *Short message service*

The amazing success of SMS in GSM/GPRS networks will be preserved in UMTS. SMS makes it possible to send a short message from a UE to the *SMS Service Centre* (SMS-SC), where this is stored and then forwarded to the final destination. This variant is referred to as SMS-MO/PP (*Mobile Originated/Point-to-Point*), to be compared with variant SMS-MT/PP (*Mobile Terminated/Point-to-Point*) where it is the UE which receives a short message forwarded by the SMS-SC. The size of the message is limited to 160 characters.

Another way to send short messages is based on point-to-multipoint approach. This service, termed CBS, enables the operator to broadcast short messages to all UEs in the area covered by a base-station. They are not acknowledged by the UEs. Each message is associated with an identifier which gives the opportunity to the UEs to get rid of unwanted broadcast messages or to identify those that have already been received [TS 22.003]. The maximum length of each cell broadcast message is 93 characters.

In the evolution path of SMS, there is also the *Enhanced Messaging Service* (EMS), where white and black images or sounds are associated to the short messages [TS 23.040, TS 22.140]. EMS relies on the concatenation of as much as 255 short messages to create a single message stream. A more important step in the enhancement of SMS is the introduction of the *Multimedia Messaging Service* (MMS). This non-real-time service ("store and forward" mechanism) is already available in GPRS networks and will be preserved in UMTS [TS 22.140] – this is a bearer independent service. MMS enables users to send and receive messages including text (ASCII, UCS2, etc.), audio (MP3, MIDI, WAV, etc.), still-images (JPEG, GIF, PNG, etc.), video (H.263, MPEG4), and optionally, streaming. It uses IP data path and IP protocols (WAP, HTML, HTTP, etc.) and supports different addressing modes (MSISDN, Email, etc.). Figure 3.8 shows a possible architecture to provide MMS services in a UMTS network.

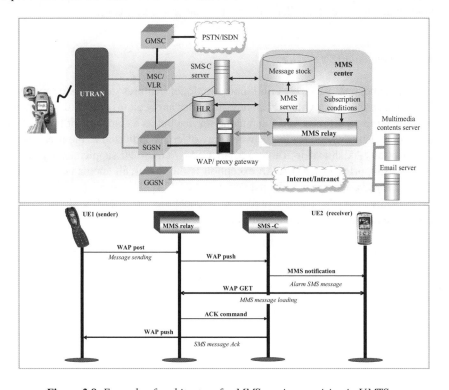

Figure 3.8. *Example of architecture for MMS service provision in UMTS*

3.3.3.4. *Facsimile transmission*

Supporting facsimile service is not required in the UMTS standards of *Release 99* [TS 22.100]. Although GSM teleservices "alternate speech and facsimile G3" and "automatic facsimile G3" could be implemented in a UMTS network, it is prescribed in [TR 22.945] to adopt two alternative (optional) solutions in UMTS which are supposed to provide a better performance and to facilitate network implementation. One solution suggests providing facsimile service based on the store-and-forward approach (e.g. as an email attachment), whereas the second is real-time and IP-based using a non-transparent UMTS bearer service.

3.3.3.5. *Voice group service*

The following voice group services are not required to be supported in UMTS networks based on *Release 99*: *Voice Group Call Service* (VGCS) that enables speech transmissions in half duplex mode; the group call that can be limited to a specific geographical area and *Voice Broadcast Service* (VBS) that supports a single voice transmitted to all available device (e.g. an emergency call).

3.3.4. Supplementary services

Supplementary services bring out additional features to the aforementioned teleservices, including video-telephony and other applications [TS 22.004]. They are not offered to a customer as a stand alone service but as a complement to a service – the supplementary services the network may offer are indicated in the user's service subscription. As shown in Table 3.5, all GSM *Release 99* supplementary services are also supported in UMTS. An exception is the multi-call supplementary service that is only available in UMTS networks [TS 22.135]. This service enables users to dynamically control parallel network connections in the CS domain – each connection using its own dedicated bearer.

Supplementary service	Supported in		
	GSM	**GPRS**	**UMTS**
Call Deflection (CD)	Yes	No	Yes
Calling Line Identification Presentation (CLIP)	Yes	No	Yes
Calling Line Identification Restriction (CLIR)			
Connected Line Identification Presentation (CoLP)			
Connected Line Identification Restriction (CoLR)			
Call Forwarding (CF) Unconditional (CFU)	Yes	No	Yes
CF on Mobile Subscriber Busy (CFB)			
CF on No Reply (CFNRy)			
CF on Mobile Subscriber not Reachable (CFNRc)			
Calling Name Presentation (CNAP)	Yes	No	Yes
Call Waiting (CW)	Yes	No	Yes
Call Hold (HOLD)			
Multi Party Service (MPTY)	Yes	No	Yes
Closed User Group (CUG)	Yes	No	Yes
Multiple Subscriber Profile (MSP)	Yes	No	Yes
Advice of Charge (Information) (AoCI)	Yes	No	Yes
Advice of Charge (Charging) (AoCC)			
User-to-User Signalling (UUS)	Yes	No	Yes
USSD/MO and USSD/MT	Yes	No	Yes
Barring of All Outgoing Calls (BAOC)	Yes	No	Yes
Barring of Outgoing International Calls (BOIC)			
Barring of Outgoing International Calls except those directed to the Home PLMN Country (BOIC-exHC)			
Barring of All Incoming Calls (BAIC)			
Barring of Incoming Calls when Roaming Outside the Home PLMN Country (BIC-Roam)			
Explicit Call Transfer (ECT)	Yes	No	Yes
Enhanced Multi-Level Precedence and Pre-emption (EMLPP)	Yes	No	Yes
Completion of Calls to Busy Subscribers (CCBS)	Yes	No	Yes
UMTS specific: Multicall (MC)	No	No	Yes

Table 3.5. *UMTS supplementary services*

3.3.5. *Operator specific services: service capabilities*

For those operators aiming at a faster creation and deployment of new services and service differentiation, the UMTS standards have specified *service capabilities*. As depicted in Figure 3.9, they comprise: bearer services; the mechanisms to maintain secure service provision across other networks; the storage of the user's service profile and billing. Apart from the bearers, new services and applications can be created by using tools and interfaces within a standardized framework [TS 22.101]. These tools were introduced in GSM Phase 2+ and concern: MExE, LCS, USAT/SAT and CAMEL.

Following a service request, the UMTS network offers the mechanisms required to establish, modify and release the associated connection as well as the bearer services with the QoS specified by the application. Mobility management procedures are also activated in order to guarantee service provision, even if the UE subscriber is moving or roaming in a visited network.

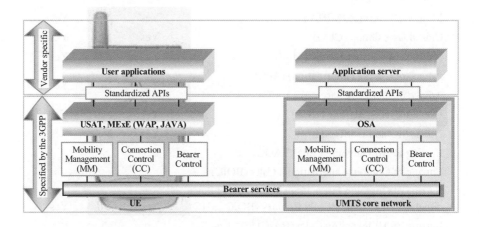

Figure 3.9. *Network and UE open architectures for service development*

MExE (*Mobile Station Application Execution Environment*) is more of a standardized framework describing the requirements of runtime environments (based either on WAP or Java) to support services than an environment itself. It has the ability to tell to a MExE service provider (not necessarily the operator), which are the current capabilities of the terminal – capabilities that can be negotiated if needed. This is achieved through a MexE classmark that identifies a category of MExE UEs supporting MExE functionality [TS 23.057]. The MExE servers can then determine the most suitable content format for the device: depending on screen and memory size, color capability, user interfaces, etc.

USAT/SAT (*USIM/SIM Application Toolkit*) is an extension of a GSM SIM application toolkit to support USIM features. USAT encompasses a number of mechanisms that enable applications, existing in the USIM, to interact and operate with any ME which supports the specific mechanisms required by the application [TS 22.038]. Such mechanisms enable for instance: the ME to tell the USIM what its capabilities are; to display text from the USIM on the ME display; to set up calls to numbers held in the USIM; to download data to the USIM; to adapt the menus in the ME display according to USIM application in use; to handle several IMSIs and to select the most appropriate as a function of the available PLMNs; etc.

LCS (*LoCation Services*) is a feature introduced in GSM *Release 98* and then enhanced and adapted in *Release 99* for UMTS. The idea behind LCS is to localize a UE by defining the mechanisms and network elements involved in the measurement of the radio signals, the calculation of the position and the dialog between the network and the external LCS client [TS 22.071].

CAMEL (*Customized Applications for Mobile network Enhanced Logic*) is a network feature that enables the use of operator specific services by a subscriber even when roaming outside the home PLMN [TS 22.078]. CAMEL is based on the *Intelligent Network* (IN) approach and relies on triggers within the visited network infrastructure to suspend call processing at a service event and communicate with a remote CAMEL service environment before proceeding to handle the service function. A typical example of application is in prepaid services (e.g. SMS and data services), when roaming in other networks.

3.3.6. *The virtual home environment*

The idea to define a standardized framework aiming at the development of new services based on *service capabilities* is part of the *Virtual Home Environment* (VHE) concept [TS 22.121]. VHE offers the service providers a set of components for flexible service creation, enabling them to develop services whose appearance adapts to the network and terminal capabilities.

VHE-enabling tools are MExE, USAT, and CAMEL. Together with OSA, these tools will enable the user to receive customized and personalized services, regardless of location, serving network or terminal type. They also offer to the network operators the flexibility to develop customized services across different networks (e.g. fixed, cellular or satellite networks), without requiring modifications of the underlying network infrastructure.

OSA (*Open Service Access*) specification is a collection of open network *Application Programming Interfaces* (APIs) for UMTS application provisioning defined by 3GPP [TS 23.127]. In Figure 3.9, the service applications are not necessarily controlled by the network operator, but can be created and deployed by any third party. However these third parties will need access to the UMTS core network capabilities, especially when it comes to call related services, location based services and services that charge for a certain content. OSA is meant to enable third party application development and deployment by means of open, secure and standardized access to core network capabilities, while preserving the integrity of the underlying network.

In the first UMTS network deployments it is very likely that the end users will not perceive the benefits offered by VHE. Even today, the number of services developed using MExE, USAT and CAMEL tools remains very limited, since most operators develop their own services using proprietary solutions. The success of VHE, including OSA and *service capabilities*, will depend on whether or not the operators will open the networks to open service creation.

3.4. Traffic classes of UMTS bearer services

As we previously mentioned, the traffic class, or QoS class, is a QoS attribute that indicates the type of application for which the UMTS bearer is optimized. There are four possible QoS classes: conversational, streaming, interactive and background [TS 23.107]. The classes are differentiated by how they support end-to-end delays and transfer errors (see Table 3.6).

	QoS class	Delay	Application	Data rate	Error sensitive
"First class"	Conversational	<< 1 s	Video-telephony	32-384 kbps	No
			Interactive games	< 1 kbps	Yes
"Business class"	Streaming	< 10 s	High-quality audio and video	32-128 kbps 32-384 kbps	No
			Still images	Not guaranteed	Yes
"Economic class"	Interactive	< 4 s	E-commerce	Not guaranteed	Yes
			WWW browsing	Not guaranteed	Yes
"Cargo"	Background	> 10 s	Fax	Not guaranteed	No
			Email arrival notif.	Not guaranteed	Yes

Table 3.6. *Service classification based on QoS and application examples*

3.4.1. *Conversational services*

Conversational traffic class services are mainly for real-time applications involving a two-way (symmetric) transport between human users. Examples of applications are speech services, video-telephony and interactive games. Typically, circuit-switched bearers are used in order to guarantee real-time bearer performance, i.e. low end-to-end delay. The human perception is a key parameter to assess the performance of a conversational service. Delay is inherent to any telecommunication system and the end-user may notice this. The recommended maximum delay for voice connections is around 400 ms. Speech communication is also sensitive to variations in delay, i.e. delay jitter, and as such, data buffering shall be used at network edges. This ensures that a constant stream of speech frames can be reproduced. Typical delay jitter buffers range between 50 and 100 ms.

3.4.2. *Streaming services*

Streaming traffic class service concerns the transferring of data, such that it can be processed as a steady and continuous stream. This class is appropriate to carry real-time traffic flows of non-interactive and very asymmetric nature as in server-to-user service schemes. With streaming, there is always a human destination that can start displaying the data before the entire file has been transmitted and the final destination is always a human. Examples of streaming applications are high-quality audio and video streaming, remote video-surveillance, radio and TV programs sent through Internet, file and still-images transfer, etc. RTSP (*Real-Time Streaming Protocol*) is recommended in UMTS specifications to provide streaming services [TS 26.234, *R4*].

3.4.3. *Interactive services*

Interactive services imply the interaction between a human or a machine with a remote equipment. Within this scheme, the entity sending a message or a command expects a response from the destination. Web browsing is an example of interactive service where a human requests information from a server and the server replies. Telemetry, where two machines interact to poll measurement records such as temperature or traffic density is another example. Location services also belong to interactive service class.

Interactive services are non-real-time, asymmetric, require a low BER $(10^{-5} - 10^{-8})$ and the data rate cannot be guaranteed (*best-effort*). Low round-trip delay is a key parameter to assess the performance of this class of services. For instance, in a web browsing application it is proposed that an HTML page appears on the UE's screen in 2-4 seconds after it has been requested. Such a delay also applies when a user retrieves its mail from a local mail server.

3.4.4. *Background services*

Background services are *almost* insensitive to the delivery time. Examples of services are: background download of emails, SMS, Fax reception, background file downloading, reception of measurement records, etc. Compared to interactive services, the destination is not expecting the data within a certain time. Instead, the key parameter characterizing the QoS of background services is the bit-error rate that shall in most applications remain very low.

Figure 3.10 illustrates the principle of the service classes described above.

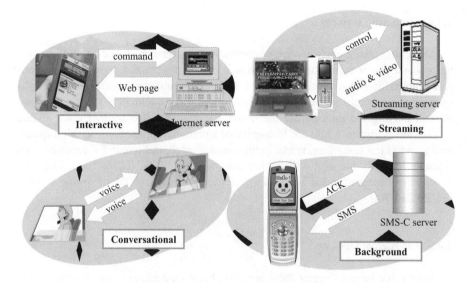

Figure 3.10. *Illustration of traffic classes in UMTS bearer services*

3.5. Service continuity across GSM and UMTS networks

3GPP specifications recommend UMTS network operators to enable continuity of service to a UE as it moves between the cells associated with GSM and UMTS radio access technologies. This functionality called "handover" requires the terminal to be Type 2 as well as the network to have appropriate resources interconnection to support the feature. Let us note that the GSM and UMTS networks involved may be either owned by the same operator or belong to different operators.

The service scenarios concern continuity of active CS, PS and multi-bearer CS/PS services when moving from UMTS to GSM radio networks and vice versa. Some examples are given in Table 3.7. Given the radical differences between UMTS and GSM radio technologies, it is impossible to guarantee service continuity for all scenarios. The 3GPP standardization philosophy is not to impose to the operators the requirements for preserving service continuity in each possible scenario. Instead, some recommendations are prescribed [TS 22.129]. For instance, it is recommended that at least active voice calls and emergency services are maintained from/to UMTS to/from GSM radio network handovers. With respect to CS data services, standards prescribe (i.e. "do your best") to maintain an active CS data call across GSM $\leftrightarrow$ UMTS handovers. Nevertheless, it is clear that a video-call, using a 64 kbps CS data bearer, cannot be preserved when handing over from UMTS to GSM, since the data rate in GSM radio interface is limited to 9.6 kbps.

As far as PS data services are concerned, there are scenarios where multiple PDP contexts (see Chapter 4) may be active at the same time either in UMTS or GPRS networks. Each PDP context represents a connection to an external network (Internet, Intranet/VPN) and may use a PS data bearer with a specific QoS. In a handover scenario, it is possible to renegotiate the QoS under the initiative of the network. In the case of unsuccessful QoS renegotiation, the connection may be terminated for that active PDP context. In the case of multi-bearer services involving active CS and PS data bearers, the handover operation seems to be much more complex and it is up to each operator to decide which service shall be maintained, which shall be terminated and for which of them the QoS can be renegotiated.

Active service before the handover	Requirement for inter-system handover	
	GSM/GPRS → UMTS	UMTS → GSM/GPRS
Voice service	Yes	Yes
Supplementary services	Yes	Yes
SMS, CBS, and fax	No	No
CS data service	Yes	Yes, for CS bearers with data rate ≤ 9.6 kbps
PS data service	Yes	Yes, provided that a possible QoS renegotiation is successful
Multi-bearer CS/PS services	Yes, for GPRS *class A* MSs	Yes, for GPRS *class A* MSs and provided that a possible QoS renegotiation is successful

Table 3.7. *3GPP requirements for selected inter-system handover scenarios*

Chapter 4

UMTS Core Network

4.1. Introduction

When a UMTS subscriber originates (or terminates) a call to access circuit- or packet-switched services, the connection is controlled by the core network. If the desired service is provided by an external network, the core network further provides interworking functionality. The core network also manages the mobility of the UE within its home network (i.e. the network where the subscription data is stored) and within a visited network in the case where the UE is roaming. Finally, the core network performs high-level security functions such as location updating and authentication, and controls charging and accounting aspects.

This chapter will give an overview of the UMTS core network as specified by *Release 99*, i.e. the 3GPP set of specifications used as a reference for the first UMTS network deployments. This chapter also outlines certain procedures handled by the core network which are studied in more detail in Chapter 8.

4.2. UMTS core network architecture

As shown in Figure 4.1, the basic core network architecture for UMTS can be seen as a combination of GSM network subsystem (NSS) and GPRS backbone. The overall UMTS architecture can be seen as the combination of a completely new radio network (UTRAN) in the "front end" of an "evolved" GSM Phase 2+ core network (see also Chapter 2).

The pragmatic choice taken for this architecture resides in the concern of 2G operators to make cost-effective the investments made on GSM and GPRS technologies. Furthermore, developing a completely new core network would have severely delayed the availability of third generation services. The core network of *Release 99* also strives to re-apply the service schemes, the roaming and charging mechanisms tried and tested in contemporary GSM and GPRS infrastructures.

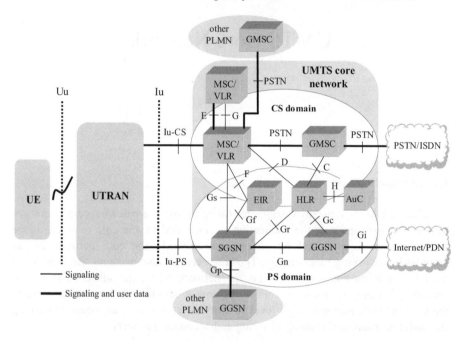

Figure 4.1. *Reference architecture of a UMTS network based on Release 99*

4.2.1. *Main features of UMTS core network based on Release 99*

UMTS standards have defined a number of requirements for the UMTS core network based on *Release 99* [TS 22.100]:

– support for circuit-switched (CS) services of at least 64 kbps per user and packet-switched (PS) data services of at least 2 Mbps. This does not prevent to supporting lower data bit rates;

– support for multiple bearers (CS and/or PS) simultaneously with different QoS attributes. During a connection, it shall be possible to add and release bearers and to modify their QoS attributes dynamically;

– support for interworking with GSM, PSTN, N-ISDN, X.25 and IP networks. Dual-mode terminals (Type 2) are required to provide seamless handover between UMTS and GSM radio access networks;

– support for operator specific services developed with the standardized tools: CAMEL, MExE, WAP and USIM/SIM toolkit and under the VHE framework.

Note that the modular architecture of a UMTS network makes it possible to interconnect its "evolved" GSM/MAP core network to other radio access architectures besides the UTRAN. Examples are the BSS of GSM, the GERAN and the HIPERLAN/2 [TS 23.121].

4.2.2. Circuit-switched and packet-switched domains

As illustrated in Figure 4.1, the core network is based on two separate domains one circuit-switched (CS) and one packet-switched (PS). The same teleservices and supplementary services as defined in GSM Phase 2+ (see Chapter 3) are supported by the UMTS network via these domains. Services with real-time constraints such as voice and video-telephony are provided by the CS domain, whereas the PS domain provides packet data services characterized by periods of alternating high and low traffic loads (bursty traffic). Examples of services offered in the PS domain by current 3G operators are streaming, web browsing, email, and SMS/MMS. Compared to the CS domain, the PS domain enables a user to be connected online to the network without being charged for the time it remains in that situation, provided that no bandwidth is used.

4.2.2.1. Simultaneous access to CS and PS services

Before having access to CS and PS services, the UE has to register with the corresponding network domain. This phase is called "IMSI attach" and "GPRS attach", when registration takes place with the CS domain and with the PS domain, respectively. If the user is authorized to access CS services a TMSI (*Temporarily Mobile Subscriber Identity*) is assigned to this user. Similarly, if authorized to access PS services, a P-TMSI (*Packet-TMSI*) is allocated. The UE registration to the CS domain and the PS domain is terminated by using, respectively, "IMSI detach" and "GPRS detach" procedures (see Chapter 8 for more details). Once registered, the UE is tracked by each domain using location management procedures. For this purpose, cells are grouped to form location areas (LA) in the CS domain and routing areas (RA) in the PS domain (see Chapter 8).

Once registered with each domain, the UMTS radio interface enables a user to have, for instance, a video-telephony conversation active in the CS domain at the same time that a web browsing session is carried out in the PS domain. Indeed, a UE can operate using one of the following configurations:

– provision of CS and PS services. The UE shall be registered to CS and PS domains so that it can access simultaneously CS and PS services. It is similar to "Class A" GPRS terminals. This is a typical configuration;

– provision of PS services. The UE is only registered to the PS domain and can only access PS services. It is similar to "Class C" GPRS terminals;

– provision of CS services. The UE is only registered to the CS domain and can only access CS services. It is similar to GSM terminals without GPRS capability.

For each successful registration to a domain, a *Mobility Management* (MM) context is activated for that domain as depicted in Figure 4.2. The context represents a logical connection with the associated domain and involves procedures such as call connection, mobility and security management. It also contains information about the temporary identifiers (TMSI or P-TMSI) and the actual area where the UE is located (LA or RA). When it is attached to both domains, two independent MM contexts are created in the UE – they are handled by the *Non-Access Stratum* (NAS) (see Chapter 2). From the UTRAN point of view, there is no a clear distinction between the two domains as far as the radio interface is concerned: CS and PS data and signaling information is conveyed via the same physical connection being under the control of the *Access Stratum* (AS). It is clear, however, that a distinction is made when the UTRAN communicates with each domain for which independent physical connections are needed.

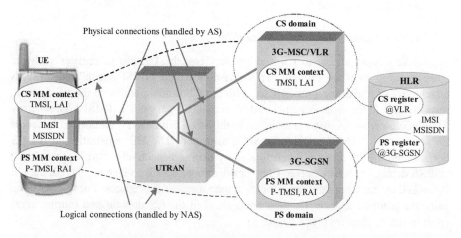

Figure 4.2. *Logical and physical interaction between the UE and the core network domains*

4.2.2.2. *Communication between domains: the Gs interface*

Procedures such as registration and location update are performed separately by the UE for CS and PS domains. Aiming at saving radio resources, combined CS/PS registration, LA/RA update and CS/PS paging can be executed via a single signaling connection towards the PS domain (see Chapter 8). This is possible due to the (optional) Gs interface. If the Gs interface is present, the UMTS network is said to be a *Network Mode I*. Otherwise, it is referred to as *Network Mode II* [TS 23.060].

4.3. Network elements and protocols of the CS and PS domains

In this section we will see that the architecture of the CS domain is no more than an evolution of the GSM NSS, whereas the PS domain is an extension of the GPRS Phase 2+ backbone.

4.3.1. *Network elements of the CS domain*

The CS domain consists of the following network elements (see Figure 4.3):

– The *Mobile-services Switching Center* (MSC) provides connection with the UTRAN and other MSCs. It manages the registration of the subscribers within the area it controls, as well as their mobility. The MSC validates the call connection requests of the UEs and allocates the necessary physical resources in combination with the UTRAN. Procedures like call routing, authentication, handovers, roaming, charging and accounting are also under the responsibility of the MSC.

– The *Visitor Location Register* (VLR) is a database that dynamically stores subscriber information when the UE is located in the LA covered by the VLR. The VLR stores MSRN (*Mobile Subscriber Roaming Number*), TMSI, the LA where the UE has been registered, supplementary services-related data, IMSI, MSISDN, HLR address, etc. The VLR is linked to one or more MSCs and may be embedded within the same MSC equipment.

– The *Gateway MSC* (GMSC) is an MSC that functions as a gateway that collects the location information to route a UE call to the MSC serving the UE at that instant. The GMSC also ensures inter-working functionality with external networks such as PSTN and N-ISDN.

– The *Home Location Register* (HLR) is a database shared by the CS and the PS domains that contains both static and dynamic information. Static information includes IMSI, MSISDN and UMTS subscription information (e.g. supplementary services). In the CS domain, dynamic information like current VLR address is used to route incoming calls towards the MSC that handled UE's registration. In the PS domain, the HLR contains the address of the serving SGSN. There is logically one

HLR per UMTS network and any modification on the subscriber data performed by the service provider is recorded in the HLR.

– The *Equipment Identity Register* (EIR) is also a database shared by CS and PS domains where a list of mobile equipments is maintained, identified by their IMEI, in order to prevent call service from stolen, unauthorized or faulty mobile terminals.

– The *Authentication Center* (AuC) is a protected database accessed by the HLR that stores a copy of the secret key contained in each subscriber's USIM, which is used for authentication, encryption and integrity procedures (see Chapter 8).

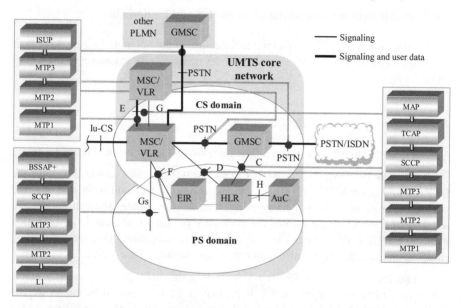

Figure 4.3. *Network elements and signaling protocols within the CS domain*

4.3.2. *Protocol architecture in the CS domain*

The protocol architecture in the CS domain is quite similar to that used in the GSM NSS. However, the fact of communicating with the UTRAN has required the definition of a new interface: the "Iu-CS" interface. This is studied in details in Chapter 6.

4.3.2.1. *Signaling System Number 7*

The internal interfaces within the CS domain are based on the *Signaling System Number 7* (SS7), which is a signaling architecture that applies out-of-band or common-channel signaling techniques – it uses a separated packet-switched network for the signaling purpose. This architecture enables the MSC to communicate with: the VLR ("B" interface), the HLR ("C" interface), other MSCs ("E" interface) and the EIR ("F" interface). There is also interface "D" between the VLR and the HLR, as well as interface "G" between VLRs.

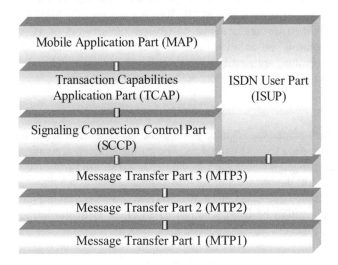

Figure 4.4. *Protocol architecture of the signaling System Number 7 in the CS domain*

In the SS7 architecture shown in Figure 4.4, the *Transaction Capabilities Application Part* (TCAP) layer performs the dialog between applications running on different nodes through a query/response scheme. The signaling transport uses the *Message Transfer Part* (MTP) layer and the *Signaling Connection Control Part* (SCCP). Error-free transport is handled by a subset of the MTP that consists of three levels, MTP1, MTP2 and MTP3, whereas logical connections are handled by a subset of the SCCP. The MTP enables a national signaling network: it realizes sequenced delivery of all SS7 message packets and provides routing, message discrimination and message distribution functions. The SCCP extends the MTP capabilities by enabling additional connectionless services as well as basic connection-oriented services. The combination SCCP and MTP provides an international signaling network for circuit-switched services.

In Figure 4.3, the interfaces requiring the establishment of a circuit to carry out user data (e.g. speech and video) implement the *ISDN User Part* (ISUP) on top of SCCP. The basic service provided by the ISUP is then the establishment and clearing of circuit-switched calls.

Mobile Application Part (MAP)

The SS7 architecture was originally defined for fixed circuit-switched networks and as such, some adaptations were needed in order to extend its use to mobile networks. The MAP layer was thus proposed for GSM and then modified to be re-used in UMTS. Let us note that MAP gives the name to the UMTS core-network: "evolved" GSM/MAP core-network. MAP uses the services of TCAP, SCCP and MTP and controls procedures such as location registration and cancellation, handover, roaming, retrieval of subscriber parameters during call set-up, authentication and security procedures [TS 29.002].

4.3.2.2. *Control plane: signaling exchange between the CS domain and the UE*

The control plane (see Chapter 2) protocol stacks used between the UE and the CS domain are depicted in Figure 4.5a. The protocol layers generating the signaling messages exchanged between the UE and the 3G-MSC are part of the NAS, whereas those involved in the transport of such messages belong to the AS. For comparison, the equivalent protocol architecture used in GSM is also shown in Figure 4.5b. Since the major innovation of UMTS compared to GSM is in the introduction of the UTRAN, the main differences between these systems relies on their transport layers. In the case of the GSM network it comprises: the *Radio Resource* protocol (RR), the *Link Access Procedures for the D channel* (LAPD), its modified version LAPDm and the *BSS Application Part* (BSSAP).

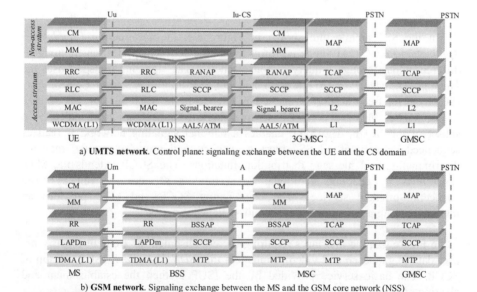

a) **UMTS network**. Control plane: signaling exchange between the UE and the CS domain

b) **GSM network**. Signaling exchange between the MS and the GSM core network (NSS)

Figure 4.5. *Control plane for CS services provision in (a) UMTS and (b) GSM*

Within the AS, the *Radio Resource Control* (RRC) and the *Radio Access Network Application Part* (RANAP) protocols replace, in some way, their homolog RR and BSSAP GSM. RRC controls radio resources and is responsible for establishing and maintaining a signaling connection between the UE and the UTRAN, whereas RANAP has a similar role by enabling communication between the UTRAN and the 3G-MSC through Iu-CS interface. As shown in Figure 4.5a, neither the LAPD nor LAPDm are supported in UMTS. Instead, reliable signaling message exchange is achieved in the radio interface due to the *Radio Link Control* (RLC) and the *Medium Access Control* (MAC) protocols, whereas RANAP relies on a broadband version of SS7 based on ATM for the same purpose but in the Iu-CS interface. The radio interface further uses the WCDMA technique in the physical layer, rather than TDMA employed in GSM. The radio protocols RRC, RLC and MAC are studied in Chapter 7, while RANAP protocol is analyzed in Chapter 6. Finally, the physical layer is largely studied in Chapters 9, 10, 11 and 12.

Within the NAS, sub-layers *Mobility Management* (MM) and *Connection Management* (CM) perform OSI layer 3 functionality. MM is responsible for all aspects concerning the user mobility as: registration (*IMSI attach* procedure), LA updating, authentication and other security functions (see Chapter 8). CM provides a point-to-point connection between two terminals and handles the call establishment, maintaining, modification and releasing. This sub-layer also controls supplementary services as well as the short message service (SMS).

4.3.2.3. *User plane: user data exchange between the CS domain and the UE*

Figure 4.6 illustrates the protocol stacks in the user plane whose major role is to materialize a *Radio Access Bearer* (RAB) with the required QoS. The lower radio layers within the AS are the same as in the control plane: RLC, MAC and the physical layer (WCDMA). The transport from/to the UTRAN to/from the 3G-MSC is realized through the Iu-CS interface, whose stack in the user plane uses ATM technology, and the *User Part* (UP) framing protocol (see Chapter 6). Within the NAS, the control protocols are replaced by the user's applications, which can be for instance, a voice or a video codec.

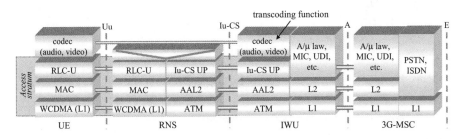

Figure 4.6. *User plane for CS service provision in UMTS*

4.3.2.4. *Example of incoming call routing in the CS domain*

The example in Figure 4.7 shows a mobile-terminated call establishment procedure with depicted utilization of involved network protocols. The case where the call is originated by the mobile is simpler and is studied in Chapter 8.

A fixed subscriber in the PSTN dials the MSISDN of a UMTS subscriber. The PSTN then establishes a signaling connection with the GMSC (1), which is the point to which a UE terminating call is initially routed, without any knowledge of the UE's location. The GMSC contacts the subscriber's HLR (2), signaling the call set-up (MSISDN contains in fact the address of the subscriber's HLR). After checking the validity of the requested number and the subscribed services, the HLR is in charge of obtaining the MSRN from the VLR (3), so that the GMSC may know the visited MSC currently serving the UE (MSC1) and forward the call to it (4). The MSRN is another temporary address number, aiming to hide identity and location of the subscriber. The MSC1 first checks the mobility context of the UE and then asks the UTRAN to start paging the LA visited by the UE (5). At this stage, the UE is identified by the TMSI assigned by the VLR. Only the UE with the correct TMSI reacts to the paging. Upon receiving the UE response, the VLR will perform authentication and some security measures and enable the MSC1 to set up a circuit connection (RAB) with the support of the UTRAN (5 and 6). The MSC also instructs the GMSC (7) so that a complete circuit path can be established between the UE and the PSTN. After this operation, the call can finally start.

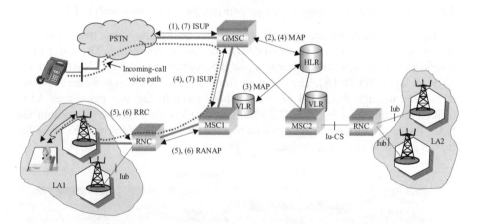

Figure 4.7. *Entities and protocols involved in the establishment of a mobile-terminated call*

4.3.2.5. *Transcoding function in the CS domain*

The *Transcoder* (TC) is a function that converts the speech from one digital format to another. Indeed, in the 3G-MSC the speech signal is transferred as a standard digital 64 kbps *Pulse Code Modulation* (PCM) signal. Over the air interface the speech signal is compressed in the range 4-13 kbps for reasons of transmission efficiency (see Appendix 1). The TC acts between these two formats. In data application the TC is disabled but a data adaptation function is still required. This is the role of the *Rate Adaptor Unit* (RAU). The TC combined with the RAU forms the TRAU. In the UMTS network, the TRAU was defined to be in the CS core network domain and logically in the NAS [TR 23.930]. This contrasts with GSM where the TRAU is located in the BSS (see Figure 4.8). In UMTS this choice is justified by the need to handle the macrodiversity (see Chapter 5). However, both in UMTS and GSM the transcoding functionality is in practice located immediately in front of the MSC. The core network architecture based on *Release 4* gives the possibility to move the transcoding function to the UMTS core network border, thus offering better transport resource efficiency (see Chapter 13).

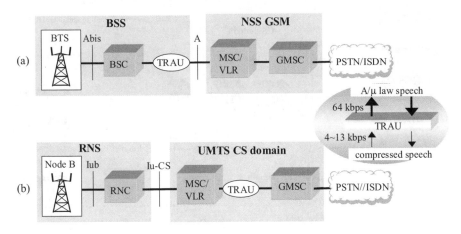

Figure 4.8. *Logical location of the TRAU in (a) GSM and (b) UMTS networks*

4.3.3. *Network elements of the PS domain*

The UMTS PS domain is an evolved version of the GPRS network. The PS domain co-exists with the CS domain and shares some common network elements, such as HLR, EIR and AuC databases (see Figure 4.9). However, in contrast with the CS domain, additional physical nodes were needed to support PS services. These are the *Serving GPRS Support Node* (SGSN) and the *Gateway GPRS Support Node* (GGSN).

The SGSN is responsible for the communication between the PS domain and all the UMTS users located within its service area (RA). Besides data transfer and routing, it handles user authentication, ciphering and integrity, charging and other mobility management functions such as attach/detach procedures of all UMTS subscribers registered within this SGSN. In the PS domain, VLR functionality is already embedded within the SGSN.

The GGSN represents the logical interface to a particular external packet data network (PDN), such as other GPRS networks (3G and 2.5G), IP and X.25 networks. The network protocol used by PDNs is referred to by the generic name of PDP. The GGSN converts user data packets coming from the SGSN into appropriate PDP format and sends them out on the corresponding external network. A similar function is performed in the other direction for incoming data packets. The GGSN can also provide dynamic allocation of network addresses (e.g. IP).

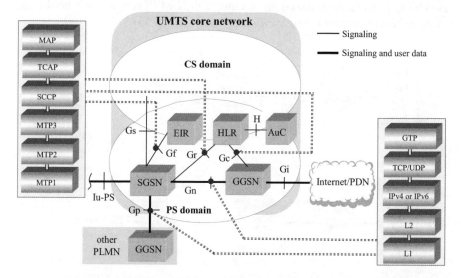

Figure 4.9. *Network elements and signaling protocols within the PS domain*

All SGSN/GGSNs are connected via an IP-based GPRS backbone network that can be intra-PLMN, if it connects the SGSN/GGSNs of the same GPRS provider, or inter-PLMN, when connecting SGSN/GGSNs of different PLMNs. In the second case, a roaming agreement between providers is necessary.

4.3.4. *Protocol architecture in the PS domain*

4.3.4.1. *Communication inside the GPRS backbone*

The names of the interfaces between the physical nodes in the PS domain identified by the "G" prefix are identical to the ones used in GPRS. However, the

Gb interface is no longer used since it has been replaced with the *Iu-PS* interface, which enables UTRAN-SGSN interaction. As in the CS domain, the MAP and SS7 architecture are used to interface the SGSN with the HLR (*Gr* interface) and the EIR (*Gf* interface). In contrast to the CS domain, protocol GTP (*GPRS Tunneling Protocol*) is used to enable communication between the SGSN and the GGSN through the *Gn* interface based on the principle of *tunneling*. A GTP tunnel is a virtual connection between the SGSN and the GGSN (or between two SGSNs and two GGSNs). Any kind of PDP (e.g. X.25 or IP) data unit is always encapsulated into IP datagrams and then tunneled through the GPRS backbone, as shown in Figure 4.10. This tunnel is identified by a *Tunnel Endpoint IDentifier* (TEID).

In the control plane, GTP-C specifies the protocols needed to create, modify and release tunnels, as well as PDP context management, location and mobility management (just as the MAP does in the CS domain for handling circuit connections). In the user plane, GTP-U provides a service for carrying user data packets. GTP is designed on top of the standard TCP/IP protocols, employing TCP (*Transport Control Protocol*) or UPD (*User Datagram Protocol*), depending on whether there is a need for reliable connection (X.25 PDN) or not (IP PDN).

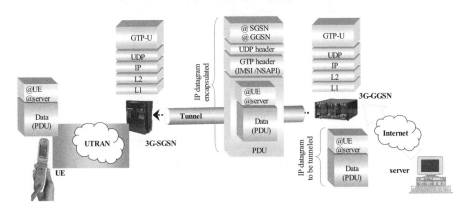

Figure 4.10. *GTP tunneling and encapsulation principle between the GGSN and the SGSN. A similar tunneling procedure is performed between the SGSN and the UTRAN*

4.3.4.2. *Control plane: signaling exchange between the PS domain and the UE*

When comparing Figure 4.5a and Figure 4.11a, there is no difference between the control plane in the CS domain and the control plane in the PS domain as far as the transport layers of the AS are concerned. With respect to NAS, the protocol layers are different. The *Session Management* (SM) layer is utilized to establish packet data sessions between the UE and the SGSN (e.g. PDP context activation, modification and deactivation), while the *GPRS Mobility Management* (GMM) layer supports mobility management procedures such as GPRS attach/detach, security and RA update. SM and GMM can respectively be considered similar to CM and MM layers in the CS domain from the functional point of view. Compared with the signaling plane in the GPRS protocol architecture (see Figure 4.11b), the *Logic Link Control* (LLC) layer that guarantees the reliable communication between the terminal and the SGSN is not supported in UMTS (part of its functionality is spread out in RRC and RANAP). This is also the case for the *Base Station System GPRS Protocol* (BSSGP) that in the GPRS network is responsible for delivering routing and QoS-specific information between the BSS and the SGSN. This employs Frame Relay as the underlying protocol. In UMTS, most BSSGP functions are covered by RANAP using ATM technology for transport purposes.

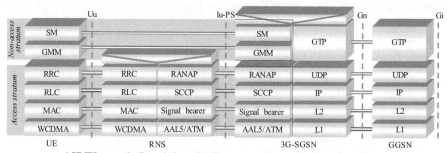

a) **UMTS network**. Control plane: signaling exchange between the UE and the PS domain

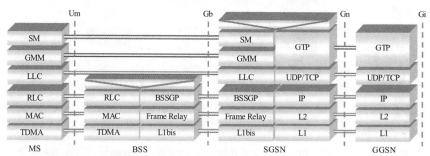

b) **GPRS network**. Signaling plane: signaling exchange between the MS and the GPRS backbone

Figure 4.11. *Control plane for PS service provision in (a) UMTS and (b) GPRS*

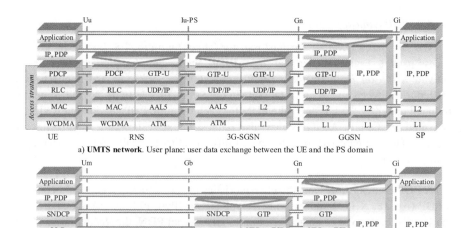

a) **UMTS network**. User plane: user data exchange between the UE and the PS domain

b) **GPRS network**. Transmission plane: user data exchange between the MS and the GPRS backbone

Figure 4.12. *User plane for PS service provision in (a) UMTS and (b) GPRS*

4.3.4.3. *User plane: user data exchange between the PS domain and the UE*

The user plane comprises the radio bearers used to transfer user data between the SGSN and the UE. The protocol stack is shown in Figure 4.12a. Transmission between the UE and the SGSN is achieved in two phases. First, the radio protocols RLC, MAC and physical layer are configured by RRC so that the radio bearers are set up between the UE and the UTRAN. On top of RLC, the *Packet Data Convergence Protocol* (PDCP) carries out the N-PDUs (*Network Packet Data Units*). Then, RANAP establishes a data bearer between the UTRAN and the SGSN based on the tunneling principle (similar to the one in SGSN-GGSN communication path). This is done by GTP-U (*GTP for the User plane*) on top of UDP/IP over ATM. Note that in the transmission (user) plane of GPRS, the transport of N-PDUs is carried out by the *Sub-Network Dependent Convergence Protocol* (SNDCP) on top of the LLC link between the terminal and the SGSN (see Figure 4.12b). SNDCP additionally performs data and header compression, while PDCP is specified to compress IP headers only.

4.3.4.4. *PDP context*

We saw that a UMTS user must register (attach) with the PS domain in order to access packet services. After that, in order to enable data transfer, the user has to activate a *Packet Data Protocol* (PDP) context. It is performed as a request-reply procedure between an UE and a GGSN via the SGSN [TS 23.060]. A PDP context is initiated either by the UE or by the PDN for each required PDP session but is always activated by the UE. A GGSN may also request the activation of a PDP context to the UE. The standard specifies the possibility of up to 15 PDP contexts existing in parallel for one user.

A PDP context represents the characteristics of the required session and contains a PDP (e.g. IP, X.121) address, a PDP type (IPv4, IPv6, PPP, X.25), the address of the GGSN that serves as an access point to an external PDN, the *Access Point Name* (APN), a quality of service (QoS) profile and other routing information including the *Network layer Service Access Point Identifier* (NSAPI). Once a PDP context is activated, the UE is visible for the associated external PDN and is able to send and receive packet data. The PDP context is stored in the UE, the SGSN and the GGSN.

The APN is the identity of the external network providing a specific service (e.g. WAP, VPN-access, "traditional" IP) that can be accessed by the UMTS subscriber. The APN consists of the network ID (mandatory), which identifies the external network accessed by the UE, with optionally the requested service, and the operator ID (optional) which identifies the PLMN enabling the access to the external network. Examples of a network ID are MYWAP.com and MYMMS.com. The format of the default operator ID is mnc<MNC>.mcc<MCC>.gprs [TS 23.003]. The list of APNs is part of the UMTS subscription data. It is stored in the HLR and is transferred to the SGSN upon activation of the PDP context.

As shown in Figure 4.13, a *secondary* PDP context may be activated. A Secondary PDP context is a PDP context with an APN/PDP address for which one or more PDP contexts have been already established. This makes it possible, for instance, to establish different flows of a multimedia communication with a specific QoS for each flow.

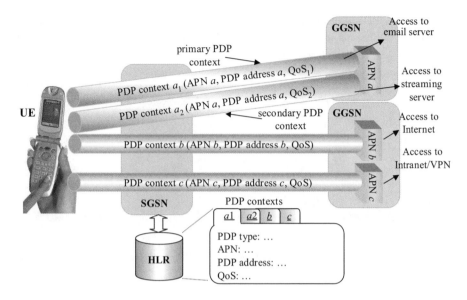

Figure 4.13. *Illustration of the concept of single and multiple PDP contexts*

PDP address allocation in the PS domain

As in GPRS, the PDP address allocation to a UE upon request of a PDP context is done either by the serving GGSN or by an external network. Depending on the network implementation, PDP address assignment to the UE can be permanent (static allocation) or different for each activated PDP context (dynamic allocation). Furthermore, the address can be public (e.g. Internet) or private (e.g. Intranet).

Relationship between NSAPI, RAB and PDP context

Successful context activation leads to the creation of two GTP tunnels, specific to the subscriber: one between the GGSN and the SGSN over the *Gn* interface and another between the SGSN and RNC over the *Iu-PS* interface. At the UE side, a PDP context is identified by an NSAPI, selected by the UE upon a PDP context activation request. The UE uses the appropriate NSAPI for subsequent data transfers in order to identify a PDN. The tunnel in the network side is identified by the TEID composed by the IMSI and the NSAPI. Once a PDP context activation request is accepted, the SGSN initiates the establishment of a RAB to communicate with the UE. The RAB is identified by a RAB ID which is in fact equivalent to that of the NSAPI. The UTRAN maintains a kind of *RAB context* to resolve the subscriber identity associated with a GTP-tunneled PDU. The UTRAN also establishes the required radio bearers (RB) associated to the RAB for which RB IDs are allocated.

For each PDP context activated, it is up to the SGSN and the UTRAN to maintain a unique relationship between the RAB, the radio bearer(s), and the GTP tunnel identities. This is illustrated in Figure 4.14.

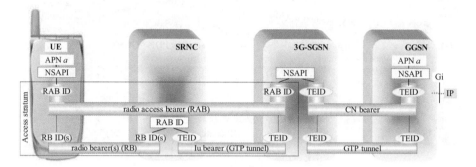

Figure 4.14. *Relationship between RB ID, RAB ID, TEID and NSAPI within a PDP context*

4.3.4.5. *Examples for PDP context activation and data packet routing*

Transparent access to Internet via WAP

In this example, the UE requests the activation of a PDP context to the SGSN with the following parameter values:

⇒ PDP type: *IP*; APN: *MYWAP.com*; PDP address: *none provided*

The SGSN receives the UE request and compares the received values with its local subscriber values. If the subscription comprises access to WAP services, it checks whether the APN=*MYWAP.com* is a legitimate APN. The SGSN then sends a query to the *Domain Name Server* (DNS) to obtain the IP address of the GGSN using the complete APN. With this address, the SGSN contacts the GGSN, which provides a dynamic public IP address to the UE from the pool of addresses owned by the operator. Let us note that this is a transparent access to the Internet, since the GGSN does not participate in user authentication; authentication is only made internally within the UMTS network using NAS procedures – the communication path between the PS domain and the external network is insecure. The transparent access mode is typically used when the UMTS operator already is an *Internet Service Provider* (ISP).

If the UE and the external network are required to communicate securely, an application such as *IP Security* (IPSec) may be used at the UE side and the remote switch of the destination network so that an end-to-end encrypted tunnel can be created (see Figure 4.15a). This is generally needed when accessing a *Virtual*

Private Network (VPN). Another possibility to provide an end-to-end secure communication is by implementing a non-transparent access mode.

Non-transparent access to a VPN

In this example, the UE requests the activation of a PDP context towards a VPN:

⇒ PDP type: *IP*; APN: *MYVPN.com*; PDP address: *none provided*

If a non-transparent access scheme is implemented, the GGSN can perform authentication and obtain a private static or dynamic IP address belonging to the VPN (or Intranet) for the UE. For this purpose, the GGSN has to send a query to a RADIUS or DHCP server. Compared to the transparent mode, the authentication and authorization are performed by the GGSN on the RADIUS server belonging to the destination network. Tunneling protocols such as IPSec or *L2 Tunneling Protocol* (L2TP) are used between the GGSN and the VPN to transmit user traffic to a final destination point (see Figure 4.15b).

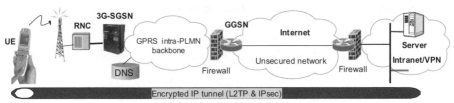

a) Transparent access to a private network which performs authentication and authorization for the user

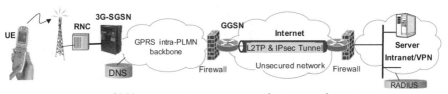

b) Non-transparent access to a private network.
Authentication and authorization are performed by the GGSN

Figure 4.15. *Principle of transparent and non-transparent access to a private network*

Mobile originated packets transmitted to an Internet Web server

In the example shown in Figure 4.16, the UE is registered with SGSN1 and has established a PDP context throughout GGSN1, offering access to the Internet. The mobile originated packets are transmitted to SGSN1, which encapsulates the IP packets, checks the PDP context and forwards them to GGSN1 through the established GTP tunnel within the intra-PLMN backbone. The GGSN1 decapsulates the packets and then sends them to the IP network in charge of their final routing towards the Web server.

Mobile terminated packets transmitted from an Internet Web server in case of roaming

Let us assume that the UE roamed from its home PLMN (PLMN1) to PLMN2. Let us suppose also that the Web server attached to the external IP network wants to communicate with the UE. It sends the packets to the IP network in charge of routing them to the GGSN1. Note that the addresses of the IP packets will have the same subnet prefix as the address of the GGSN1 (located in the home PLMN of the UE). GGSN1 asks the HLR to obtain the information on where the UE is located, i.e. to find out which SGSN the UE is registered to. When informed that this is SGSN2, GGSN1 encapsulates the IP packets and then tunnels them throughout the inter-PLMN backbone. SGSN2 decapsulates the GTP-encapsulated IP packets and delivers them to the UTRAN. If the UE is in an idle state, i.e. monitoring paging indications of arriving packets, it is paged in the appropriate RA and it responds to the page. After some AS and NAS procedures, the UE can finally receive the packets from the external Web server (see Figure 4.16).

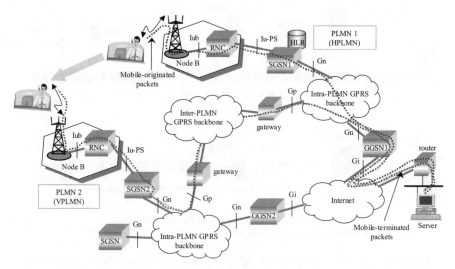

Figure 4.16. *Routing principle for mobile originated/terminated packets in the PS domain*

4.3.5. *Integrated UMTS core network*

In UMTS, the integration of the network elements of the CS and the PS domains within a single equipment referred to as UMTS MSC (UMSC) is optional. It may enable the reduction of the initial cost on network elements and optimize operation and maintenance tasks. Since both CS and PS domains may share the same physical Iu interface, additional investment savings are possible. This may also facilitate and optimize the coordination between the two domains for certain procedures without the need of a physical *Gs* interface.

4.4. Network elements not included in UMTS reference architecture

Figure 4.1 showed the elements of the UMTS core network used as the reference architecture for supporting CS and PS services. There are however other network entities that can be integrated to this basic architecture in order to provided additional teleservices and/or applications [TS 23.002]. They are optional and their implementation shall not affect the functioning of the elements within the reference core network architecture. Examples of these entities are:

– the *InterWorking Function* (IWF) typically located in the MSC which provides a data bridge between the CS domain and fixed networks such as ISDN, PSTN or packet-data networks. Its functionality comprises signaling control and protocol conversion [TS 29.004];

– the *SMS service centre* (SC), which is in charge of SMS message delivery and is common to CS and PS domains. In the case of mobile-terminated SMS messages, an SMS *Gateway MSC* (SMS-GMSC) is required to enable the communication with the serving MSC or SGSN. Similarly, an SMS-*interworking* MSC function is needed to route SMS messages from the core network to the SC;

– the *Cell Broadcast Centre* (CBC), which is in charge to deliver broadcast short messages within a given service area (see Chapter 3). The CBC is part of the core network and communicates with the UTRAN using the "Iu-BC" interface [TS 23.041];

– the entities for *location service* (LCS) provision, which enable the system to determine the geographical position of a UMTS subscriber;

– the entities related to *CAMEL*, which propose a set of functions and procedures based on the *Intelligent Network* concept that make operator-specific services available to UMTS (and GSM) subscribers who roam outside their home network (e.g. prepaid roaming services) [TS 22.078]. The key entity within CAMEL architecture is the *GSM Control Function* (gsmSCF);

Figure 4.17 gives a more complete view of the network elements within the UMTS core network architecture. In the figure is also represented the *Charging Gateway Function* (CGF) that in the PS domain collects and validates recorded traffic data information. The results are formatted and then sent for processing to the network billing system.

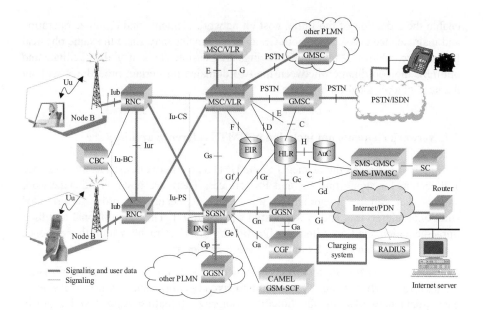

Figure 4.17. *Global view of UMTS network architecture including optional entities*

4.5. Interoperability between UMTS and GSM core networks

With UMTS, new network operators entered in the European mobile telecommunications panorama (e.g. in the UK and Italy). These operators have only UMTS licenses and do not possess GSM/GPRS networks. They have been the first to offer UMTS services and are ahead of "traditional" operators in terms of UMTS coverage in the countries they are installed.

For traditional operators who already have GSM/GPRS networks there are two basic choices for the UMTS launch: a common CN solution or an independent CN solution. As shown in Figure 4.18, the common CN solution reuses the current GSM/GPRS core network with the appropriate upgrades. The same MSC and SGSN/GGSN elements are reused for both GSM and UMTS radio access networks. The common CN approach facilitates operation and maintenance (O&M) tasks as well as intersystem handover (UMTS ↔ GSM). However, the network elements need to be re-dimensioned to take into account the new UMTS subscribers. This is in fact a risky solution for the launching phase of the UMTS.

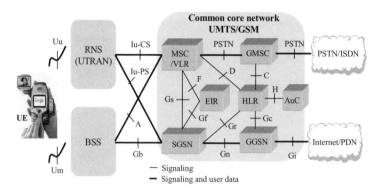

Figure 4.18. *Architecture of a common CN approach UMTS/GSM*

The independent CN solution uses different MSCs and SGSN/GGSNs to support the UMTS radio access network (see Figure 4.19). Within this solution, operators can test and validate UMTS services in relative isolation from the live, revenue-earning GSM/GPRS network. UMTS and GSM/GPRS core networks share only HLR, EIR and AuC databases. The O&M tasks are less cost-effective in an independent CN architecture and the implementation of procedures like intersystem handover become more complex. A possible UMTS roll-out may consist of launching UMTS based on independent CN solution on limited areas. Then, deploy UMTS nationwide and integrate it with GSM by using a common CN architecture.

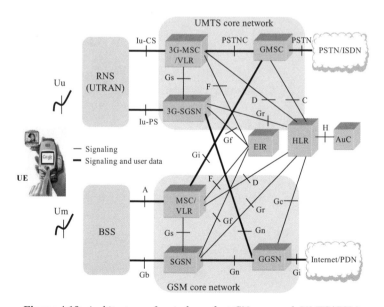

Figure 4.19. *Architecture of an independent CN approach UMTS/GSM*

Chapter 5

Spread Spectrum and WCDMA

5.1. Introduction

In the 1980s, the co-founder of the American firm Qualcomm, Andrew Viterbi, revealed that the capacity of a CDMA (*Code Division Multiple Access*) system was up to 20 times greater than that obtained in analog mobile communication systems. This announcement surprised the scientific community as well as the telecommunication manufacturers: they believed that a new and revolutionary age in the mobile communication arena had just begun. At first they were disappointed since in theory that was proved to be true but in real-life networks the performance of CDMA was quite limited – the first commercial mobile communications network based on CDMA was deployed in 1995, i.e. five years after the first deployment of GSM in Europe.

Despite its late entrance in the mobile communications scene, CDMA technology seduced numerous operators and manufacturers. Supported by the USA, Europe and Japan, CDMA was adopted by most 3G systems in the IMT-2000 framework, including UTRA technology. This chapter studies the key ideas behind CDMA as well as the technical background of the spreading techniques for use in direct sequence CDMA cellular networks.

5.2. Spread spectrum principles

CDMA can be seen as an application of *spread spectrum*, enabling multiple access communication within the same frequency band – the users are divided by different spreading codes. Spread spectrum is defined as a transmission technique in which a code sequence, independent of the information data, is employed as a modulation waveform to "spread" the signal energy over a bandwidth much greater than the signal information. At the receiver, the signal is "despread" using synchronized replica of the code sequence. Since the 1940s spread spectrum systems were developed for military use including *antijam* and *low probability of intercept* applications [SIM 94]. Commercial applications started from the 1980s when the American government expressed a desire to extend spread spectrum technology outside of the military-only realm. It followed an amazing race aiming at the development of applications for wireless communication systems. This led to CDMA technology and related standards (see Table 5.1).

1942	H. Markey and G. Antheil propose patent 2,292,387 called *Secret Communications System* outlining a spread spectrum military application.
1948/1949	The theoretical principles of spread spectrum are formalized by C.E. Shannon in the articles *Mathematical Theory of Communication* and *Communications in the Presence of Noise*.
1950s	Development of the first electronic (military) system based on spread spectrum by the American firm Sylvania Electronic Systems Division.
1956	R. Price and P.E. Green submit patent 2,982,853 called *anti-multipath receiving system* giving the fundamentals of the Rake receiver.
1978	G.R. Cooper and R.W. Nettleton suggest the possibility of applying the spread spectrum principles to cellular commutations systems.
1980s	The American government allows the usage of spread spectrum for non-military applications – this is the beginning of the first commercial applications.
1989	American firm Qualcomm makes a first demonstration in San Diego of a mobile telecommunications system based on CDMA.
1993	The results obtained by Qualcomm are used to come up with the standard IS-95A, also known as "cdmaOne".
1995	Commercial launch of cdmaOne in Hong Kong.
1999	Four out of six radio access technologies in the IMT-2000 networks are based on CDMA: UTRA/FDD, UTRA/TDD, cdma2000 and TD-SCDMA.

Table 5.1. *Key milestones in the evolution of spread spectrum towards CDMA*

5.2.1. *Processing gain*

The main importance parameter in the discussion of spread spectrum systems is the so-call processing gain, defined as the ratio of the transmission bandwidth to the information bandwidth. By denoting with B_{inf} the bandwidth of the information signal spread over a much larger bandwidth B_{spr}, the processing gain G_p is defined as:

$$G_p = \frac{B_{spr}}{B_{inf}}$$ [5.1]

The processing gain determines the number of users that can be allowed in a system, the amount of multipath effect reduction, the difficulty to jam or detect the signal, etc. – in CDMA based systems it is advantageous to have a processing gain as high as possible. The principle of spread spectrum is illustrated in Figure 5.1. In this figure S denotes the average modulating signal power satisfying $S = A_0 B_{inf} = A_1 B_{spr}$, where A_0 and A_1 represent the spectral density of the signal before and after the spreading process, respectively. It can easily be shown that $A_0/A_1 = B_{spr}/B_{inf} = G_p$.

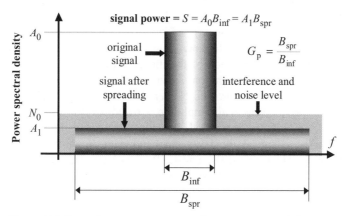

Figure 5.1. *Spread spectrum concept and processing gain definition*

5.2.2. *Advantages of spread spectrum*

The interest of spread spectrum is better appreciated when considering the famous Shannon's information-rate theorem:

$$C = B_{RF} \log_2\left(1 + \left(\frac{S}{N}\right)_{IN}\right)$$ [5.2]

where C is the channel capacity in bits per second (bps); B_{RF} is the RF bandwidth of the transmitted signal; S is the signal power; N is the noise power; and $\log_2$ is the 2-base logarithm. By developing in series equation [5.1], it can easily be shown that:

$$C \approx \frac{B_{RF}}{\ln(2)}\left(\frac{S}{N}\right)_{IN} \Rightarrow \left(\frac{S}{N}\right)_{IN} \approx \frac{\ln(2).C}{B_{RF}} \qquad [5.3]$$

In equation [5.2] is appreciated the inverse relationship between SNR and the bandwidth: if B_{RF} is increased, a lower SNR is needed for a fixed channel capacity C.

5.3. Direct sequence CDMA

The technique used in cdmaOne, cdma2000, UTRA/FDD and UTRA/TDD to perform spread spectrum is known as *Direct sequence* (DS) – for this reason these systems are known as *DS-CDMA* systems.

In DS-CDMA systems, bandwidth expansion is achieved via symbol-rate extension, as shown in Figure 5.2a. Each $d_k^{(n)}$ data symbol of user k is directly multiplied by a sequence or code $c_k(p)$ of length M, where $p = 1, 2, 3\ldots M$ ($M = 4$ in this example). The codes used for spreading are unique to every user. This enables the receiver who knows the code of the intended transmitter to select the desired signal. The symbols are of duration T_s with symbol rate $B_s = 1 / T_s$. The code is independent of the binary data. The elements of the code are referred to as *chips* consisting of square pulses with amplitude $+ 1$ and $- 1$ and time duration T_c. The chip-rate $B_{spr} = 1/T_c$ is expressed in chips-per-second (cps). The effect of multiplication is to spread the baseband bandwidth B_s of $d_k^{(n)}$ to a baseband bandwidth of B_{spr}. In UTRA/FDD and UTRA/TDD the data symbol B_s is variable depending on the service (voice, video, data), whereas the chip rate is constant: $B_{spr} = 3.84$ Mcps.

In the receiver (see Figure 5.2b), the despreading operation is basically the same as the spreading operation: the received baseband signal is multiplied again by the same (synchronized) code sequence $c_k(p)$ of length M. Since the code is made of $+ 1$ and $- 1$, this operation completely removes the code from the signal and the original data signal is left provided that orthogonal codes are used.

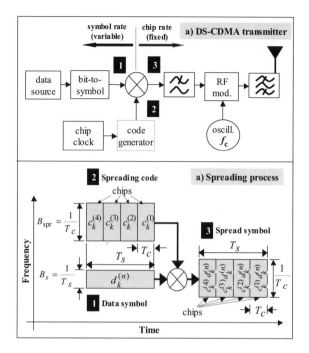

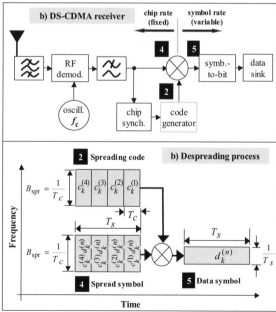

Figure 5.2. *DS-CDMA (a) transmitter and (b) receiver principle*

Spreading factor

In UTRA/FDD and UTRA/TDD the spreading operation is applied on the data signal comprising not only the user data but also the redundancy added by the channel encoder and upper layer headers. The spreading factor *SF* is defined as $SF = B_{spr}/B_s$. The definition of *SF* can be compared with that of the processing gain G_p given in expression [5.1]. Figure 5.3 illustrates the difference between these two parameters.

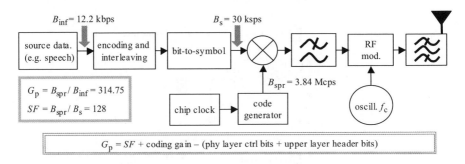

Figure 5.3. *Example showing the difference between spreading factor and processing gain*

5.4. Multiple access based on spread spectrum

A key metric in digital communications is the energy per bit per noise power density, i.e. E_b/N_0. It can easily be shown that this quantity is related to the SNR by equation:

$$\frac{E_b}{N_0} = \frac{S/B_{inf}}{N/B_{spr}} = \frac{B_{spr}}{B_{inf}}\left(\frac{S}{N}\right) \qquad [5.4]$$

If we denote the SNR before the spreading process by $(S/N)_{in}$ and replace E_b/N_0 in expression [5.4] by the SNR after the spreading process denoted $(S/N)_{out}$, we have:

$$\left(\frac{S}{N}\right)_{out} = \frac{B_{spr}}{B_{inf}}\left(\frac{S}{N}\right)_{in} = G_p\left(\frac{S}{N}\right)_{in} \Rightarrow \left(\frac{S}{N}\right)_{out,dB} = G_{p,dB} + \left(\frac{S}{N}\right)_{in,dB} \qquad [5.5]$$

EXAMPLE.– let us consider the UTRA/FDD case where the chip rate is 3.84 Mcps. Let us assume that a single user in the uplink transmits speech at a bit rate of 12.2

kbps. Finally we know that in order to achieve a *Bit-Error-Rate* (BER) of 10^{-3}, a $(S/N)_{out,dB} = 5$ dB is needed. From expression [5.5], the required $(S/N)_{in,dB}$ is:

$$\left(\frac{S}{N}\right)_{in,dB} = 5 - 10\log_{10}\left(\frac{3.84 \times 10^6}{12.2 \times 10^3}\right) = 5 - 25 = -20 \text{ dB} \qquad [5.6]$$

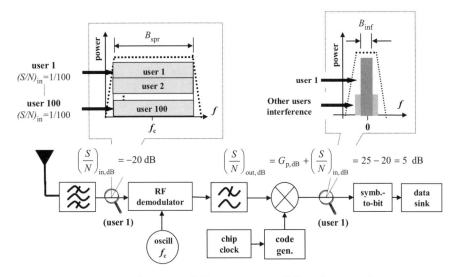

Figure 5.4. *Receiver showing multiple access principle based on spread spectrum*

From the above example it can be inferred that a $G_{p,dB} \approx 25$ dB provides the receiver with a margin of about $10^2 = 100$ times the required SNR to achieve a BER of 10^{-3}. As this will be studied in the next section, such processing gain may be used in a multiple access communication scheme enabling $10^2 \approx 100$ users to communicate with the network within the same frequency carrier and at the same time as shown in Figure 5.4. This property of spread spectrum technology is the basis of CDMA.

5.5. Maximum capacity of CDMA

The CDMA capacity depends on different factors including receiving demodulation, power-control accuracy, thermal noise and power interference generated by other users in the same cell and in the neighboring cells [GIL 91]. In what follows, the number of users is estimated in the uplink by assuming perfect power control (i.e. equal to the received power by the base station for all mobile users) and ignoring other sources of interference. With these assumptions, the SNR of one user at the base station receiver is related to the sum of powers from K individual users present in the band such that:

$$\left(\frac{S}{N}\right) = \frac{S}{(K-1)S} = \frac{1}{(K-1)} \qquad [5.7]$$

Substituting [5.7] in [5.4] yields:

$$\frac{E_b}{N_0} = \frac{S/B_{inf}}{(K-1)S/B_{spr}} = \frac{B_{spr}}{B_{inf}(K-1)} = \frac{G_p}{(K-1)} \qquad [5.8]$$

5.5.1. *Effect of background noise and interference*

The above equation does not consider background noise N, caused by spurious interference and thermal noise. Incorporating N in [5.8] yields:

$$\frac{E_b}{N_0} = \frac{S.G_p}{(K-1)S+N} = \frac{G_p}{(K-1)+N/S} \qquad [5.9]$$

Solving [5.9] for K gives the number of users that a single CDMA cell can support. This means that the cell is omnidirectional and has no neighboring cells, and the users are transmitting all the time. In practice, there are many cells in a UMTS network. Although these other users are power-controlled by their respective Node Bs, their signal powers constitute inter-cell interference. The ratio between the interference generated by the users in the serving cell and the interference from other cells satisfies $1/(1 + \beta)$. Where factor β varies from 0 to 100% ($\beta = 0$ in the single-cell case). Equation [5.9] is modified to account for the effect of interference:

$$\frac{E_b}{N_0} = \left(\frac{1}{1+\beta}\right)\left(\frac{G_p}{(K-1)+N/S}\right) \qquad [5.10]$$

This implies that the capacity K in terms of number of users is given by:

$$K = \left(\frac{1}{(1+\beta)}\right)\left(\frac{G_p}{E_b/N_0} - \left[\frac{S}{N}\right]^{-1}\right) + 1 \qquad [5.11]$$

Expression [5.11] shows that the capacity in a CDMA system increases as the processing gain G_p increases. In the case where the G_p cannot be increased, other techniques shall be implemented so that E_b/N_0 can be decreased, while keeping the target BER. Modulation and channel coding schemes are used in UTRA to reduce E_b/N_0 and increase capacity. However, beyond a particular limit, these methods reach a point of diminishing returns for increasing complexity. The other way to increase capacity is by reducing the interference. This is discussed in the following sections.

5.5.2. *Antenna sectorization*

Rather than using omnidirectional antennae, which have a pattern radiating over 360°, sectorized antennae covering three sectors so that each sector is only receiving signals over of 120° is an attractive possibility to mitigate interference from other users. Indeed, a sectorized antenna rejects interference from users that are not within its antenna pattern. The amount of interference that can be rejected is relative to the number of sectors and is quantified by γ, i.e. the *sectorization gain*.

5.5.3. *Voice activity detection*

Speech statistics show that a user in a conversation typically speaks between 40 and 60% of the time. When users assigned to a cell are not talking, *Voice Activity Detection* (VAD) algorithms can be used to reduce the total interference power by this voice activity factor denoted by α. In effect, the codec used in UMTS is variable rate (see Appendix 1) which means that the output rate of the voice codec is adjusted according to a user's actual speech pattern. And if the user is not speaking during part of the conversation, the transmitted data rate is lowered to prevent power from being transmitted unnecessarily. Finally, with VAD and sectorization, E_b/N_0 now becomes:

$$\frac{E_b}{N_0} = \left(\frac{1}{1+\beta}\right)\left(\frac{\gamma.G_p}{\alpha(K-1)+N/S}\right) \qquad [5.12]$$

The number of users per cell works out to be:

$$K = \left(\frac{1}{\alpha(1+\beta)}\right)\left(\frac{\gamma.G_p}{E_b/N_0} - \left[\frac{S}{N}\right]^{-1}\right) + 1 \qquad [5.13]$$

EXAMPLE.– determine the maximum uplink capacity in a UTRA/FDD cell where the chip rate is 3.84 Mcps. Consider a voice service at 12.2 kbps for which $E_b/N_0 = 5$ dB is needed to get a BER of 10^{-3}. Neighboring cells make the noise rise in the serving cell by up to 60%, which means that $\beta = 0.6$ and $1/(1+\beta) = 0.625$. Thermal noise is negligible ($N = 0$).

Without VAD and sectorization, the number of users for this example is from equation [5.11]:

$$K = \frac{1}{(1+\beta)}\left(\frac{G_p}{E_b/N_0}\right) + 1 = \frac{1}{(1+0.6)}\left(\frac{3.84.10^6/12.2\times10^3}{10^{0.5}}\right) + 1 \approx 63 \quad [5.14]$$

Let us now assume that VAD is used such that $\alpha = 1/2$ and that Node B covers three different sectors ($\gamma = 3$). The number of users for this example works out to be from equation [5.13]:

$$K = \frac{1}{\alpha(1+\beta)}\left(\frac{\gamma.G_p}{E_b/N_0}\right) + 1 = \frac{1}{0.5(1+0.6)}\left(\frac{3\times\left(3.84.10^6/12.2\times10^3\right)}{10^{0.5}}\right) + 1 \approx 374 \quad [5.15]$$

The result in equation [5.15] shows a 6-fold capacity increase when compared to the case without VAD and sectoring. However, it must be pointed out that this is only a gross estimate and the estimation of the real UTRAN system capacity requires much more detailed and sophisticated analysis. This analysis must take into account issues like imperfect power control, imperfect interleaving, multipath fading and non-uniform distribution of users (see, for instance, [AKH 99]).

5.6. Spreading code sequences

The fundamental problem of spread spectrum in CDMA is that each user causes *Multiple Access Interference* (MAI) affecting all the other users. In previous sections it was assumed that spreading and despreading procedures were ideal. However, in practice the correlation properties of the spreading code sequences may degrade the performance of the CDMA system by increasing the level of MAI. We are interested in codes with low cross-correlation and high auto-correlation properties. Cross-correlation refers to the comparison between two sequences from different sources, whereas auto-correlation compares the same sequence with a shifted copy of itself.

5.6.1. *Orthogonal code sequences*

Orthogonal codes are characterized by zero cross-correlation and so MAI from other users is cancelled. Orthogonal codes are usually generated from a Walsh-Hadamard matrix built by:

$$H_M = \begin{bmatrix} H_{M/2} & H_{M/2} \\ H_{M/2} & -H_{M/2} \end{bmatrix} \qquad [5.16]$$

where $-H_M$ is a square matrix that contains the same elements as H_M with the sign inverted. For example, four orthogonal codes c_1, c_2, c_3 and c_4 of length $M = 4$ are generated from the rows of the Walsh-Hadamard matrix as follows:

$$H_1 = [+1], \; H_2 = \begin{bmatrix} +1 & +1 \\ +1 & -1 \end{bmatrix}, \; H_4 = \begin{bmatrix} +1 & +1 & +1 & +1 \\ +1 & -1 & +1 & -1 \\ +1 & +1 & -1 & -1 \\ +1 & -1 & -1 & +1 \end{bmatrix} \qquad [5.17]$$

By calculating the orthogonality (i.e. multiplying element by element) of codes $c_1 = [+1 \; +1 \; +1 \; +1]$ and $c_2 = [+1 \; -1 \; +1 \; -1]$, we get: $1 \times 1 + 1 \times (-1) + 1 \times 1 + 1 \times (-1) = 0$. Hence, c_1 and c_2 are orthogonal.

5.6.1.1. *Orthogonal multiple access in a CDMA system*

In a CDMA system, each user is assigned one or many orthogonal codes, thus enabling users with different codes not to interfere with each other. This communication scheme requires synchronization among the users since the codes are orthogonal only if they are aligned in time.

EXAMPLE.– Figure 5.5 illustrates an example of a CDMA system with two users in the downlink. It is assumed that Node B transmits the symbols $d_1^{(1)} = +0.9$ and $d_1^{(2)} = -0.7$ to *user 1* and the symbols $d_2^{(1)} = -0.8$ and $d_2^{(2)} = +0.6$ to *user 2* within the same carrier and at the same time. Orthogonal codes $c_1 = [+1 \quad -1 \quad +1 \quad -1]$ and $c_2 = [+1 \quad +1 \quad -1 \quad -1]$ are allocated respectively to *user 1* and *user 2*. Note that this is a didactic example where the waveform of the information symbols is ignored as well as the modulator/demodulator impairments and the effect of the propagation channel. Figures 5.6a and 5.6b show the despreading process performed respectively by the receivers of *user 1* and *user 2*.

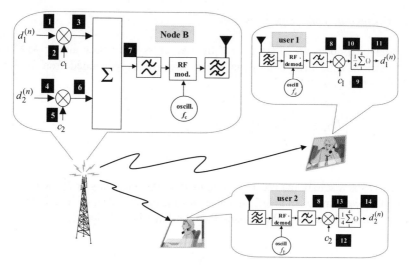

Figure 5.5. *Example of orthogonal DS-CDMA multiple access in the downlink with two users*

5.6.1.2. *Disadvantages of orthogonal codes*

The cross-correlation function of orthogonal codes varies remarkably as a function of the timeshift of the sequences. In the uplink such a timeshift is naturally produced when the users' transmission is not synchronized. In the downlink this is caused by the multipath phenomenon. Moreover, the auto-correlation function of Walsh-Hadamard codes does not have good characteristics: it can have more than one peak and therefore it is not possible for the receiver to detect the beginning of the code sequence.

5.6.2. *Pseudo-noise code sequences: Gold codes*

Compared to orthogonal sequences, pseudo-noise (PN) or *random* sequences have better auto- and cross-correlation properties. This is due to the fact that PN sequences behave like noise [VIT 95]. The family of PN sequences used in UTRA is called Gold sequences. Gold sequences (codes) have only three cross-correlation peaks, which tend to decrease as the length of the code increases [GOL 68]. As a consequence, intra- and inter-cell interference is mitigated. On the other hand, they have a single auto-correlation peak at zero, just like ordinary PN sequences. This is beneficial for the receiver synchronization: bit information detection is improved. Gold sequences are constructed from the modulo-2 addition of two maximum length *preferred* PN sequences. By shifting one of the two PN sequence, we get a different Gold sequence [GOL 68]. This property can be used to generate a large number of codes, which will enable multiple access on the network. Figure 5.7 gives an example of a Gold generator and the associated correlation function.

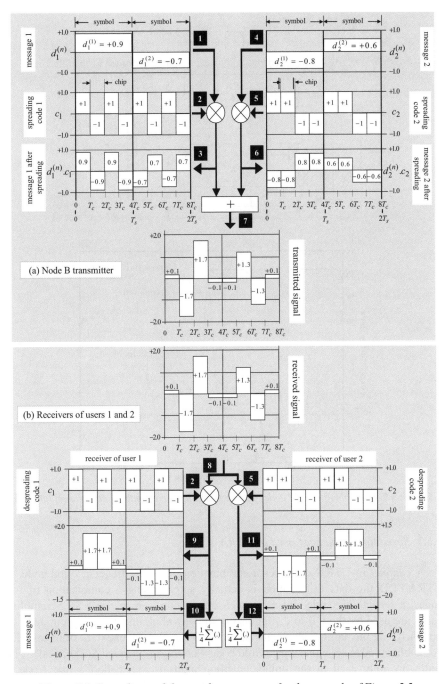

Figure 5.6. *Spreading and despreading processes for the example of Figure 5.5*

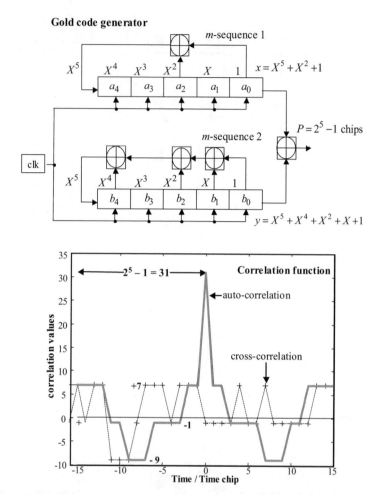

Figure 5.7. *Example of a Gold code generator and its corresponding correlation function*

5.6.3. *Spreading sequences used in UTRA*

A *channel* in UTRA describes a combination of carrier frequency and code. In UTRA/FDD the data information is actually spread using a combination of two codes: a *scrambling* code and a *channelization* code (see Figure 5.8). Scrambling codes are used by the UE to distinguish in the downlink one Node B from another. Similarly, in the uplink, they enable Node B to discern one UE from another. Scrambling codes are either *long* (Gold codes) or *short* (extended S(2) codes). Short codes are only used in the uplink enabling multi-user detection at the Node B receiver: their bit-periodic property may be exploited for MAI canceling [PAR 00].

Channelization codes are orthogonal Walsh-Hadamard codes. In downlink they are used by each UE receiver to distinguish each of its own channels from all other channels transmitted by Node B. In uplink, they enable Node B receiver to discern the physical channels of a same UE.

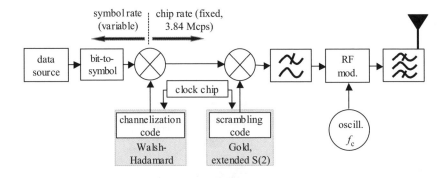

Figure 5.8. *Spreading codes used in the UTRA transmitter*

5.7. Principles of wideband code division multiple access

As a function of the bandwidth occupied by the spread signal, CDMA systems are known as *narrow band CDMA* or *wideband CDMA* (WCDMA). The 2G cdmaOne system is a narrow band radio system since bandwidth is limited to 1.25 MHz with a chip rate of 1.2288 Mcps. This contrasts with UTRA technology, also referred to as "WCDMA", where the bandwidth is 5 MHz and where the chip rate is 3.84 Mcps. Inherent advantages of WCDMA are:

– wide range of bit rates (8 kbps to 2 Mbps) with one 5 MHz carrier;

– fading is less deep than in the narrowband systems;

– large bandwidth makes it possible to use many multipath components (Rake receiver);

– improved narrowband interference rejection;

– facilitated frequency planning; reuse of the same frequency in adjacent cells;

– improved capacity due to the principle of *macrodiversity*.

The following sections discuss the aforementioned advantages of WCDMA as well as inherent drawbacks.

5.7.1. *Effects of the propagation channel*

The propagation channel is the bottleneck in any radio-communication system. The signal transmitted throughout the radio interface is affected by the radio channel impairments including: thermal noise, path loss, fading at low rates, *inter-symbol interference* (ISI) at high rates, shadowing, intra-cell interference and inter-cell interference. These phenomena are illustrated in Figure 5.9.

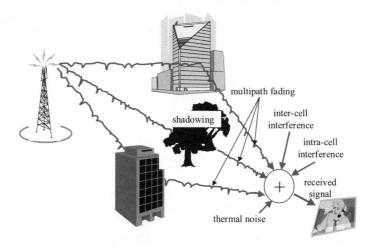

Figure 5.9. *Radio impairments affecting received signal quality*

5.7.1.1. *Path loss*

Path loss is part of the so-called large-scale fading and refers to the reduction in signal power at the receiver relative to the transmitted power. This loss is proportional to the distance d (raised to some appropriate power n) between the transmitter and the receiver and carrier frequency. There exist many models that are used to predict path loss. The difference lies in the way they consider other effects that come into play in addition to distance, i.e. on whether the signal propagates in a free or in an obstructed space [RAP 96]. For instance, in a free-space model, the signal strength diminishes inversely to the square of the distance.

5.7.1.2. *Shadowing*

Shadowing is the loss of signal strength, which is typically due to a diffracted wave emanating from an obstacle between transmitter and receiver such as walls, floors, motion of objects, etc. Shadowing is sometimes called "slow fading", since, from a mobile viewpoint, passing through a shadow region takes considerable time.

The statistics of shadowing are usually modeled with a log-normal distribution of mean signal power and a standard deviation varying from 6 to 12 dB [RAP 96].

5.7.1.3. Multipath propagation

Multipath propagation occurs when a UE is completely out of sight from Node B transmitter. As a consequence, the received signals are made up of a group of reflections from hills, buildings, vehicles and other obstacles; and none of the reflected paths is any more dominant than the other ones. The multipath creates one of the most difficult problems in the mobile radio environment: fading. In radio systems the envelope of the received carrier signal varies according to a Rayleigh distribution and therefore, this sort of fading is called *Rayleigh fading*. Figure 5.10 shows an example of the signal-power variation caused by Rayleigh fading.

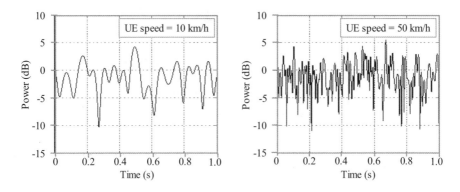

Figure 5.10. *Rayleigh fading effects on the UE received signal for different speeds*

Flat fading refers to the case where the latest copy of the signal arrives at the receiver after a time duration that is smaller than the symbol period. The combined sum of several paths that arrive from all directions at the receive antenna at the same time may add up constructively or destructively. Destructive interference can cause the received signal to fall more than 30 dB below the average. The depth, number and duration of dips are a function of *Doppler frequency*, which is proportional to mobile speed and the carrier frequency [RAP 96].

Frequency-selective fading occurs when the transmitted signal follows several paths, each arriving at the receiver antenna at different times. The result is the dispersion of the received signal in time called *multipath delay spread*. If the delay spread is significant compared to one data symbol period, we are in the case of a frequency-selective channel and ISI can occur. That means that symbols arriving significantly earlier or later than their own symbol periods can corrupt preceding symbols. In the frequency domain, the signal could see large variations in power over its bandwidth.

Figure 5.11 shows the combined effect of path loss, shadowing and Rayleigh fading within the received signal strength.

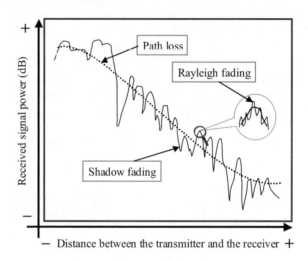

Figure 5.11. *Effect of path loss, shadowing and Rayleigh fading on the received signal*

5.7.2. Techniques used in WCDMA for propagation impairment mitigation

5.7.2.1. Multiple forms of diversity in UTRA/WCDMA

Diversity is a favored approach to mitigate fading. There are basically three major types of diversity: frequency, time and space diversity. All these approaches are utilized in UTRA by the following way.

Frequency diversity. Wideband signal itself produces frequency diversity. The wide signal bandwidth of UTRA (5 MHz) provides robustness against fading (via frequency diversity), as the channel is unlikely to fade as a whole at any given time. In the example of Figure 5.12, the strength signal fades up to 15 dB over a bandwidth of 500 kHz. Thus, only 1/10 of the signal spectrum will be affected within a 5 MHz bandwidth: the average attenuation of the signal will be 1 dB.

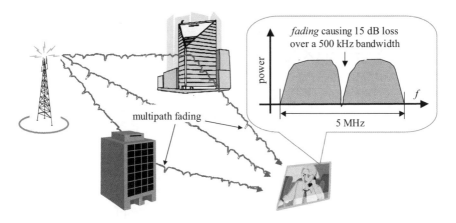

Figure 5.12. *Example showing the robustness of UTRA within a frequency selective channel*

Time diversity. Channel coding combined with interleaving produces time diversity. Channel coding aims at fighting errors resulting from fades and, at the same time, keep the signal power at a reasonable level. Most error-correcting codes perform well in correcting random errors. However, during periods of deep-fades, long streams of successive errors may render the error-correction process useless. The purpose of interleaving techniques is precisely that of randomizing the bits in a message stream so that successive errors introduced by the channel can be converted to random errors. Channel coding and interleaving schemes used in UTRA are described in Chapter 9.

Multipath diversity: the Rake receiver. Receiving different multipaths separately (fading independently) and then combining them produces multipath diversity. A special receiver, called a Rake receiver, has been devised for this purpose. The principle of the Rake receiver is shown in Figure 5.13. It consists of correlators called "fingers". The Rake locks onto the different multipath components that are separately identified as distinct echoes. After despreading by the fingers, the signals are then brought in phase and combined to yield a final composite received signal. The technique *Maximal Ratio Combining* (MRC) can be used for this purpose [PRO 95].

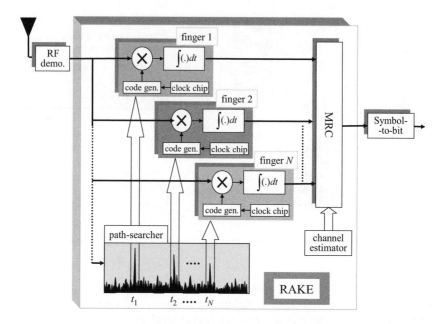

Figure 5.13. *Rake receiver principle*

The channel impulse response $h(t - t_i)$ estimates are required by the MRC approach in order to perform a coherent combination of the different paths arriving at the receiver at time t_i. Figure 5.14 shows the scatter diagram at the output of each finger in the case of a static channel with three echoes, as well as the output of the Rake receiver (sum of the finger outputs). Clearly the Rake receiver is able to add the energy from each finger, thus improving the output SNR. One important limitation of the Rake receiver that may impact its performance is that its search window is of limited size and Rake receivers can only exploit multipath components within one symbol period. Other techniques such as multi-user detection can be implemented to overcome such a limitation.

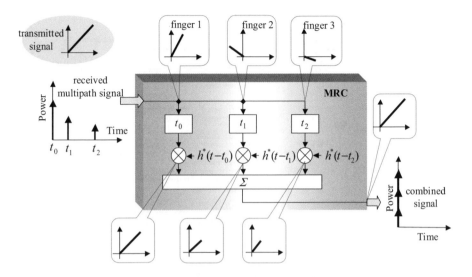

Figure 5.14. *Maximum ratio combining (MRC) operation with three multipaths*

Antenna diversity

Besides frequency, time and multipath diversity, antenna diversity adds a fourth dimension of diversity: space diversity.

Receiver antenna diversity. Two antennae that are sufficiently far apart experience uncorrelated transmission channels. This means that it is less likely that both antennae will receive a destructive fading at any given time. Moreover, having receiver antenna arrays gives the possibility to reduce co-channel interference due to antenna selectivity [TAN 00]. The principle of the Rake receiver can be exploited by implementing the MRC concept within all space diversity branches, as shown in Figure 5.15. It should be noted, however, that receiver antenna diversity is envisioned to be only in the base station; indeed, that is impractical to have two antennae in the UE separated far apart to consider that the transmission channels are uncorrelated.

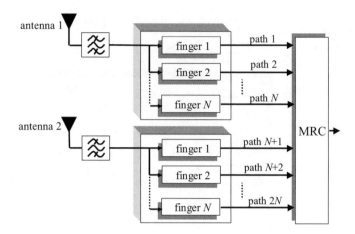

Figure 5.15. *Space diversity principle using two receiver antennae*

Transmit diversity. Transmitter (Tx) diversity techniques in the downlink are intended to avoid the need of having multiple antennae at the mobile receiver, while still having the benefit of receiver antenna diversity. Different solutions have been proposed by suggesting that multiple antennae at the base station will increase the downlink capacity with only a minor increase in terminal implementation [ROH 99]. Figure 5.16 illustrates the principle of Tx diversity with two antennae. We can see that the information follows a specific Tx diversity processing before transmission (see Chapter 10).

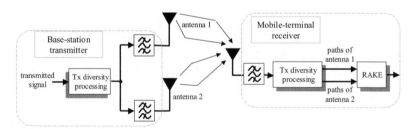

Figure 5.16. *Tx antenna principle based on two antennae*

Macrodiversity: soft and softer handover

The state where the terminal receives the same information transmitted simultaneously from different cells produces space diversity. Such a state is called macrodiversity.

Soft handover is the procedure that consists of adding and releasing cells during the macrodiversity scenario. Each cell is covered by a different Node B and all of them transmit the same data signal using a different spreading code. At the UE receiver, the multipaths of each transmitted signal are combined based on the MRC principle (see Figure 5.17). *Softer handover* is a special case of soft handover where the radio links that are added and removed belong to the same Node B (see Figure 5.18).

Macrodiversity mitigates the slow fading caused by terrain variations as well as fast fading. It is also beneficial for reducing inter-cell interference since the cells involved in macrodiversity control the power of the UE in cell-to-cell transitions.

The disadvantage of soft handover is that it creates more interference to the system in the downlink since the new Node B now transmits an additional signal for the UE. It is possible that the UE cannot catch all the energy that Node B transmits due to a limited number of Rake fingers. Thus, the gain of soft handover in the downlink depends on the gain of macrodiversity and the loss of performance due to increased interference. In the uplink, the soft-handover needs more radio resources to be allocated given extra transmission across the interface between the Node Bs and the RNC (i.e. Iub interface). Indeed, after decoding a frame, Node B sends it to the RNC that chooses Node B with the best signal quality on a frame-to-frame basis and sends the data further, as shown in Figure 5.17.

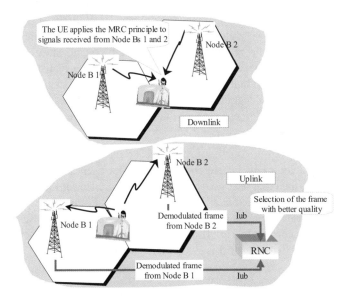

Figure 5.17. *Soft handover concept*

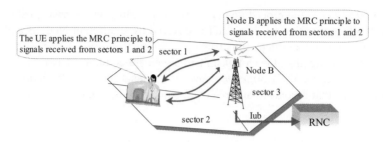

Figure 5.18. *Softer handover concept*

5.7.2.2. *Power control*

Power control is an essential feature in any CDMA radio system because it:

– reduces the interference caused by the near-far effect;

– mitigates fast fading effects;

– maintains radio link quality by keeping a target BER/BLER;

– prolongs battery autonomy thereby saving power in the UE.

Power control for the UTRA uplink is the single most important system requirement because of the *near-far effect*. In this case, the target for the power control means very large path loss dynamic range requirement of the order of 80 dB [GIL 91], because mobiles may be very close or far away from Node B. For the downlink, no power control needs to be required in a single cell system, since all signals are transmitted together and hence vary together. However, in a realistic multiple cells environment, interference from neighboring cell sites fades independently from the given cell site and thereby degrades performance. Thus, it is necessary to apply power control also in this case, in order to reduce inter-cell interference.

Near-far effect

Each user is a source of interference for the other users and if one is receiving with more power, then that user generates more interference for the other users. This phenomenon known as *near-far effect* is illustrated by Figure 5.19. Let us assume that the signal carrier strengths of *user A* (C_A) and *user B* (C_B) decrease as a function of the distance d according to $1/d^4$. If the distance of user A from the serving Node B is $d_A = 1,000$ m and that of user B is $d_B = 100$ m, we get:

$$\frac{C_B}{C_A} = \left(\frac{d_A}{d_B}\right)^4 = \left(\frac{1,000 \text{ m}}{100 \text{ m}}\right)^4 = 10^4 \ (\approx 40 \text{ dB}) \tag{5.18}$$

From the above example, it appears clearly that user B may "mask" user A from the perspective of Node B receiver. The use of power control ensures that all users arrive at about the same power at the receiver and thus no user is unfairly disadvantaged compared to the others.

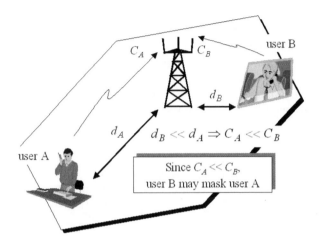

Figure 5.19. *Near-far effect in CDMA with two users transmit simultaneously in the same carrier bandwidth without power control*

Open loop power control

Open loop power control is used to provide a wide dynamic range and thereby compensate the large abrupt variations in the path loss attenuation. The initial UE transmitted signal is also defined from this process. The UE measures the received signal strength and determines an estimate of the path loss with respect to a given Node B: the smaller the received power, the larger the propagation loss and vice versa. The transmitted power is then changed accordingly.

Closed loop power control

Open loop power control can compensate for path loss and large-scale variations such as shadowing, but it cannot compensate for multipath fading because uplink and downlink are not correlated. Thus, closed loop power control is used as illustrated by Figure 5.20. The UTRA standard controls the power level of each mobile and Node B by using power control commands. On both link directions, the level of interference is measured through dedicated pilot symbols. The UE/Node B then sends a command back to Node B/UE, depending on the ratio between this estimate and the target SIR (*Signal-to-Interference Ratio*). This command is transmitted at a rate of once every 1.6 kHz. The target SIR is typically calculated as a function of the required signal quality (BER/BLER).

Errors on the power control are possible because the received power may be wrongly estimated and a closed-power command may suffer from noise, interference or multipath propagation. Also, at high speeds, the channel significantly changes within one 1/1.6 ms period and the closed loop power control cannot cope with the fast fading any more, thus leading to performance degradation after the Rake receiver. Fortunately, this loss in performance is somewhat compensated by a higher coding gain, due to a better efficiency of the interleaving scheme. The actual power control algorithms used in UTRA are studied in-depth in Chapters 10 and 12 for UTRA/FDD and UTRA/TDD modes, respectively.

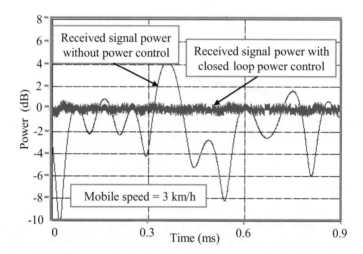

Figure 5.20. *Effect of closed loop power control to compensate for Rayleigh fading*

5.7.2.3. *Multi-user detection*

The simplest way to despread a CDMA signal is to apply the Rake principle and correlate the received signal with the known spreading sequence. The results will be good as long as the influence of other users can be neglected. In practice, however, the Rake receiver does not take into account the existence of this interference, i.e. the MAI (see also Chapter 12). The signal quality can be improved when the knowledge of the cross-correlation between the different spreading codes is taken into account – instead of assuming that the correlations are zero. This leads to a more complex class of DS-CDMA receivers, based on the multi-user detection principle also referred to as joint-detection or interference cancellation.

A number of multi-user receivers have been proposed [MOS 96, KLE 96]. One example is to multiply the received despread signal vector with the inverse of the cross-correlation matrix of the spreading sequences [VER 86]. However, this "optimal" approach is computationally expensive. There has been a great deal of interest in finding suboptimum detectors with acceptable complexity and marginal performance degradation compared with the optimum detector. Various suboptimum detectors have been proposed, most of which can be classified into linear multi-user detector and subtractive interference canceller. In linear multi-user detectors, a linear transformation is performed to the outputs of the conventional detector to produce a new set of decision variables with MAI greatly decoupled. In subtractive interference cancellation, estimates of the interference are generated and subtracted out. The cancellation can be carried out either successively or in a parallel manner.

Chapter 6

UTRAN Access Network

6.1. Introduction

Compared to the GSM network architecture, the radio access network of UMTS, called the UTRAN (*Universal Terrestrial Radio Access Network*), constitutes the main innovation. The UTRAN is responsible for the control and the handling of radio resources, and allows data and signaling traffic exchange between user equipment (UE) and core network (CN). It handles the allocation and withdrawal of radio bearers required for the traffic support on the radio interface and controls some functions related to UE mobility and network access.

The UTRAN architecture and functioning are determined by the characteristics of the new radio access technology based on WCDMA with its two modes UTRA/FDD and UTRA/TDD. We will see for instance that with the use of macrodiversity in UTRA/FDD mode, the mobility handling is partly located in the UTRAN. All functional entities composing the UTRAN and their different roles are described in this chapter. Furthermore, the chapter will also detail the communication protocols used for information exchange on the one hand between UTRAN internal nodes and on the other hand between the UTRAN and the CN. As ATM is the transfer mode chosen for data transport through internal and external interfaces, its principle is also described in this chapter.

6.2. UTRAN architecture

As illustrated by Figure 6.1, the UTRAN is the link between the UE and the CN domains through, respectively, the Uu and Iu interfaces. The UE and the CN are described respectively in Chapters 3 and 4. We therefore only focus here on UTRAN components and its internal and external interfaces.

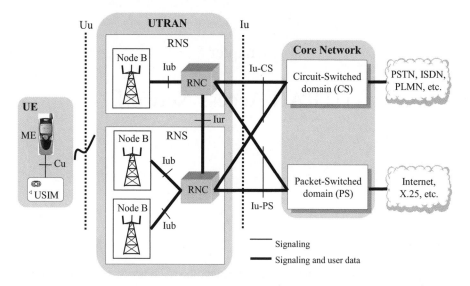

Figure 6.1. *UTRAN architecture*

Compared with GSM radio access network, the UTRAN brings the following innovations:

– *specification of four new interfaces*: "Uu", "Iub", "Iur" and "Iu", the latter being declined into two interfaces "Iu-CS" and "Iu-PS" with respectively the CS and the PS core network domains. These interfaces presented in Table 6.1 are *open* and should allow interworking between equipments from different manufacturers, and therefore offer more flexibility to operators on the choice of equipment providers;

– *use of CDMA access mode*. The use of (wideband) CDMA as multiple-access technique in the UTRAN requires supporting the macrodiversity procedure, non-existent in radio access systems based on TDMA access mode;

– *use of Asynchronous Transfer Mode* (ATM) for the transport layer of "Iu", "Iub" and "Iur" interfaces. This transfer mode is particularly appropriate for transport in the network of information with variable bit rate while respecting the requested transmission delay;

– *handling of the mobility in the UTRAN*. The UTRAN handles mobility at cell and URA (*UTRAN Registration Area*) levels independently from the mobility management handled by the CN. This enables efficient radio resources management and less signaling exchange between the mobile station and the CN. Temporary identities for UTRAN mobility management purposes were also specified.

Interface	Location		Equivalent in GSM
Uu	UE ↔UTRAN		Um
Iu	UTRAN↔CN	Iu-CS: RNC↔MSC	A
		Iu-PS: RNC↔SGSN	Gb
Iur	RNC↔RNC		None
Iub	Node B↔RNC		Abis

Table 6.1. *UTRAN interfaces and their equivalent in GSM*

6.2.1. *The radio network sub-system (RNS)*

The UTRAN is composed of one or several radio sub-systems called *Radio Network Sub-system* (RNS) and equivalent to the GSM BSS functional sub-system. An RNS is in its turn composed of one *Radio Network Controller* (RNC) and one or several Node Bs controlled via Iub interface.

6.2.1.1. *Node B*

Node B[1] is the exchange node between the UTRAN and all the UEs located in the cell or sectors covered by Node B. It mainly assures physical layer functions such as interleaving, channel coding and decoding, rate matching, spreading, QPSK modulation, etc. The UTRA physical functions are studied in Chapters 9, 10, 11 and 12.

Node B and power control

The transmit power of the UE is systematically controlled by Node B in order to maintain the QoS level regardless of the position of the UE. The purpose of power control is also to enable more efficient battery saving in the UE and to avoid useless interference level increasing in the cell. The interference level in a cell is a key issue for CDMA since it is highly linked with the access network capacity (see Chapter 5).

From measurements performed on the received uplink signal, Node B sends to the UE a command for the latter to adjust its transmit power. On the UE side measurements on downlink are also used to determine a power control command to

---

1 Throughout the book the term "base station" is also used to designate Node B.

send to Node B. This mechanism consisting of a mutual transmit power control between the UE and Node B is called in CDMA *inner loop power control*.

The commands exchanged between UE and Node B to increase or decrease transmit power are based on a quality threshold established by the RNC, and the updating procedure of this threshold is called *outer loop power control* (see Chapter 10).

Node B and handover

In GSM, the handover implies replacement of the receiving and transmit carrier frequencies with new ones allocated to the mobile station. This type of handover is called *hard handover* as the communication is interrupted during the channel switching (see Figure 6.2). In UTRA it is possible to avoid the radio link interruption. Indeed, with CDMA, a same carrier frequency could be used in several adjacent cells and this enables a UE to use multiple and simultaneous data paths involving different Node Bs for a same communication service.

The macrodiversity is the mechanism consisting of the support of multiple paths in order to increase performance of the radio link. The *soft handover* is the procedure in which the UE exploits macrodiversity to perform a handover procedure with no interruption of the radio link, as the current link will be only released when a new one has been established (see Figure 6.2). Two cases of soft handover where respectively one or several Node Bs are involved could be distinguished:

– *First case*: UE is moving through sectors controlled by the same Node B. This is a special case of soft handover and is referred to as *softer handover* procedure. For downlink traffic, the support of macrodiversity requires that in the sectors the UE is moving through, Node B transmits the same signal using the same carrier frequency with specific spreading code for each sector. The UE combines the received signal to increase the performances of the radio link. The *Maximal Ratio Combining* (MRC) described in Chapter 5 could be used to combine CDMA signals. In the case of uplink traffic, it is up to Node B to combine signals received by the antennas affected by sectors. The combined signal is then conveyed to the RNC via the Iub interface.

– *Second case*: UE is moving through cells served by different Node Bs. This case is treated in the next section.

NOTE.– the soft handover procedure is only applicable if the radio access technology UTRA/FDD is used by the UTRAN. It does not exist for UTRA/TDD.

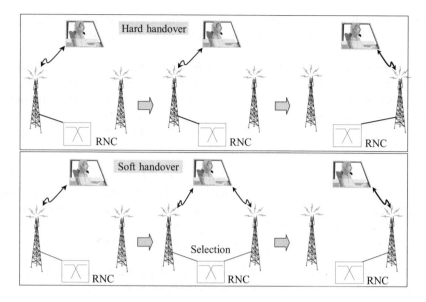

Figure 6.2. *Differences between hard handover and soft handover*

6.2.1.2. *The radio network controller (RNC)*

The RNC enables radio resource management in the UTRAN. Its main functions are:

– outer loop power control;

– handover control;

– mobile stations admission and traffic load control;

– channelization and scrambling code allocation handling;

– data transmission scheduling in packet transfer mode;

– combining/distribution of signals from/to different node Bs in a macrodiversity situation.

Depending on the role it plays for a given UE, the RNC is called *Controlling RNC* (CRNC), *Serving RNC* (SRNC) or *Drift RNC* (DRNC). One RNC equipment is generally able to play each of these three logical roles.

The CRNC role is not dependent on whether the UE has established or not an *RRC connection*. It handles radio resources for all Node Bs under its control – each Node B is controlled by a unique CRNC. The CRNC is also responsible for admission control of UEs in the cells it covers, and control of resources allocation during for instance a handover execution towards its controlling area.

The SRNC is the RNC that handles radio resources for a given UE that has already established an *RRC connection*. It is thus responsible for the handling of radio connection with the UE and some associated procedures such as handover, radio bearer allocation, SRNS relocation and outer loop power control. The SRNC is also responsible for selecting the best frame out of those received from different Node Bs involved in a macrodiversity scenario.

The DRNC is any RNC different from the SRNC but involved in a connection between a UE and the UTRAN. It is linked to the SRNC by the Iur interface. When an RNS runs out of radio resources or if a soft handover procedure has been decided, the SRNC could request an RNC belonging to another RNS to support it in terms of radio resources (spreading codes). This other RNC is called DRNC.

The RNS containing the SRNC is designated as the *Serving RNS* (SRNS), whereas the one containing the DRNC is called *Drift RNS* (DRNS).

Differences between the DRNC and the SRNC

In a macrodiversity context, when the involved Node B belongs to different RNSs, one of the RNC plays the drift role and the other the serving role as illustrated in Figure 6.3.

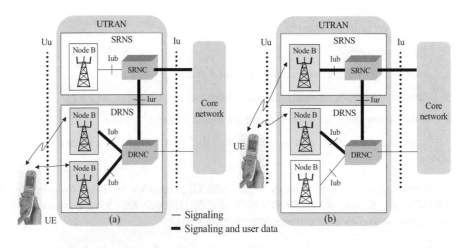

Figure 6.3. *Examples illustrating the drift and serving RNC roles in a macrodiversity context*

For uplink traffic, the DRNC conveys to the current SRNC of the UE the signal received from Node B after MRC and demodulation operations. In the case where several Node Bs controlled by the DRNC are involved in the macrodiversity, the signal with less bit error rate is selected and conveyed to the SRNC. The decoded data is finally transferred transparently from the DRNC to the SRNC without analyzing the contents.

For downlink traffic, the SRNC sends the signal to the DRNC which in turn distributes it to each Node B under its control involved in the macrodiversity (see Figure 6.3).

NOTE.– the examples in Figure 6.3 are only applicable when a dedicated physical channel is used for the traffic and if the Iur interface supports user data traffic.

6.2.2. Handling of the mobility in the UTRAN

As described in detail in Chapter 7, the *Radio Resources Control* (RRC) protocol layer could be in idle mode or in one of the four connected modes: CELL_DCH, CELL_FACH, CELL_PCH or URA_PCH. In connected mode, an RRC connection has been established between the UE and the UTRAN, and the UE mobility management changes from core network to UTRAN responsibility. The UE position is followed at cell level by the SRNC in CELL_DCH, CELL_FACH and CELL_PCH states. However, when cell change notification (CELL UPDATE) procedure performed by the UE in CELL_FACH and CELL_PCH becomes very frequent, the SRNC could move the UE to the state URA_PCH where the mobility is handled at the URA level. In the URA_PCH state the UE will signal its position to the SRNC only when it enters a new URA which is a registration area composed of a group of cells and unknown at core network level. As a consequence of the mobility management in the UTRAN level, new temporary identifiers have been introduced and a new procedure called *SRNC relocation* has been specified. The Iur interface also plays a major role as further described in this section.

6.2.2.1. UE temporary identifiers in the UTRAN

Beside the temporary identifiers TMSI and P-TMSI allocated to the UE by the CN for security purpose, a set of temporary identifiers are also used in the UTRAN for identification of a UE related data in common channel and mobility management in the UTRAN level. The following four types of these identifiers called *Radio Network Temporary Identifiers* (RNTI) have been specified (see also Chapter 7):

– S-RNTI (*Serving RNC RNTI*) which is used to associate a UE connected to the UTRAN to a unique SRNC at a given time. It is allocated by the serving RNC during the RRC connection establishment;

– D-RNTI (*Drift RNC RNTI*) which is associated to a UE by a DRNC but is neither used in Uu interface nor communicated to the UE. It is possible to have at the same time both an S-RNTI and a D-RNTI allocated to a given UE, and in this case the related SRNC and DRNC should be capable of handling both these identifiers;

– C-RNTI (*Cell RNTI*) which is only valid in a given cell and allocated by a CRNC to a UE entering in a new cell under its control;

– U-RNTI (*UTRAN RNTI*) used to identify a UE in the whole UTRAN. It is composed of the SRNC identifier (SRNC-id) and the S-RNTI described above.

6.2.2.2. *SRNS relocation*

As illustrated in Figure 6.4, when the mobile is out of cells controlled by the SRNC, the Iur interface is used to maintain the link between the UE and its SRNC. For radio resources saving purpose or when data traffic is not supported on this interface, it could be decided to remove Iur from the traffic path by transferring the role of serving RNS to the DRNS. This transfer of RNS role, called "SRNS relocation", is completely transparent to the end user (see also Chapter 8). The resources rationalization is also associated to the concept of *streamlining*.

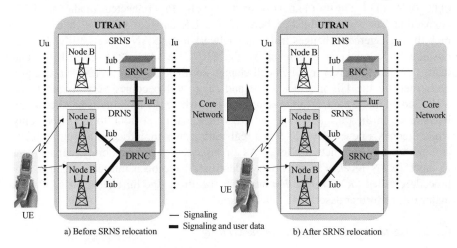

Figure 6.4. *Illustration of the SRNS relocation procedure*

6.2.3. *Summary of functions provided by the UTRAN*

The main functions assured by the UTRAN are presented in the non-exhaustive list below:

– *transfer of user data*. This is the main function of the UTRAN: serving as the bridge between the UE and the CN through Uu and Iu interfaces;

– functions related to network access. These are network access control, congestion control and system information broadcast. The UTRAN can authorize or deny access to the network to new users or allocate or refuse new resources in order to avoid an overload situation with the already connected users. The UTRAN is also responsible for system information broadcast in each cell, providing UEs with AS and NAS information.

– functions related to security. The UTRAN implements the ciphering and integrity algorithms also present in the UE and used for information protection and confidentiality in the interface between the UE and the UTRAN. The security management function is described in Chapter 8;

– functions related to mobility. These functions include the handover management, the SRNS relocation, the paging and notification, and the UE position estimation in case a location service (LCS) is offered;

– functions related to radio resources management. They regroup functions associated to the allocation and maintenance of resources required for radio communication: measurement for channel quality evaluation based on bit error rate, received power strength, interference level in the cell, etc. Radio resources management includes also allocation and de-allocation of radio bearers, procedures related to power control, channel coding and decoding, etc.;

– synchronization. It is up to the UTRAN to maintain in each cell the time reference on which any UE must be synchronized in order to be able to transmit or receive information. In the UTRA/TDD radio access case, the UTRAN shall in addition assure the time synchronization between all Node Bs composing the radio network. The UTRAN is also in charge of the synchronization between Node Bs and the RNC, and between the RNC and the CN during data transfer.

6.3. General model of protocols used in UTRAN interfaces

The previous section has been devoted to the study of equipments that constitute the UTRAN and their different roles. In what follows we will describe how these equipments exchange signaling and data information through specified interfaces.

Open interfaces

All UTRAN interfaces (both external and internal) are open interfaces as their functional characteristics and protocol stacks are completely standardized. That means that there is in principle no ambiguity in what the different manufacturers have to implement on these interfaces and therefore equipments compliant with the specified standard could be interconnected even if provided by different manufacturers. The generic model for the description of protocols used in UTRAN interfaces is presented in Figure 6.5 [TS 24.401]. It is inspired by the OSI model and protocol stacks used by the ISDN.

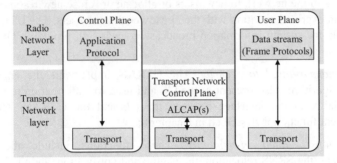

Figure 6.5. *Generic model in the protocol architecture of UTRAN interfaces*

Independent transport network and radio network layers

As illustrated by Figure 6.5, the protocol architecture in UTRAN interfaces is composed of horizontal layers and vertical planes that are independent logical components. This enables the evolution of each interface independently from the other. For example, one of the elements of the protocol stack in the vertical plane could be modified without having any impact on the other vertical plane.

6.3.1. *Horizontal layers*

Horizontal layers separate functionalities related to radio access network (*Radio Network Layer*) and those related to the technology used for transport of signaling and user data (*Transport Network Layer*). In UMTS networks, the *Radio Network Layer* implements UTRAN-specific protocols. Let us note that such layer-based architecture makes it possible, in theory, to use other access technologies (e.g. GERAN, HIPERLAN/2, etc.) on top of the *Transport Network Layer*. The *Transport Network Layer* implements transport bearers used to convey signaling and user data. It also includes physical layer protocols used for the physical media chosen by the network operator: optical fiber, radio waves, coaxial cable, etc.

6.3.2. *Vertical planes*

Usually, signaling information is used in addition to user traffic data for the related service handling purpose. In Figure 6.5, we can distinguish a *Control Plane* and a *UserPlane*.

6.3.2.1. *Control plane*

The control plane includes the application protocol in charge of the allocation and the handling of RABs. Signaling messages generated by this application protocol are conveyed on one or several signaling bearers.

For each of the different UTRAN interfaces there is a specific application protocol:

– RANAP (*Radio Access Network Application Part*) for the Iu interface;

– RNSAP (*Radio Network Subsystem Application Part*) for the Iur interface;

– NBAP (*Node B Application Part*) for the Iub interface.

Figure 6.6 represents a global view that illustrates the interaction between the different application protocols during exchanges between the UE, Node B, the RNC and the CN. The radio interface protocols (RRC, RLC, MAC, physical layer) are studied in Chapter 7, while application protocols are studied further in this chapter.

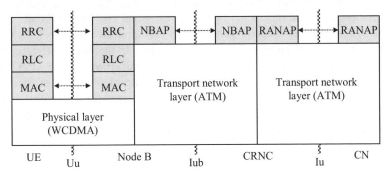

a) Control plane in idle mode: the UE is monitoring possible incoming calls

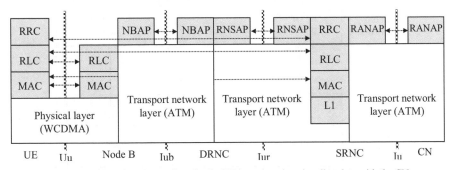

b) Control plane in connected mode: the UE is exchanging signaling data with the CN

Figure 6.6. *Example of signaling exchange between the UE, the UTRAN and the CN*

6.3.2.2. *User plane*

User traffic associated to a given service (voice, video or data) is conveyed on the *user plane* either in circuit or in packet mode. As it will be illustrated further on, it could also happen that non-user traffic (e.g. signaling traffic) is conveyed on the user plane. The *Radio Network Layer* in the user plane includes the *Frame Protocols* that are physical level protocols in charge of the scheduling of the transmission of data flows submitted by upper layers. These data flows are then conveyed by the *Transport Network Layer* on the data bearers.

6.3.3. *Control plane of the transport network*

The control plane of the transport network does not contain information related to the used radio access technology. ALCAP (*Access Link Control Application Protocol*) protocols are located in this plane. ALCAP is the generic name for signaling protocols used in the user plane for data bearer establishment, maintenance and release. As shown in Figure 6.7, ALCAP takes charge of requests from the application protocol and includes a signaling bearer that could be different from the one conveying signaling message generated by the application protocol. ALCAP is present in the UTRAN interfaces Iu-CS, Iur and Iub. In the Iu-PS interface, where the control plane of the transport network is not present, the data bearer characteristics are pre-established. This illustrates one of the goals of the UMTS standard which was to specify independent protocol stacks for the control and user planes – stacks interworking through the control plane of the transport network. This architecture enables the independent definition of signaling and data bearers for which different QoS are usually required.

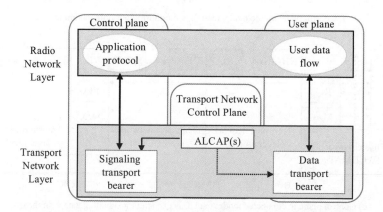

Figure 6.7. *Interaction between control and user planes through ALCAP*

6.4. Use of ATM in the UTRAN network transport layer

ATM (*Asynchronous Transfer Mode*) is a transfer mode characterized by data transmission in packet switched mode with an asynchronous time division multiplex [PUJ 00]. The transmission data unit in ATM is a *cell* consisting of a 53 bytes fixed-size unit. The term "asynchronous" in ATM means that a cell is not exclusively dedicated to a given active service, but is instead allocated alternately to several services on readiness of data to transfer. One could compare the ATM mode with the synchronous transfer mode used in GSM. The latter is based on the exclusive allocation to a service of a time slot in a TDM (*Time Division Multiplex*) frame during all the call duration. Though suitable for service with constant bit rate, the synchronous mode is not efficient for a service with variable bit rate because in this case resources remain reserved to the service even when there is no data to transfer. ATM associates an efficient use of network resources and the support of real-time traffic. It offers for example a better QoS than the IP-based transport.

In the UMTS network, ATM is used in the network transport layer. This choice relies on two key properties of ATM:

– possibility to convey variable bit rate traffic both for PS and CS services. This is particularly useful for UMTS given the large variety of offered services. The use of ATM cells enables the optimization of the transport of speech traffic in the network, particularly for silence periods during the communication;

– possibility to keep required QoS for conveyed media. Indeed, with the use of ATM for the transport layer, the QoS needed for each service class will remain insignificantly modified, especially for delay sensitive services like telephony and video telephony.

6.4.1. *ATM cell format*

An ATM cell is composed of a header field followed by a payload containing the upper layer data (see Figure 6.8). Cells are allocated according to the data sources activity and transport resources availability. There are two types of header: one used for ATM cells exchanged on a *User-to-Network Interface* (UNI) and another used for ATM cells exchanged on a *Network Node Interface* (NNI). The header is composed of 5 bytes and begins with the VPI (*Virtual Path Identifier*) field. The length of the VPI field could be either 8 bits (UNI format) or 12 bits (NNI format). As for the VCI (*Virtual Channel Identifier*) field, its length is fixed and equal to 16 bits.

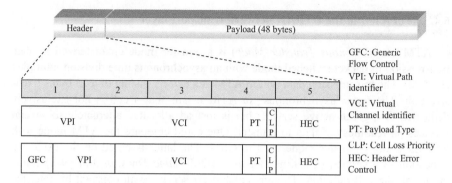

Figure 6.8. *ATM cell structure*

The GFC (*Generic Flow Control*) field is only present in headers of type UNI and could be used to identify several stations sharing the same ATM interface. The PT (*Payload Type*) field is a 3-bits size used to indicate the type of information (control or data) transported in the payload field, and also as congestion indicator. A one bit length field CLP (*Cell Loss Priority*) is used to identify ATM cells of low priority that could be discarded in network congestion situation. The payload field following the header is composed of 48 bytes. Regarding the header, an error control field HEC (*Header Error Control*) of one byte length is used to detect and correct any single erroneous bit or to detect several erroneous bits. No cell retransmission is performed since non-recoverable erroneous cells are simply discarded.

6.4.2. *ATM and virtual connections*

The transfer mode used by ATM is connection-oriented, meaning that it requires the establishment of a logical (virtual) connection prior to any data transfer: ATM cells do not include identifiers of the related source and destination end-points contrary, for example, to IP datagrams. Instead, cells contain VCI and VPI fields, the role of which is to identify the connection [PUJ 00]. The ATM connection is established by means of a supervisor cell (connection-signaling request message) propagated towards the destination end-point and leaving a mark at each crossed node. The routing of the supervisor cell consists of determining at each network node an association between the input port and an output port, resulting thus in a routing table indicating to which node an input cell shall be routed.

6.4.3. *ATM reference model*

From a functional point of view, the ATM reference model is quite similar to the generic model of UTRAN interfaces previously described (see Figure 6.9).

The *physical layer*, equivalent to the physical layer of OSI reference model, manages the transmission/reception of bits composing ATM cells on the chosen physical medium (optical fiber, radio channel, coaxial cable, etc.).

The *ATM layer* is independent from the underlying physical medium. It is responsible for cell multiplexing over virtual circuits (enabling several media/upper layers to share a same virtual circuit) and cell relay through ATM network nodes. VPI and VCI fields of cell header are used to enable these functions.

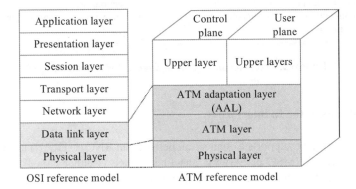

Figure 6.9. *ATM layered reference model in comparison with OSI model*

The *ATM adaptation layer* (AAL), combined with the *ATM layer*, is equivalent to the *data link layer* of the OSI reference model. It is responsible for ATM signaling according to QoS requested by applications. There are different types of ATM adaptation layers, each being associated with a specific service/application (see Table 6.2). For UMTS, we are only interested in AAL type 2 (AAL2) and AAL type 5 (AAL5) that are used in UTRAN interfaces.

AAL2 is suitable for the transport of low bit rate (up to 64 kbps) and bursty real-time traffic. It enables the adaptation of several "micro cells" into one ATM cell. Given its characteristics, AAL2 is particularly suitable for the transport of variable bit rate voice or video within the network: several voice calls could be supported by a single ATM connection resulting in bandwidth saving. AAL5 is used for variable bit rate and non-real-time traffic, both for connection-oriented and connectionless mode at network layer. It is typically appropriate for transport of IP packets and signaling message between cellular network nodes.

	AAL-1	AAL-2	AAL-3/4	AAL-5
Used in the UTRAN	No	Iu-CS, Iur and Iub	No	Iu-CS, Iu-PS, Iur and Iub
Delay constrain	Real-time	Real-time	Non-real-time	Non-real-time
Bit rate mode	Constant	Variable	Variable	Variable
Network layer connection mode	Connection-oriented		Connectionless and connection-oriented	Connectionless

Table 6.2. *Characteristics of the different ATM adaptation layers*

6.5. Protocols in the Iu interface

The Iu interface supports a connection between an RNC (UTRAN) and a node of the core network (MSC or SGSN). The Iu interface is either called Iu-CS if the UTRAN is connected to the CS core network domain or Iu-PS if the UTRAN is connected to the PS core network domain. There is also another type of Iu interface called Iu-BC which is not addressed here and that links an RNC to the core network domain providing *Cell Broadcast Services* (CBS) (see also [TS 25.410]).

6.5.1. *Protocol architecture in Iu-CS and Iu-PS interfaces*

Figure 6.10a shows the architecture of the protocols involved in the Iu-CS interface, while Figure 6.10b shows the protocols used in the Iu-PS interface. These interfaces are quite similar – the main difference being located on the control plane of the transport network.

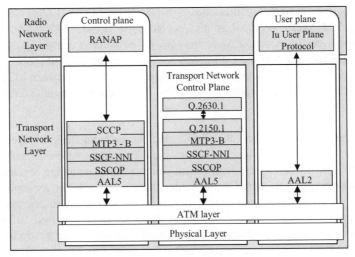

a) Iu-CS protocol architecture

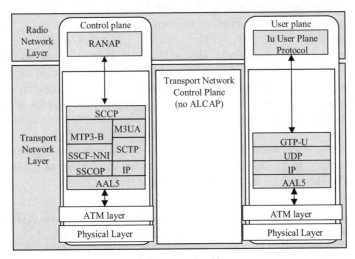

b) Iu-PS protocol architecture

Figure 6.10. *Protocol architecture of Iu-CS and Iu-PS interfaces*

6.5.1.1. *Control plane protocol stack for Iu-CS and Iu-PS interfaces*

On top of the protocol stack in Iu-CS and Iu-PS, there is the *Radio Access Network Application Part* (RANAP) protocol using as signaling bearer the *Broadband Signaling System* #7 (BB-SS7). Compared with the Narrowband SS7 used in the CN *Release 99*, the *Message Transfer Part Level 3* (MTP3) enabling reliable routing of signaling messages is replaced by MTP3-B (*Broadband MTP Level 3*). The MTP2 protocol is also replaced by SAAL-NNI (*Signaling ATM Adaptation Layer for Network Node Interface*). The latter is composed of two sub-layers: SSCF (*Service Specific Coordination Function*) and SSCOP (*Service Specific Connection Oriented Protocol*). SSCF coordinates exchanges between MTP3-B and SSCOP, while the latter provides mechanisms for the establishment and release of a connection (see Figure 6.11).

SCCP is common to broadband SS7 and narrowband SS7. It represents the control sub-system of signaling connections and offers two supplementary services in comparison to MTP3-B: support of international signaling exchange and connection-oriented services. Connectionless mode is used, for example, for the transfer of location updating or paging messages, while connection-oriented mode is more appropriate when several messages have to be transferred to the same destination or for the transfer of longer messages (e.g. radio bearer allocation/release, handover execution). For the Iu-PS interface, UMTS specifications enable operators to choose an IP over the ATM approach for the transport of signaling messages. In that case M3UA (*MTP3-user adaptation*) and SCTP (*Stream Control Transmission Protocol*) are used on top of IP. M3UA and SCTP have been developed by the working group SIGTRAM in the IETF organization (see www.ietf.org). The M3UA sub-layer is used to adapt SS7 signaling messages to/from IP network-based signaling messages. As for SCTP sub-layer, it assures a reliable transfer of signaling messages generated in the control plane. One of the reason for which SCTP has been preferred to TCP is the shorter delay for the message transfer.

NOTE.– it should be noted that for *Release 5* of the standard, a UTRAN transport network entirely based on IP is considered as an alternative to ATM transport [TR 25.933]. This is valid for both the control and the user plane.

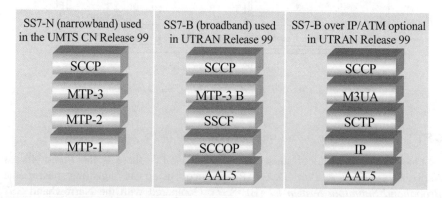

Figure 6.11. *Comparison of the different SS7 architectures in UMTS*

6.5.1.2. *Transport network control plane protocol stack for Iu-CS and Iu-PS interfaces*

The transport network control plane protocol stack in the Iu-CS interface is based on a technical recommendation [ITU Rec. Q.2630.1] specifying the management of an AAL2 connection in the user plane. For the Iu-PS interface, it clearly appears in Figure 6.10b that a control plane of the transport network does not exist. That is why the signaling protocol ALCAP used for establishment of data bearers in the user plane is not necessary – the characteristics of these bearers are pre-established by the operator.

6.5.1.3. *User plane protocol stack for Iu-CS and Iu-PS interfaces*

In the user plane of Iu-CS and Iu-PS interfaces, at the radio network layer, we have the protocol Iu UP (*User Plane*) ensuring the transport of user data (voice, video, IP datagrams, etc.) associated to a RAB. User traffic is conveyed on one or several *Iu bearers*. Each protocol is associated to a unique RAB and if several RABs are involved in a communication, several Iu UP instances shall be used. The Iu UP protocol could operate either in *transparent mode* or in *support mode*. The mode to use is chosen by the core network when specifying the QoS of the RAB(s) according to the service/application requirements.

In *transparent mode*, information is transferred without any processing. In *support mode*, Iu UP could perform user data segmentation and associated functions such as transmission error handling if required by the RAB. The segment (SDU[2]) size is typically the size of a voice frame from the AMR codec or a data frame exchanged during the establishment of a circuit-switched communication.

For data transfer in the user plane, Iu-CS and Iu-PS are slightly different. For Iu-CS, user data blocks are directly submitted to AAL2 and transported in ATM cells. On the contrary, Iu-PS uses the same principle as the one used for the Gn interface between SGSN and GGSN of the UMTS core network (see Chapter 4). Indeed, the user plane of Iu-PS uses the GTP protocol (called here GTP-U for *GPRS Tunneling Protocol User-plane*), which enables the encapsulation of user data in IP datagrams. GTP uses UDP (*User Datagram Protocol*) as transport protocol over IP meaning connectionless and unacknowledged transfer mode. AAL5 is then used as ATM adaptation protocol. In *Release 5*, an IP-only based transport (no ATM) is also considered [TS 25.410, *R5*].

---

2 SDU (*Service Data Unit*) is the generic name of a data unit exchanged between adjacent layers of a protocol stack.

6.5.2. RANAP

As defined in the previous sections, RANAP is the application protocol providing the signaling service between the UTRAN and the CN, both for Iu-CS and Iu-PS interfaces. From a functional point of view, RANAP is equivalent to its counterpart in GSM called BSSMAP (*Base Station Subsystem Management Application Part*).

RANAP functions are defined in [TS 25.413] and comprise:

– SRNS relocation, consisting of transferring the role of serving RNC from one RNC to another RNC with and without (hard handover) Iur interface;

– transport in transparent mode of NAS signaling messages between the UE and the CN;

– overall RAB management (set-up, modification and release);

– release of all Iu resources involved in a connection, for both user and control planes;

– handling of the CN originated paging towards the UE;

– transfer of security mode commands to the UTRAN for activation/de-activation of ciphering and integrity protection procedures (see Chapter 7);

– transfer of UE actual location information from RNC to CN.

Transparent transport of signaling messages between the UE and the CN

NAS signaling exchange requires the establishment of two connections: an *RRC connection* and an *Iu connection*. The purpose is to setup a complete signaling path (signaling bearers) between the UE and the CN. At the radio interface level, the transport of signaling messages between the UE and the RNC is made possible by the RRC connection establishment procedure described in Chapter 7. Upon receipt of the *RRC INITIAL DIRECT TRANSFER* message containing the first NAS message, the RNC establishes the Iu signaling connection between the RNC and the CN to complete the signaling path set-up (*SCCP CONNECTION REQUEST* and *SCCP CONNECTION CONFIRM* messages in Figure 6.12 [TS 25.410]). Then the RNC uses a message *RANAP INITIAL UE MESSAGE* to forward to the CN the NAS signaling information received from the UE [TS 25.413].

RAB establishment

A RAB is a service offered by the UMTS network for transport of user data between the UE and the CN. It could be seen as a "pipe" offering a given QoS. The RAB establishment process is always controlled by the core network which keeps the subscription data of each user. The RAB establishment is initiated by the CN and executed by the UTRAN. The CN uses the RANAP message *RAB ASSIGNMENT REQUEST*, including the unique identifier of the RAB (RAB ID) and QoS parameters (traffic type, bit rate, error rate, delay class, etc.) associated with the RAB (see Figure 6.13). The CN indicates also how the user plane and the transport layer of the Iu interface should be configured. After a successful execution of the CN request, the RNC sends back to the CN the RANAP message *RAB ASSIGNMENT COMPLETE* as acknowledgement [TS 25.413]. After the establishment of an Iu bearer, the RNC proceeds to the establishment of one or several radio bearers (RB) to complete the "pipe" between the UE and the CN. The RB establishment is studied in Chapter 7.

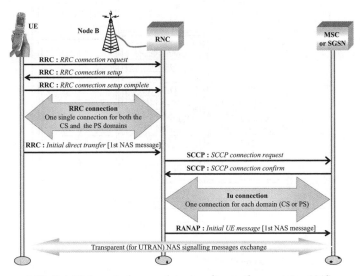

Figure 6.12. *Establishment of a complete signaling path supporting NAS messages*

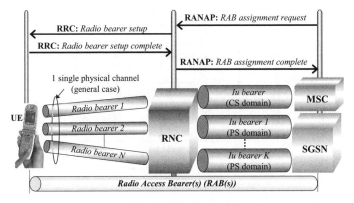

Figure 6.13. *Signaling exchange for the RAB set-up*

6.6. Protocols in internal UTRAN interfaces

6.6.1. *Iur interface (RNC-RNC)*

The Iur interface, for which the equivalent does not exist in GSM, supports the information exchange between the SRNC and the DRNC. The protocols involved in the Iur interface are presented in Figure 6.14a.

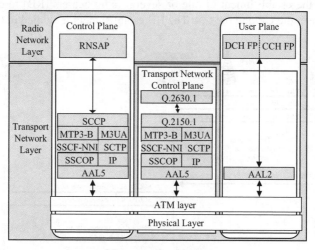

a) Iur protocol architecture

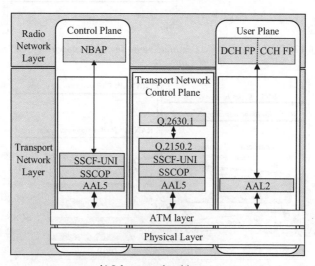

b) Iub protocol architecture

Figure 6.14. *Protocol architecture of Iur and Iub interfaces*

6.6.1.1. *Control plane protocol stack of the Iur interface*

RNSAP is the application protocol in charge of signaling exchange on the Iur interface. It relies on SCCP protocol to provide this service. For the transport of signaling messages, there are two possibilities under SCCP: one using SS7 signaling system and another relying on IP, and for both AAL5 is used as an ATM adaptation layer.

6.6.1.2. *RNSAP*

RNSAP ensures the signaling exchange between two RNCs [TS 25 423]. From a functional point of view, it is composed of four modules: a basic module used to handle the mobility within the UTRAN (*RNSAP Basic Mobility Procedures*), one module dedicated to the handling of dedicated and shared transport channels (*RNSAP DCH Procedures*), one for the handling of common transport channels (*RNSAP Common Transport Channel Procedures*) and finally the module for global procedures (*RNSAP Global Procedures*) not related to a specific UE. A network operator could implement all modules or only some of them as needed. For example, to support macrodiversity, the module *RNSAP DCH procedures* shall be implemented. This means that the support of macrodiversity requires the presence of the user plane, whereas for a procedure such as *URA updating* only signaling messages of the control plane are needed.

RNSAP Basic Mobility Procedures

This module comprises the minimum number of procedures an Iur interface shall support to exist. It is composed of procedures enabling exchange of uplink and downlink signaling messages between an SRNC and a DRNC. Only signaling traffic (not user traffic) is handled by procedures of this module. Among basic mobility procedures we can mention the paging procedure for which the SRNC sends to the CRNC a request to page the UE in a given cell or URA under its control (see Figure 6.15). This *basic module* contains also procedures enabling two RNCs to exchange cell/URA change notification information sent from the UE (see Chapter 7). Since the *basic module* does not support user traffic, the user plane and the transport network control plane are not needed.

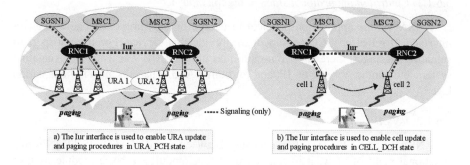

Figure 6.15. *Examples of RNSAP basic mobility procedures in Iur interface*

RNSAP Dedicated Transport Channel procedures

These procedures are used to handle DCHs, DSCHs and USCHs (TDD only) transport channels between two RNSs. They enable the establishment, modification and release of these channels between the DRNS and the UE. The UTRAN relies, for example, on this module to handle macrodiversity in which different RNSs are involved in a communication. This module also contains the following elementary procedures between SRNC and DRNC: radio link handling (set-up, addition, reconfiguration, failure/restoration notification by the DRNC, pre-emption indication, deletion); dedicated measurement handling (initiation/termination requested by the SRNC, initiation/reporting executed by the DRNC); downlink power control used by the SRNC to control the level of downlink transmitted power and to request a DRNC to balance downlink transmission powers of the radio links of one UE; compressed mode commands used to activate/deactivate the compressed mode in the DRNS for one UE-UTRAN connection.

RNSAP Common Transport Channel Procedures

This module enables information exchange via the common transport channels. It is used to coordinate the establishment and the release of common transport channels in the user plane which are involved in a connection between the DRNC and the UE (i.e. for traffic on RACH and FACH channels). If this module is not present in the Iur interface between the SRNC and the DRNC, SRNS relocation will be required when the UE performs a cell change towards the DRNC, while user data traffic is being conveyed on common channels. The complexity of the MAC protocol layer increases with the support of this module, as it shall distinguish functions at SRNC and DRNC levels. Elementary procedures of this module are common transport channel resources initialization and common transport channel resources release initiated by the SRNC.

RNSAP Global Procedures

This module contains procedures which are not related to a specific UE. One example is the *Error indication* procedure used by the RNC to signal to its peer the node errors detected in a received message, when there is no failure response message that could be used to report the detected errors.

6.6.1.3. *Control plane protocol stack in the transport network of the Iur interface*

This protocol stack is used when modules used to handle dedicated and common transport channels are integrated in the application protocol RNSAP. The technical recommendation [ITU Rec. Q.2630.1] specifies the control of AAL2 connections in the user plane. The signaling bearer over ATM could be based either on wideband SS7 or IP stacks (see Figure 6.14a).

6.6.1.4. *User plane protocol stack of Iur interface*

On top of this plane are located the DCH-FP and CCH-FP frame protocols for the transfer of, respectively, dedicated and common transport channels data frames. Functions assured by frame protocols are equivalent for Iur and Iub interfaces and are described in the following section.

6.6.2. *Iub interface (RNC-Node B)*

This is a logical interface through which a Node B communicates with its unique controlling RNC (CRNC). For example, when RNC needs to request a Node B to release a radio link previously established with a UE, this is done throughout the Iub interface. The protocol stacks at this interface is shown in Figure 6.14b.

6.6.2.1. *Control plane protocol stack of the Iub interface*

In the control plane of the Iub interface, NBAP is used as radio network layer signaling protocol. The signaling bearer (transport layer) used by NBAP in the control plane is reduced to SAAL-UNI (SSCF-UNI and SSCOP) over ATM, with AAL5 as ATM adaptation layer.

6.6.2.2. *NBAP*

The NBAP application protocol consists of two families of signaling procedures [TS 25.433]: *common procedures* and *dedicated procedures*.

The NBAP *common procedures* contain procedures related to the handling of common resources: common transport channel set-up, modification and release; common measurement initiation, reporting, termination, and failure notification; cell set-up, reconfiguration, release and status indication; system information update, etc.

The NBAP *dedicated procedures* contain the following elementary procedures between the CRNC and Node B: radio link handling (set-up, addition, reconfiguration, failure/restoration notification by Node B, pre-emption indication, release); dedicated measurement handling (initiation/termination requested by the CRNC, initiation/reporting executed by Node B); downlink power control used by the CRNC to request a Node B to balance downlink transmission powers of the radio links of one UE; compressed mode command used to activate/deactivate the compressed mode in Node B for one communication between Node B and a specific UE.

In addition to the above two procedures, there is the *error indication* procedure which is used by the CRNC/Node B to signal to Node B/CRNC the errors detected in a received message, when there is no failure response message that could be used to report the detected errors.

6.6.2.3. *Transport network control plane protocol stack of the Iub interface*

The transport network control plane protocol stack of the Iub interface is quite similar to the one used in the same plane in the Iur interface (see Figure 6.14). The only difference is the non-support of the SCTP/IP option in the Iub interface.

6.6.2.4. *Protocol stack at the user plane of the Iub interface: Frame Protocols*

In both Iur and Iub interfaces, the *Frame Protocols* (FP) are physical layer protocols. As illustrated by Figure 6.16, the downlink data flow (already processed at RLC and MAC layers) arrives at the user plane containing the frame protocols. They enable synchronization functionality between L2 radio protocols (RLC/MAC) and the physical layer of the radio interface: data can only be transmitted on these interfaces at pre-established transmission time intervals of 10, 20, 40 or 80 ms.

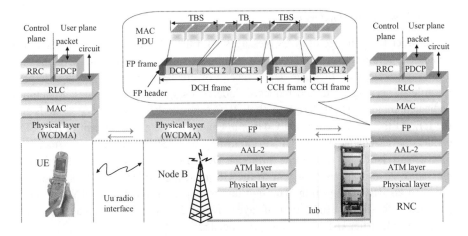

Figure 6.16. *Illustration of the frame protocol role in the RNC and the UE*

The DCH-FPs are in charge of the transport of downlink and uplink MAC PDUs, also called TBs (*Transport Blocks*), between Node B and the RNC. These blocks are associated to the DCH transport channels and contain [TS 25.427]: data generated by or destined to the UE; outer loop power control information between the RNC and Node B; parameters for synchronization of Node B and RNC, etc.

The CCH-FPs are in charge of the transport of downlink and uplink MAC PDUs between Node B and the CRNC. These blocks are generated by common transport channels RACH, CPCH, FACH, PCH, DSCH and USCH (mode TDD), and contain [TS 25.435]: user data generated by or destined to the UE; measurement results; parameters for synchronization of Node B and RNC, etc.

6.7. Data exchange in the UTRAN: example of call establishment

The example here described shows how protocols of the UTRAN interfaces described above interact. It is assumed that the UE is in RRC idle state, is already registered in the network, and needs to exchange user data in packet or circuit switched mode with another user. For that, it must perform the following (see Figure 6.17):

– establish an RRC connection;

– request the establishment of an Iu connection;

– establish RABs for user data transfer between the UE and the CN.

RRC connection establishment

In order to exchange of NAS messages with the CN, the UE needs to establish dedicated radio resources in the access stratum. For that it sends to the RNC the *RRC CONNECTION REQUEST* message. Upon receipt of this request, the RNC starts to play the role of serving RNC and requests Node B, using NBAP messages, to setup a dedicated channel (DCH) between Node B and the SRNC, and between Node B and the UE. At the Iub interface, ALCAP is in charge of the establishment of the ATM connection associated to the UE. Prior to any transfer on the radio interface, the SRNC and Node B use frame protocol DCH-FP to exchange synchronization messages. A message is then sent to the UE as acknowledgement of its request to establish an RRC connection. Through this response message, the SRNC provides parameters for the configuration of layers 1 and 2 of the UE and may allocate a temporary identifier U-RNTI to the UE if needed.

Establishment of an Iu connection and NAS signaling messages exchange

After the RRC connection is established, the UE is, in our example, moved to CELL_DCH state. Radio resources requested for the communication between the UE and the SRNC are setup, and the UE must request the SRNC to setup the other resources required for a complete signaling path up to the CN (CS and/or PS domains). This is done by using the RRC message *INITIAL DIRECT TRANSFER* encapsulating the NAS message that could be, for example, a voice service request (*CM SERVICE REQUEST*). In this example, the message container indicates the CS domain.

The SRNC uses the RANAP message *INITIAL UE MESSAGE* to indicate to the CN that the mobile has a NAS message to transfer and that a SCCP connection is requested to be established. After the establishment of this connection, the NAS message is transferred to the CN (e.g. the MSC in the case of a voice service request). Subsequent NAS messages will use the established SCCP connection and will be transferred in a RANAP message *DIRECT TRANSFER*. Similarly, subsequent NAS messages will be transferred between the UE and the SRNC on the existing RRC connection and will be encapsulated in the RRC messages *DL DIRECT TRANSFER* or *UL DIRECT TRANSFER* for, respectively, the downlink and the uplink. Messages following the initial transfer are transparently transferred by the UTRAN, which is only a relay towards the CN.

RAB establishment

After establishing the signaling connection (RRC connection + Iu connection) between the UE and the CN, several NAS messages can be exchanged before the CN decides to allocate the resources required for the transport of user data. Upon reception of the service request from the UE, the CN first performs access control. Then if the UE has a valid subscription, the CN requests the UTRAN to allocate to the UE a RAB with the required QoS parameters. The RANAP RAB ASSIGNMENT REQUEST message is used by the CN to initiate the RAB establishment. Upon reception of this message, the SRNC will setup the *Iu bearer* with the CN using ALCAP signaling, and the *radio bearer(s)* with the UE using RRC signaling. At the Iu-PS interface, the Iu bearer is pre-established and therefore ALCAP protocol stack is not needed.

The SRNC then requests Node B to reconfigure the radio link that has been setup for the transport of NAS signaling messages, in order to support in addition the transport of user data in a dedicated channel DCH. Upon the ATM path between Node B and the SRNC established and after the synchronization procedure, the SRNC sends the RRC message RADIO BEARER SETUP to the UE. In this message, the SRNC provides the UE with the parameters for the configuration of physical layer, MAC and RLC layers, according to the QoS instructed by the CN. The UE notifies the SRNC of the completion of the bearer set-up by using the RRC message RADIO BEARER SETUP COMPLETE. The SRNC in turn uses the RANAP message RAB ASSIGNMENT RESPONSE to indicate to the CN that a complete path for user data transport has been successfully setup. It is only after this step that the UE can start sending and receiving user data on the established RAB.

6.8. Summary of the UTRAN protocol stack

Figure 6.18 shows the protocol stacks the RNC must support in order to be able to communicate with the CN, Node B, the UE and other RNCs. The radio protocols are described in Chapter 7.

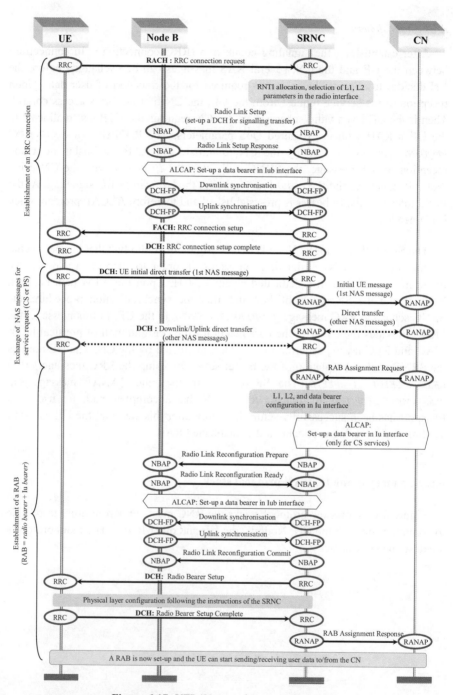

Figure 6.17. *UTRAN procedures to setup a RAB*

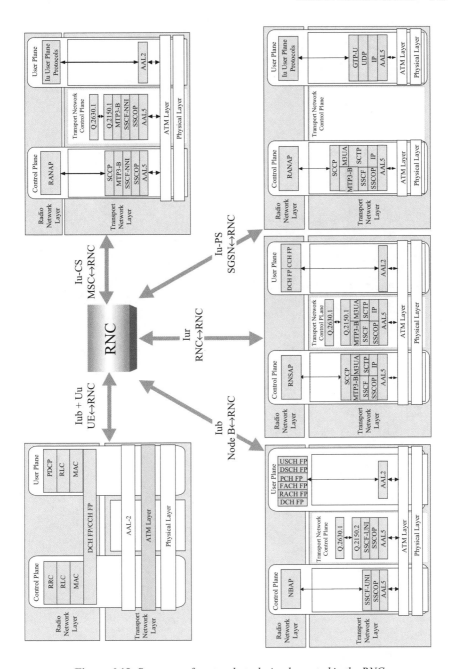

Figure 6.18. *Summary of protocol stacks implemented in the RNC*

Chapter 7

UTRA Radio Protocols

7.1. Introduction

This chapter is devoted to the description of protocols implemented in the UTRA/FDD radio interface, whose main role is the set-up, reconfiguration and release of *radio bearer* services in the *Access Stratum*. It consists of three protocol layers:

– the physical layer (layer 1);

– the data link layer (layer 2);

– the network layer (layer 3).

Layer 2 is composed of four sub-layers:

– MAC (*Medium Access Control*);

– RLC (*Radio Link Control*);

– PDCP (*Packet Data Convergence Protocol*);

– BMC (*Broadcast/Multicast Control*).

Layer 3 consists of one protocol named RRC (*Radio Resource Control*). The other components, MM (*Mobility Management*) and CM (*Connection Management*) of layer 3 are part of the *Non-Access Stratum* and are then described in Chapter 8. Figure 7.1 presents the layered architecture of UTRA protocols.

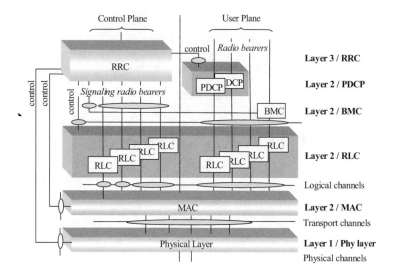

Figure 7.1. *Layered architecture of UTRA protocols*

We notice in Figure 7.1 that the UTRA protocol stack is divided into control (C) plane and user (U) plane, supporting respectively the signaling messages between the UE and the network and the user data traffic. While MAC and RLC are present in the C- and U-planes, RRC only exists in the C-plane and PDCP and BMC are present in the U-plane exclusively. Signaling messages generated by RRC or submitted to RRC by NAS are transmitted on *signaling radio bearers*, whereas user data are conveyed over one or several *radio bearers* with a negotiated QoS that should match the type of service. The "control" lines between RRC and each of the other layers or sub-layers illustrate that the configuration of lower layers is controlled by RRC. We can also note that the service access points consist of transport channels for access to layer 1 and logical channels for access to MAC sub-layer.

7.2. Channel typology and description

Three types of channels are defined in the UTRA radio interface:

– logical channels;

– transport channels;

– physical channels.

The focus in this chapter is on the UTRA/FDD radio interface, while bearing in mind that Chapter 12 gives more details about the UTRA/TDD mode.

7.2.1. *Logical channels*

A logical channel is defined by the type of information conveyed on, and we can distinguish two sub-types: those that carry control messages and those used for the transport of user traffic.

Logical control channels

Logical control channels are used to convey information in the C-plane. There are four types:

– BCCH (*Broadcast Control Channel*): unidirectional downlink channel supporting the system information broadcast in a cell. The system information consists of network access parameters, identity and type of the serving PLMN, information on measurement to be performed by the UE, parameters for cell selection and reselection, etc.;

– PCCH (*Paging Control Channel*): unidirectional downlink channel transporting paging information broadcast in a cell;

– CCCH (*Common Control Channel*): bidirectional common channel used for transfer of signaling messages exchanged during the RRC connection establishment, cell update and URA update procedures;

– DCCH (*Dedicated Control Channel*): bidirectional channel conveying signaling messages dedicated to a given UE. A DCCH is only used when the RRC connection has been already setup and the conveyed messages could be either generated by RRC, or originated from NAS for signaling between UE and the Core Network (CN).

Logical traffic channels

Logical traffic channels are used to convey information in the U-plane. Two types have been specified:

– DTCH (*Dedicated Traffic Channel*): point-to-point bidirectional channel dedicated to one UE for the transfer of user data after establishment of a communication;

– CTCH (*Common Traffic Channel*): point-to-multipoint unidirectional downlink channel used for transfer of user information dedicated to a group of UEs.

7.2.2. *Transport channels*

A transport channel is defined by the way the binary data is processed at the physical layer and can be seen as a service provided by the physical layer to MAC for data transfer. They are classified in three groups: common, shared and dedicated channels.

Common transport channels

A common transport channel is a point-to-multipoint channel that could be used for the transfer of information related to one or several UEs:

– BCH (*Broadcast Control Channel*). This is a downlink channel used for the transport of system information broadcast in a cell;

– PCH (*Paging Channel*). This is a downlink channel used for the transport of *paging* messages in one or several cells;

– RACH (*Random Access Channel*). This is an uplink random access channel with collision risk used for the transport of both signaling and user data from several UEs. A major difference between GSM RACH and UTRA RACH channels is that, in addition to the transport of the initial UE request to access to the network, UTRA RACH is also used to transport non-real-time user data;

– CPCH (*Common Packet Channel*). This is an uplink random access channel with collision detection only used for transport of non-real-time user data. Let us note that this channel is optional and is not used in real networks. It is removed from *Release 5* onwards;

– FACH (*Forward Access Channel*). This is a downlink channel used for the transport of signaling messages and user data. As the channel is common to several UEs, it includes the identity of the addressed user.

Shared transport channels

Only one shared transport channel, DSCH (*Downlink Shared Channel*), has been defined. DSCH is a downlink channel used in combination with one or several dedicated channels. It is dynamically shared by different users and transports control and traffic data. DSCH is optional and not used so far in real networks. It is removed from *Release 5* onwards for FDD mode.

Dedicated transport channels

A dedicated channel DCH is a point-to-point channel dedicated to only one UE. This type of channel exists in both uplink and downlink and transports control and traffic data.

Table 7.1 shows possible values of transport format attributes for each of UTRA/FDD transport channels presented above.

	BCH	PCH	FACH, DCH, DSCH, CPCH	RACH
TB size (bits)	246	1 to 5,000	0 to 5,000	0 to 5,000
TBS size (bits)	246	1 to 200,000	0 to 200,000	0 to 200,000
TTI (ms)	20	10	10, 20, 40, 80	10, 20
Type of channel coding	convolutional	convolutional	Turbo, convolutional	convolutional
Coding rate	1/2	1/2	1/2, 1/3	1/2
CRC size	16	0, 8, 12, 16, 24	0, 8, 12, 16, 24	0, 8, 12, 16, 24

Table 7.1. *Possible values of transport format attributes for UTRA/FDD transport channels*

Transfer attributes of transport channels

The characteristics or attributes of the transport channels [TS 25.302] are defined in the so-called *Transport Format* (TF). A TF is the format used for data exchange between physical and MAC layers on a given transport channel within a *Transmission Time Interval* (TTI). The TF is composed of a dynamic part that can be autonomously modified by the transmitting part at every TTI, and a semi-static part that can be modified only by reconfiguration of the channel via RRC messages.

The dynamic part further consists of the following attributes:

– *Transport Block* (TB) representing the smallest exchange unit between layer 1 and MAC on the transport channel;

– *Transport Block Set* (TBS) defined as a set of TBs within the same transport channel;

– *Transport Block Set Size* defined as the number of bits in a TBS. All TBs of a TBS have the same size.

The semi-static part comprises the following attributes:

– *Transmission Time Interval* (TTI) defined as the inter-arrival time of TBSs. A TTI can have the value of 10, 20, 40 or 80 ms;

– type of channel coding applied for error protection (convolutional coding or turbo coding);

– coding rate applied to the channel coding (1/3 for turbo code, 1/2 or 1/3 for convolutional code);

– static rate matching parameter;

– size of CRC (*Cyclic Redundancy Check*), whose result is delivered to RLC.

A *Transport Format Set* (TFS) is a set of TFs associated to a transport channel with identical semi-static parts and different dynamic parts. Following well defined TF selection rules, one of the TFs associated to the transport channel is selected at each TTI, thus allowing a fast adaptation of the transmission rate to the source data rate. A *Transport Format Combination* (TFC) is an authorized combination of currently valid TFs that can be simultaneously submitted to layer 1 on a CCTrCH of a UE. A *Transport Format Combination Set* (TFCS) is a set of TFCS given to MAC by RRC. MAC chooses between the different TFCs specified in the TFCS. MAC has control over the dynamic part of the TFC while the semi-static part related to the service QoS (error rate, transfer delay) is managed by the admission control in the RNC.

Table 7.2 gives an example of TFS and TFCS definition on a configuration with three active dedicated transport channels. Every 10 ms corresponding to the smallest TTI, MAC can select one out of the three authorized TFC defined in the example.

Transport channels	TFS	Dynamic part			Semi-static part		
		TB (bits)	TBS (bits)	TTI (ms)	Channel coding	RM	CRC
DCH1	$TF0_1$	20	20				
	$TF1_1$	40	40	10	convolutional	1	16
	$TF2_1$	160	160				
DCH2	$TF0_2$	320	320	10	convolutional	1	8
	$TF1_2$	320	1,280				
DCH3	$TF0_3$	320	320	20	turbo	2	0
The TFs above are grouped to constitute the TFCs below							
Possible TFCs (TFCS)		DCH1		DCH2		DCH3	
TFC1		$TF0_1$		$TF0_2$		$TF0_3$	
TFC2		$TF1_1$		$TF0_2$		$TF0_3$	
TFC3		$TF2_1$		$TF1_2$		$TF0_3$	

Table 7.2. *Example of TFS and TFCS*

7.2.3. *Physical channels*

On the UTRA/FDD air interface, a physical channel is defined by a carrier frequency, a channelisation code and a scrambling code. For uplink, the relative phase is also taken into account in the definition of a physical channel. In this chapter we will only present the physical channels onto which are mapped transport channels presented above. A detailed description of all UTRA/FDD physical channels is given in Chapter 9.

Uplink physical channels

Four types of UTRA/FDD uplink physical channel are defined for *Release 99*:

– PRACH (*Physical Random Access Channel*). This is the physical channel on which the RACH transport channel is mapped. It uses a random access procedure described in detail in Chapter 10;

– PCPCH (*Physical Common Packet Channel*). This is the physical channel on which the CPCH transport channel is mapped. It uses a random access procedure with collision detection described in detail in Chapter 10;

– DPDCH (*Dedicated Physical Data Channel*). This type of channel is used to convey information transported by an uplink DCH. Several DPDCH could be simultaneously used for one physical channel of one user. In this case the global bit rate is distributed on the different DPDCH;

– DPCCH (*Dedicated Physical Control Channel*). It is a physical channel used to convey control information generated by the physical layer and associated to one or several uplink DPDCH. Only one DPCCH is used regardless of the number of DPDCH it is combined with (see Chapter 10).

Downlink physical channels

Downlink physical channels may or may not support downlink transport channels, since they convey physical layer internal control information:

– DPCH (*Dedicated Physical Channel*). This is a physical channel supporting a downlink transport channel of type DCH. Associated control information (DPCCH) is time multiplexed with user data (DPDCH). Its structure is then different from the structure of a dedicated uplink physical channel for which the DPCCH and the DPDCH are basically two different physical channels;

– SCH (*Synchronization Channel*). It consists of transmitting in parallel two synchronization codes: *Primary Synchronization Code* (PSC) and *Secondary Synchronization Code* (SSC). It enables UEs to time synchronize with a given cell; to decode information carried by the downlink channels in that cell; and perform power measurements on the serving and neighbor cells (see Chapter 10);

– CPICH (*Common Pilot Channel*). This channel conveys a pre-defined sequence of pilot bits. It is a phase reference for the other downlink channels and serves for channel estimation and quality measurement of active and neighboring cells (see Chapter 11);

– P-CCPCH (*Primary Common Control Physical Channel*). This is the physical channel supporting the BCH transport channel. One and only one P-CCPCH channel is associated to each UTRA cell;

– S-CCPCH (*Secondary Common Control Physical Channel*). This type of physical channel supports a transport channel of type PCH and/or one or several FACH transport channels;

– PICH (*Paging Indicator Channel*). It is a physical channel always associated to an S-CCPCH supporting a PCH. It conveys *Paging Indicator* (PI) bits. PICH monitoring mechanism is studied in Chapter 10;

– PDSCH (*Physical Downlink Shared Channel*). A PDSCH supports transport channels of type DSCH. It is always associated to a dedicated physical channel DPCH that conveys in the DPCCH sub-channel the PDSCH control information;

– AICH (*Acquisition Indicator Channel*). This is a physical channel associated to PRACH and used to carry *Acquisition Indicators* (AI) which are used to acknowledge or unacknowledge the reception of the access preamble in the PRACH;

– AP-AICH (*CPCH Access Preamble Acquisition Indicator Channel*). This channel is quite similar to AICH and is associated to PCPCH to acknowledge or unacknowledge the access preamble on the PCPCH;

– CD/CA-ICH (*CPCH Collision Detection/Channel Assignment Indicator Channel*). This is a signaling channel also associated to PCPCH. It carries collision detection indicator and channel assignment indicator information sent by Node B to acknowledge a collision detection preamble on PCPCH;

– CSICH (*CPCH Status Indicator Channel*). This is rather a sub-channel time multiplexed with an AICH, an AP-AICH or a CD/CA-ICH channel. It conveys information on the availability of PCPCH channels in a given cell.

7.3. Physical layer

The physical layer provides transport service to MAC through transport channel as illustrated in Figure 7.2. In uplink, a channel coding is applied to data received from MAC before their transmission on the air interface. To each transport block delivered to the physical layer, MAC associates a TFI chosen in a TFS attributed to the corresponding transport channel. Then the physical layer combines the different TFIs associated to current active transport channels and indicates to the receiver this combination using the corresponding TFCI (*Transport Format Combination Indicator*). This TFCI allows the receiver to decode and demultiplex the information it receives. When several transport channels of the same type are simultaneously supported by the same physical channel, the channel coding is followed by a multiplexing chain and the channel decoding is preceded by demultiplexing chain. Operations performed in the coding/decoding and multiplexing/demultiplexing chains are described in Chapter 9.

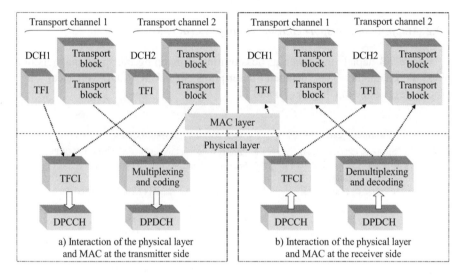

Figure 7.2. *Interaction between the physical layer and MAC in the UE*

7.3.1. *Physical layer functions*

The physical layer performs the following functions:

– channel coding/decoding for error protection on transport channels;

– multiplexing of several transport channels of the same type on a CCTrCH (*Code Composite Transport Channel*) and transmission of this CCTrCH on one or several physical channels. In the receiving side, the CCTrCH is demultiplexed towards the different transport channels;

– rate matching consisting of adding or retrieving bits after channel coding in order to adapt the data size to the physical channel capacity;

– modulation and spreading of physical channels and the corresponding reverse functions: demodulation and despreading;

– frequency and time synchronization;

– closed loop power control. Fast power control is performed at slot level by using *Transmit Power Control* (TPC) bits transmitted in the dedicated physical control channels (see Chapter 10);

– measurement and indication of measurement result to higher layers;

– support of macrodiversity in a soft handover situation.

7.3.2. *Mapping of transport channels onto physical channels*

Figure 7.3 depicts the different possibilities of mapping transport channels onto physical channels. The detailed description of UTRA/FDD physical layer procedures is done in Chapters 10 and 11.

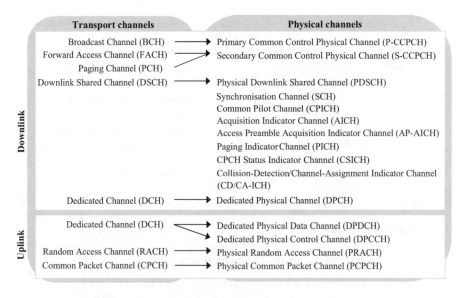

Figure 7.3. *Mapping of transport channels onto physical channels*

Models of UE's physical layer for RACH and DCH

The RACH (see Figure 7.4a) is supported by one specific physical channel, the PRACH. When several data flows are multiplexed on the RACH, the multiplexing is performed in MAC layer. As shown by Figure 7.4b, several DCHs could be simultaneously active and multiplexed in the physical layer. Channel coding is applied to each of the active DCHs and the outputs are multiplexed to form a

CCTrCH which then could be transmitted by using one or several dedicated physical channels (multicode transmission). To these physical channels conveying user data is associated a DPCCH used to transfer pilot bits, the TFCI used in the TTI, the feedback information (FBI) for the control of transmit diversity and power control bits TPC.

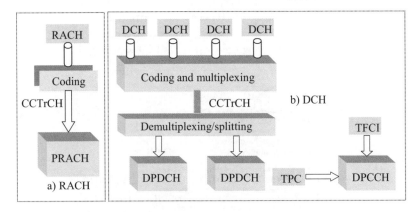

Figure 7.4. *Models of the UE's physical layer in the uplink*

Models of UE's physical layer for PCH, FACH and DCH

The BCH is exclusively supported by the P-CCPCH (see Figure 7.5a). It is characterized by fixed values of transfer attributes summarized in Table 7.1. One transport channel of type PCH and one or several transport channels of type FACH could be multiplexed to constitute a code composite CCTrCH supported by an S-CCPCH (see Figure 7.5b). When user data are transported by a FACH, the S-CCPCH could also carry a TFCI and pilot bits. A physical channel of type PICH carrying paging indicator bits (PI) is associated to any S-CCPCH supporting a PCH.

As for the uplink case, several downlink dedicated transport channels could be multiplexed in the physical layer to constitute a CCTrCH, which could be transmitted using one or several dedicated physical channels. In a macrodiversity situation (see Chapter 5), the associated DPDCHs are supported by as many links as there are cells involved in the macrodiversity (see Figure 7.5c).

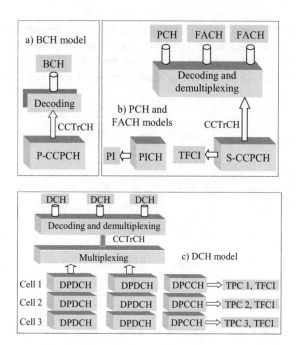

Figure 7.5. *Models of the UE's physical layer in the downlink*

7.4. MAC

The functional model of MAC sub-layer is illustrated in Figure 7.6. MAC-d controls access to dedicated channels, while the MAC-c/sh entity controls common and shared channels. With regards to MAC-b, it controls the BCH. In the network side, the MAC-d and MAC-c/sh entities are located in the SRNC, while MAC-b is located in Node B.

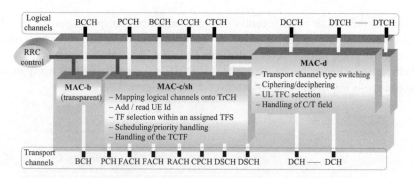

Figure 7.6. *Functional model of MAC in the UTRAN side*

7.4.1. *Main functions of MAC*

MAC controls access to the transmission medium through the following functions [TS 25.321]:

– mapping of logical channels onto transport channels; this function makes it possible to make transparent, to higher layers using MAC services, the type of transport channel on which a logical channel is mapped;

– transport channel switching on the request from RRC; this is a flexibility offered by UTRA on resource management by making it possible to dynamically adapt allocated radio resources. For example, a DTCH logical channel could be mapped on a DCH at the beginning of an Internet session. Then, upon detection of a given period of inactivity, RRC could proceed to a reconfiguration by requesting MAC to change the mapping of DTCH from a DCH to a RACH;

– control of traffic volume on each active transport channel with the help of RLC providing information on transmission and retransmission buffers. Traffic measurement results are notified to RRC which could use them for radio resources reconfiguration. MAC could be configured to deliver traffic measurement results to RRC periodically or upon detection of thresholds (low or high) of buffer occupancy;

– TFC selection of each TTI, based on priority associated to the different active logical channels and current data rate on each of these logical channels. The shorter TTI is used for TFC selection;

– priority handling between different data flows of one user and scheduling of traffic from different users on common and shared channels;

– multiplexing/demultiplexing of several logical channels on one transport channel;

– ciphering/deciphering of data when RLC transparent mode is used;

– identification of UEs on common transport channels.

7.4.2. *Mapping of logical channels onto transport channels*

Figure 7.7 shows the possible associations between logical channels and transport channels. Correspondence between transport and physical channels is also shown. We could note that for PCCH, CCCH and CTCH there is only one possible association, whereas for BCCH, DCCH and DTCH there are multiple possible associations. However, some channel associations are exclusive. For instance, BCCH is either mapped on BCH or FACH depending on the state of the connection between the UE and the network. Similarly, DCCH and DTCH are exclusively mapped either on common transport channels or on dedicated transport channels.

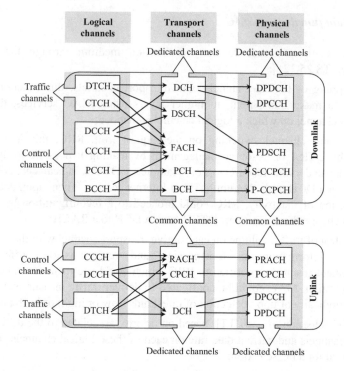

Figure 7.7. *Mapping of logical channels onto transport and physical channels*

7.4.3. *MAC PDU*

As illustrated in Figure 7.8, the MAC *Protocol Data Unit* (PDU) is composed of the *Service Data Unit* (SDU) and an optional header. The SDU field represents data submitted by a higher layer using the MAC data transmission service. This is the RLC PDU on which only ciphering/deciphering could be applied by MAC layer in the case of RLC transparent mode. MAC transfers data submitted by RLC without segmentation or concatenation.

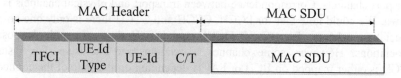

Figure 7.8. *MAC PDU structure*

The presence of the different fields of the MAC PDU header will depend on the type of transport channel and whether several logical channels are multiplexed:

– C/T (*Control/Traffic*) field is used when several dedicated logical channels are multiplexed onto the same transport channel. The name of this field is misleading since one could assume that it only indicates the type of data (control or user traffic) conveyed on the channel, whereas identifies instead the instance of the logical channel. The field is coded on four bits enabling the coding of fifteen different values of dedicated logical channel identities and one reserved value;

– UE-Id field is used to identify a UE on a common transport channel. Three types of this identifier are used in the UTRAN:

- C-RNTI (*Cell Radio Network Temporary Identity*) coded in 16 bits is used to identify the UE on a DTCH or DCCH logical channel mapped on a common transport channel (FACH, RACH or CPCH). This identifier is valid in a cell basis,

- DSCH-RNTI (*DSCH Radio Network Temporary Identity*) coded in 16 bits is used to identify the UE in the DSCH,

- U-RNTI (*UTRAN Radio Network Temporary Identity*) coded in 32 bits is used in a DCCH mapped on a common transport channel when it is not possible to identify the UE with C-RNTI. U-RNTI is composed of two parts: S-RNCID (*Serving RNC Identity*) identifying the SRNC within the UTRAN and S-RNTI (*SRNC RNTI*) identifying the UE within the SRNC. This identifier is useful in case of cell change where C-RNTI allocated to the old cell can no longer be used in the new cell. U-RNTI is exclusively used in downlink;

– UE-Id type indicates the type of identifier used and makes it possible to decode the UE-Id field;

– TCTF (*Target Channel Type Field*) is used to indicate the type of logical channel supported by FACH or RACH. For common logical channels (BCCH, CCCH, CTCH), a different value of TCTF is used to identify each of them, whereas for dedicated logical channels, a unique value of TCTF is associated to the entire group of channels – the multiplexing/demultiplexing being performed thanks to the channel instance identifier C/T.

Table 7.3 indicates, for each possible association between logical and transport channels, the presence or lack of each field in the MAC header.

	BCCH		PCCH	CCCH	CTCH	DCCH or DTCH	
	BCH	FACH	PCH	RACH FACH	FACH	DCH	RACH FACH
C/T	N	N	N	N	N	N/Y	Y
UE-Id type and UE-Id	N	N	N	N	N	N	Y
TCTF	N	Y	N	Y	Y	N	Y

Table 7.3. *Presence of optional fields in the MAC header; N: not present, Y: present, N/Y: present if several logical channels are multiplexed on the transport channel*

7.5. RLC

The RLC sub-layer provides transfer service to upper layers in three different modes [TS 25.322]:

– *transparent mode* (TM) in which data submitted by the upper layer is transferred by RLC without any additional control information and without any error control, and only segmentation/reassembly being possible to be applied. This mode is used for conversational services like voice call and video telephony;

– *unacknowledged mode* (UM) providing transfer service without any guarantee of delivery to the receiving side. This mode is suitable for packet service like streaming;

– *acknowledged mode* (AM) providing a transfer service with guarantee of delivery to the receiver, or in case of impossible delivery, notification of the non-delivery to the upper layer. It also ensures notification to the upper layers of non-recoverable error. AM is used for packet transfer without real-time constraint.

Table 7.4 presents RLC transfer mode applicable to the different logical channels.

Transfer mode	BCCH	PCCH	CCCH	CTCH	DCCH	DTCH
TM	yes	yes	yes (UL)	no	yes	yes
UM	no	no	yes (DL)	yes	yes	yes
AM	no	no	no	no	yes	yes

Table 7.4. *RLC transfer modes applicable to the different logical channels*

7.5.1. *Main functions of RLC*

In addition to user data transfer, RLC assures the following functions:

– segmentation of SDU submitted by the upper layer if its size is greater than the maximum PDU size allowed by the related RLC entity. In case of transparent transfer mode the way segmentation is performed is defined at the establishment of the RLC entity;

– reassembly of RLC SDU from received segments in a receiving RLC entity;

– concatenation of several RLC SDUs, when allowed by their size, to form a RLC PDU;

– padding of RLC PDU when it is not completely filled and concatenation is no longer possible;

– error correction by retransmission in acknowledged transfer mode;

– in-sequence delivery of upper layer PDUs in acknowledged transfer mode;

– duplicate detection in acknowledged mode;

– flow control by transmission "suspend and resume" mechanism;

– sequence number check in unacknowledged transfer with discard of incomplete RLC SDUs;

– protocol error detection and recovery: erroneous PDU sequence number, invalid PDU format, etc.;

– ciphering/deciphering for acknowledged and unacknowledged RLC transfer mode;

– suspension and resumption of transmission on request of RRC layer for non-transparent RLC transfer mode.

7.5.2. RLC PDU

RLC layer uses two types of PDUs: data PDU containing user data and control PDU used in acknowledged mode.

RLC data PDUs

For user data transfer, RLC uses three types of PDUs corresponding to the three types of transfer mode. For transparent mode, the RLC PDU contains only user data. For non-transparent mode, the RLC PDU contains, in addition to user data, a header of variable length and also a padding field when the PDU is not fully filled with user data. In the case of acknowledged transfer mode, the data PDU could be also used to convey control PDUs (*Piggybacked Status* PDU). Figures 7.9a, 7.9b and 7.9c represent the data PDU formats for the three different RLC transfer modes.

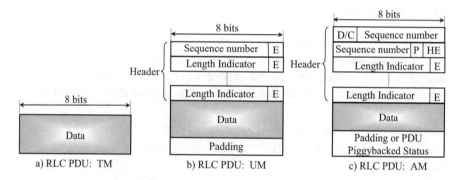

Figure 7.9. *Format types of RLC data PDUs*

In Figure 7.9 the following control fields are distinguished:

– *Sequence Number* is coded in 7 bits for unacknowledged mode or 12 bits in acknowledged mode;

– *Length Indicator* is coded in 7 or 15 bits depending on the size of the PDU. It is used to indicate boundaries of RLC SDUs in a RLC data PDU and to define whether padding or piggybacked status PDU is included in the RLC data PDU;

– *E* (*Extension*) indicates whether the subsequent byte is a payload or header data;

– *HE* (*Header Extension*) is coded in 2 bits and present in the second byte of an acknowledged mode PDU to indicate, like E bit field, whether the subsequent byte contains payload or header data;

– *P* (*Polling*) is coded in 1 bit and used to request acknowledgement to a receiving RLC entity;

– *D/C* (*Data/Control*) is coded in 1 bit. When is set to 1 it indicates that the PDU contains data.

RLC control PDUs

Control PDUs are only used in acknowledged transfer mode for handling the acknowledgement and retransmission mechanisms. We can distinguish the following four types of control PDUs:

– *Status PDU*, used by the RLC transmitting entity to request its peer receiving entity to move the reception window. It is also used by the RLC receiving entity to notify the peer transmitting entity about missing and received PDUs, the allowed size of transmission window, or to acknowledge a request from the transmitting entity to update the receiving window;

– *Piggybacked Status PDU*, which is a status PDU conveyed within a data PDU;

– *Reset PDU*, used to reset the protocol (state, variables and timers) between two peer RLC entities;

– *Reset Ack PDU*, used to acknowledge reception of a *Reset PDU*.

Figures 7.10a, 7.10b and 7.10c show the different control PDU formats, and the fields used in control PDUs:

– *D/C* is coded in 1 bit. When is set to 0 it indicates a control PDU;

– *PDU Type* is coded in 3 bits and it indicates the type of control PDU (*Status*, *Reset* or *Reset Ack*);

– *SUFI* (*Super Field*) using a TLV (type, length, value) structure to code various types of status information. Details on the different type of SUFI and the coding of each one could be found in [TS 25.322];

– *R2* (*Reserved 2*) is 1 reserved bit used for octet alignment and always set to 0;

– *R1* (*Reserved 1*) corresponds to three reserved bits always set to 0 and only used for octet alignment;

– *RSN* (*Reset Sequence Number*) is coded in one bit and represents a sequence number used to differentiate two subsequent *Reset PDUs*;

– *HFNI* (*Hyper Frame Number Indicator*) is coded in 20 bits used to synchronize HFN values in the UE and the network. HFN is an input parameter for ciphering.

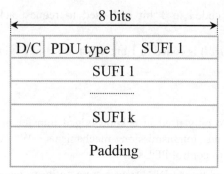

a) Control RLC PDU: *Status*

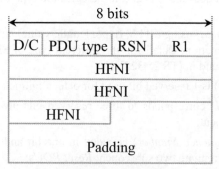

b) Control RLC PDU:
Piggybacked Status

8 bits

D/C	PDU type	RSN	R1
HFNI			
HFNI			
HFNI			
Padding			

c) Control RLC PDU:
Reset or *Reset Ack*

Figure 7.10. *Different formats of control RLC PDUs (acknowledged mode)*

7.5.3. *RLC transmission and reception model*

Transparent mode

As illustrated in Figure 7.11a, when the RLC TM entity receives an RLC SDU from the upper layers, it segments it if it is larger than the PDU size used by MAC and if segmentation has been configured by RRC. Then, the resulting RLC PDUs are submitted to MAC for transmission on the TTI. If segmentation is not configured, several RLC SDUs of the same size and for which the total size fits in the PDU size allowed in the TTI, could be submitted to MAC for transmission in the TTI. When RLC PDUs are received, either the segmentation is configured and the received PDUs are reassembled within the TTI to form the RLC SDU delivered to the upper layer, or the segmentation is not configured and the RLC SDU is equal to the RLC PDU.

Unacknowledged mode

A UM RLC transmitting entity forms a UM RLC PDU by adding a header to a payload corresponding to the upper layer data (see Figure 7.11b). The payload is constituted either by an entire SDU or by a segment of SDU or a concatenation of SDUs depending on the size of received SDUs and the maximum PDU size allowed by MAC for the current TTI. Padding bits are added if needed and if configured, ciphering is applied to the payload part of the PDU before it is submitted to MAC. In the receiving side, the RLC PDU header is used to infer how the PDU has been constituted. Then, if segmentation, concatenation or ciphering have been applied by the transmitter, the reverse operations (reassembling, separation or deciphering) are performed by the receiver before delivery to the upper layers. If missing PDUs are detected with the sequence number check, the SDU is considered as invalid and therefore discarded.

Acknowledged mode

When an AM RLC entity receives RLC SDUs from the upper layers, it segments and/or concatenates them (if necessary) to form the payload of the RLC PDU to be sent (see Figure 7.11c). *Padding field* and/or *Piggybacked Status* could be also included in this payload. Then, a header is added to the payload to constitute the PDU that is submitted for transmission and buffered for possible retransmissions. The MUX unit performs multiplexing of new PDUs and PDUs to be retransmitted. If ciphering is activated, the payload part of the PDU is ciphered and then the resulting RLC PDU is submitted to MAC for transmission. In the receiving side, reverse operations are performed. Indeed, received PDUs are first deciphered if the function has been activated. Then, deciphered PDUs are put in the receiver buffer where they are controlled for potential retransmission requests. After the reassembling is performed (if needed), the resulting RLC SDUs are delivered to the upper layers.

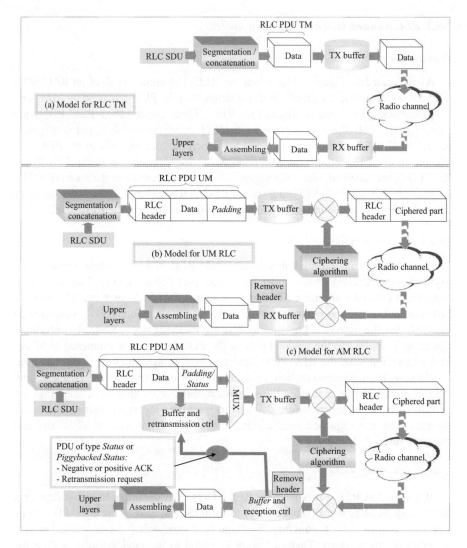

Figure 7.11. *Transmission and reception models for RLC PDUs*

7.6. PDCP

PDCP only exists in the user plane and only for services in the PS domain [TS 25.323]. PDCP functions are:

– header compression/decompression. This consists roughly of removing redundancies on control information, which is done by avoiding retransmitting information that do not vary from frame to frame (e.g. source and destination IP addresses) and using more compact coding (e.g. using delta of subsequent values rather than actual values) for some fields of the header, the value of which varies from frame to frame. Header compression enables the optimization of the use of radio resources by reducing the header size by 10 for IPv4 and even more for IPv6. In principle, a PDCP entity could support zero, one or several header compression algorithm. In *Release 99*, only the algorithm specified by IETF in RFC2507 for compression of TCP/IP header is defined;

– RLC lossless transfer of PDCP SDUs for bearer services configured for *Lossless SRNS relocation*. This service can only be provided by a PDCP entity relying on an AM RLC entity assuring in-sequence delivery of RLC SDUs. The mechanism used for lossless SRNS relocation consists of maintaining locally a sending and a receiving sequence number (*Send PDCP SN* and *Receive PDCP SN*) both in the UE and the SRNC. The *Send PDCP SN* is increased each time a PDCP PDU is submitted to RLC and the *Receive PDCP SN* is increased each time a PDCP PDU is received.

During SRNS relocation (see Chapter 8):

– the SRNC sends to the target RNC the sequence number of the next PDCP SDU expected and the sequence number of the next PDCP SDU to be sent, as well as all subsequent PDCP SDUs to be sent to the UE;

– the target RNC sends to the UE the sequence number of the next expected uplink PDCP SDU;

– the UE sends to the target RNC the sequence number of the next expected downlink PDCP SDU.

The purpose of the exchanges listed above is to acknowledge in each side the PDCP SDUs successfully transferred, thus avoiding losses and duplications of SDUs.

PDCP PDUs

As illustrated in Figure 7.12, three types of PDUs are used by the PDCP protocol:

– *No-Header* PDU (see Figure 7.12a) which is used when no header compression is applied. It corresponds to the SDU received from the upper layer;

– *Data* PDU (see Figure 7.12b) which contains a header and a data field corresponding to the compressed or non-compressed PDCP SDU, or header compression signaling information;

– *SeqNum* PDU (see Figure 7.12c) which contains a header followed by a sequence number and then a data field corresponding to the compressed or non-compressed PDCP SDU or header compression related signaling information. This type of PDU is only used during SRNS relocation.

In Figures 7.12b and 7.12c, *PDU type* field coded in 3 bits indicates whether the PDU is of type *Data* or *SeqNum*, while the *PID* field coded in 5 bits indicates the applied header compression algorithm and the type of packet. In Figure 7.12c, the sequence number field is coded in 16 bits.

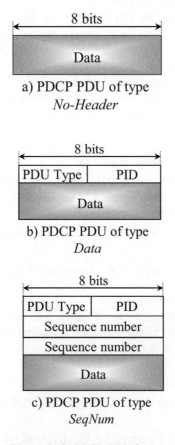

a) PDCP PDU of type
No-Header

b) PDCP PDU of type
Data

c) PDCP PDU of type
SeqNum

Figure 7.12. *PDCP PDU formats*

Example of IP header compression

Figure 7.13 shows an example where PDCP performs header compression on an IP datagram. The PDCP PDU submitted to RLC is segmented and a header is added to each segment. MAC then adds a header if required and the resulting MAC PDU (transport block) is submitted to layer1 which will add the error control bits (CRC).

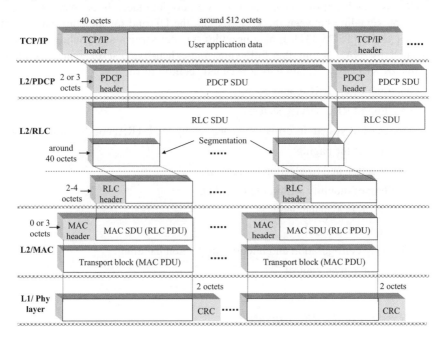

Figure 7.13. *Example of the interaction of PDCP with RLC, MAC and the physical layer*

7.7. BMC

In the UTRAN side, BMC provides broadcast service of user messages originated from the *Cell Broadcast Center* (CBC) connected to the RNC. BMC only exists in the user plane and relies on an RLC entity using unacknowledged transfer mode and mapped on the CTCH. Two types of messages are broadcast by BMC in the air interface: messages addressed to the end user and messages containing scheduling information. In *Release 99*, only user messages of type SMS-CB (*Short Message Service-Cell Broadcast*) are specified.

In the UTRAN, the BMC entity assures the reception and the storage of scheduling information and user messages sent by the CBC. In the UE, when BMC receives a scheduling message, it analyses it and provides to RRC the scheduling parameters. These parameters are used by RRC to configure lower layers for discontinuous reception. When user messages are received by the UE BMC entity, they are delivered to the related application layer when corresponding to the expected message.

7.8. RRC

RRC is the "control tower" in UTRA radio interface. It is responsible for generating signaling between the UTRAN and the UE and for the configuration of Layers 1 and 2 [TS 25.331]. It also provides transfer service of signaling messages generated by NAS. From the following non-exhaustive list of functions made possible by RRC may explain why it is called "control tower":

– handling of the RRC connection;

– handling of RRC service states;

– broadcast of system information generated by the UTRAN or the core network;

– handling of the paging;

– cell selection and reselection processes;

– handling of mobility within the UTRAN;

– radio bearers management;

– measurement control;

– ciphering configuration and integrity protection handling;

– outer loop power control.

7.8.1. *Handling of the RRC connection*

The RRC connection is a signaling connection between the UE and the UTRAN. The RRC connection establishment is always initiated by the UE when its RRC entity receives from NAS a request to send a message to the core network while no RRC connection has been established. As shown in Figure 7.14, the connection request is performed by using the *RRC CONNECTION REQUEST* message relying on the RACH procedure described in Chapter 10. The current UE identity (IMSI, TMSI, P-TMSI, IMEI, etc.) and the connection establishment cause (outgoing call, incoming call, signaling transmission, etc.) are indicated to the RNC in the *RRC CONNECTION REQUEST* message. The RNC then replies with the *RRC CONNECTION SETUP* message sent on a FACH channel and containing information about the *Signaling Radio Bearers* (SRB) to be used on the connection and the indication of the RRC state. RRC entity in the UTRAN also indicates to the UE the transport and physical channel configuration, including the uplink scrambling code, all transport formats and the carrier frequency. Depending on the indicated RRC state, the network could also allocate to the UE an RNTI for mobility handling. The UE acknowledges a successful RRC connection establishment using the *RRC CONNECTION COMPLETE* message which is sent either on a DCH in CELL_DCH state or on a RACH in CELL_FACH state. In this message the UE also indicates to the network its L1/L2 capabilities.

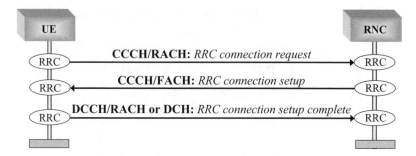

Figure 7.14. *Establishment of an RRC connection*

The RRC connection release procedure is only requested by the RRC layer in the SRNC in order to release the RRC connection, including the signaling link and all RBs between the UE and the UTRAN.

7.8.2. *Handling of RRC service states*

The UTRA radio access interface has been specified in a way that enables a flexible usage of radio resources. The principle consists of adapting at any time resources allocated to a UE to its traffic needs. Figure 7.15 presents the different service states of the RRC protocol. According to whether an RRC connection is established or not, the mobile station operates in two different modes:

– *idle mode* in which there is no RRC connection established between the UE and the UTRAN;

– *connected mode* in which an RRC connection has been established. This mode is subdivided in four states: CELL_DCH, CELL_FACH, CELL_PCH and URA_PCH.

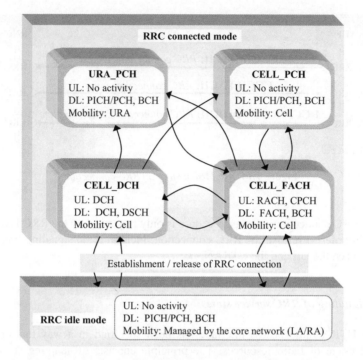

Figure 7.15. *RRC service states: mobility and transport channels involved*

CELL_DCH state

UE enters the CELL_DCH state from idle mode or CELL_FACH state when dedicated radio resources (one or several transport channels of type DCH or DSCH) are allocated to it. Dedicated resources are used for real-time traffic or transfer of a huge amount of data in packet mode. In CELL_DCH the UE receives RRC messages sent on DCCH, and with regard to the UE capabilities, system information could be received on BCH or FACH. Transitions to states CELL_FACH, CELL_PCH or URA_PCH are triggered by using signaling messages between the UE and the UTRAN. For instance, in a packet mode session, if there is no user data exchange for a while, UTRAN could ask UE to move to CELL_PCH or URA_PCH state. Upon release of the RRC connection, the UE is automatically in idle mode.

CELL_FACH state

In CELL_FACH state, no dedicated resources are allocated to the UE. Common transport channels RACH, FACH or CPCH are used for the exchanges between the UE and the network, and the UE mobility is handled at the cell level. This state is suitable for transfer of small amount of non-real-time user data and signaling messages. In CELL_FACH the UE listens to RRC messages sent on BCCH, CCCH or DCCH. Transition to CELL_PCH or URA_PCH is done when explicitly requested by the UTRAN. Upon release of the RRC connection, the UE moves to idle mode.

CELL_PCH and URA_PCH states

CELL_PCH and URA_PCH are states with very low activity in the radio interface and without transmission/reception of user traffic. In these states, UE is in DRX mode: its activity being reduced to PICH/PCH monitoring and cell reselection process. Transitions to these states are ordered by the UTRAN, for instance, when there is a long period without any user traffic exchange in CELL_DCH or CELL_FACH.

Before resuming user traffic, RRC shall move to CELL_FACH state and perform the UTRAN location update process (*Cell Update* or *URA Update*). Indeed, in CELL_PCH or URA_PCH:

– when the UTRAN receives downlink user traffic, it sends to the UE a paging message indicating traffic resumption. The UE then moves to the CELL_FACH state and performs a *Cell Update* procedure with a cause set to *paging response*. Upon completion of this procedure user traffic can then be resumed;

– for uplink traffic, the RRC entity in UE moves to the CELL_FACH state and performs a *Cell Update* procedure with a cause set to *uplink traffic resumption*. After successful completion of the procedure, the user traffic is resumed.

The main difference between CELL_PCH and URA_PCH is that the UE position is known for the former at cell level and for the latter at URA level. Moving from CELL_PCH to URA_PCH state will reduce the location update activity by performing the procedure upon URA change instead of cell change. The importance of this URA_PCH state can be easily understood if we consider for example a fast moving UE in CELL_PCH without traffic in an area composed of several micro cells.

7.8.3. *System information broadcast*

System information comprises core network or UTRAN originated control information that is broadcast in each cell. This information enables UEs to identify cells in the network coverage, know their radio environment and receive parameters needed for the use of common radio resources.

SIB and MIB definition

System information is broadcast on *System Information Blocks* (SIB), each SIB being composed of system information elements of the same nature. SIBs are transmitted in a message *RRC SYSTEM INFORMATION* which is of fixed length and that can convey one SIB segment or several concatenated SIBs. Segmentation and concatenation of SIBs are performed by the UTRAN RRC layer according to SIB size and PDU size of the related transport channel. At the receiving side (RRC layer of UE) the reverse operation (reassembling or de-concatenation of SIBs) is performed. The *RRC SYSTEM INFORMATION* message is transmitted on the BCCH, which is mapped for most of SIBs on a BCH itself supported by a P-CCPCH.

System information elements are periodically and continuously broadcast by the UTRAN in order to enable any UE that has been just switched on to rapidly get information on its radio environment. After this first acquisition of system information elements, the UE reads them again only when required, for the purpose of power saving. Indeed, the UTRAN indicates any change on broadcast system information and a UE that has already camped on a cell needs only to monitor change indications. For a UE in idle mode or CELL_PCH or URA_PCH states, the change information is a paging message, whereas for a UE in CELL_FACH a message *RRC SYSTEM INFORMATION CHANGE INDICATION* sent on FACH is used.

Figure 7.16 shows the structure of system information in the UTRAN as well as the contents of the different information blocks. A *Master Information Block* (MIB) contains change notification information (tag) and scheduling information. The MIB is essential since it shall be decoded before any SIB can be read. Scheduling information in the MIB is related to SIBs or *Scheduling Blocks* (SB). Up to two SBs can be included in the MIB containing only references to other SIBs.

In addition to the references to SIBs and SBs, the MIB contains an information element *MIB value tag* used to indicate change on system information, and an information element *PLMN type* that could take the value of GSM MAP or ANSI-41. A reference to a SIB or a SB is composed of the type of block (SIB or SB), the block changed indication and information on the scheduling of the block. The scheduling information element comprises the number of segments SEG_COUNT; the position of the first segment SIG_POS(0) within one cycle of the *Cell System Frame Number* (SFN); the repetition period SIB_REP in the number of frames and the offset SIB_OFF giving its distance in the number of frames to the previous segment.

The system information block scheduling is based on the SFN. This is the cell clock system and has a period of 4,096 frames, with SFN varying from 0 to 4,095. In order to enable the easy decoding of MIB, predefined values have been attributed to its scheduling parameters: SIB_POS, SIB_REP and SIB_OFF of MIB take, respectively, values 0, 8 and 2. This means that a MIB is broadcast from SFN 0 with a repetition period of 80 ms.

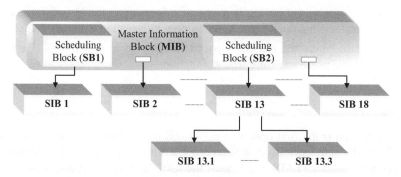

SIB1: NAS-related information and timers used by the UE in idle and connected states
SIB2: URA identity
SIB3 and SIB4: Cell selection and reselection parameters
SIB5 and SIB6: Common and shared physical channel configuration (PRACH, PDSCH…)
SIB7 and SIB8/SIB9: RACH/CPCH-related parameters
SIB10: DRAC-related parameters
SIB11 and SIB12: Intrafrequency, interfrequency and inter-RAT measurement parameters
SIB13: ANSI-41 core network information
SIB14 and SIB17: UTRA/TDD information
SIB 15: Location services parameters
SIB 16: Preconfigured radio bearers used during handover to UTRAN
SIB 18: PLMN and equivalent PLMN identities

Figure 7.16. *System information structure and SIB contents*

7.8.4. *Handling of the paging*

When the UTRAN wants to alert a UE, it sends to it an RRC paging message that could be *paging type 1* or *paging type 2*, depending on the current RRC serving state.

The message *paging type 1*, transmitted on PCCH, is used in idle mode and CELL_PCH and URA_PCH states of connected mode to notify a UE of an incoming call or change on system information broadcast in the cell. This type of paging is also used by the UTRAN when it receives downlink user traffic addressed to a UE in CELL_PCH or URA_PCH state, as an indication to move to CELL_FACH state for traffic resumption. The message *paging type 2*, also called *Dedicated paging*, is used in CELL_FACH and CELL_DCH states to notify a mobile of an incoming call. It is transmitted on a dedicated logical channel supported by a transport channel of type FACH or DCH.

When the RRC layer in the UE receives a paging message:

– if the paging has been triggered by the core network, it informs upper layers with indication of the paging cause and the relevant core network service domain;

– if the purpose of the paging is a system information change notification, the mobile will start reading the BCH to update stored information with new value;

– in the case where the mobile is paged for moving to CELL_FACH state, it starts *Cell Update* procedure for traffic resumption.

7.8.5. Cell selection and reselection

Cell selection is the first operation performed by the RRC layer of a UE at the start of its radio activity. Indeed, whenever a UE with a valid USIM is switched on, the NAS will select a PLMN and request the RRC layer to select a suitable cell of the selected PLMN. Cell selection is also performed when a UE moves back to RRC idle mode after RRC connection release. After cell selection, a UE in RRC idle mode or CELL_FACH, CELL_PCH or URA_PCH state will continue searching the best cell by regularly monitoring radio environment in accordance with defined criteria. When a more suitable cell is found, it is selected instead of the previously selected one. This procedure is called cell reselection. Cell selection and reselection procedures are discussed in detail in Chapter 11.

7.8.6. UTRAN mobility handling

Procedures that are performed for UTRAN mobility management will depend on the RRC mode.

Mobility in idle mode

When a UE is in RRC idle mode, its presence is ignored by the UTRAN. If it is attached to CS and/or PS service domain, its mobility will be managed by the core network using TMSI and/or P-TMSI temporary identities. Once attached, the UE in RRC idle mode will monitor the PCH (via PICH) of the current cell to detect any paging message notification. DRX mechanism is used by the UE for power saving. In this mode the UTRAN mobility management is limited to neighboring cells monitoring and cell reselection. As previously mentioned, it is up to the UTRAN to determine the reselection rules by broadcasting in each cell the system information elements related to cell reselection (list of neighbor cells to monitor, measurement to perform, reselection criteria, etc.). The cell reselection procedure is, in this case, autonomously executed by the UE.

Mobility management in URA_PCH, CELL_PCH and CELL_FACH states

In these states, the UE is connected to the UTRAN and has been attributed a temporary identifier which is U-RNTI in URA_PCH state and C-RNTI in CELL_PCH and CELL_FACH states. Similarly to idle mode case, also here the mobility related procedures are initiated by the UE and cell reselection is based on system information broadcast by the UTRAN in the cell.

In addition to cell reselection, the UE also performs in these states location update procedures as follows:

– after cell reselection in CELL_FACH, the *Cell Update* procedure is performed by the UE RRC layer to enable the localization of the terminal at cell level;

– if cell reselection occurs in CELL_PCH, the UE RRC will first move to CELL_FACH, perform the *Cell Update* procedure and then move back to CELL_PCH if neither the UE nor the UTRAN have user data to transfer;

– in the case of cell reselection in URA_PCH resulting in a new cell that belongs to a new URA, the *URA Update* procedure is performed by the UE RRC layer to enable the localization of the UE at URA level. In order to perform the *URA Update* procedure, UE RRC shall move first to CELL_FACH and then, if after completion of the procedure there is no user data to transfer, RRC will move back to URA_PCH.

In all the above cases, if the new selected cell belongs to a new RNS, the UTRAN could reallocate a new U-RNTI identifier to the UE, in particular if the location update procedure has resulted in an SRNS relocation procedure (see Chapter 8).

Mobility in CELL_DCH

In CELL_DCH, the position of the UE is known at cell level. Mobility within the UTRAN is controlled by RRC in the SRNC by using the measurement results provided by the UE and the knowledge of the network topology to perform soft or hard handover. *Soft handover* is executed by means of the ACTIVE SET UPDATE procedure used to update active set cells in a macrodiversity situation. This type of handover consists of removing and/or adding radio links, while assuring that at any time there is at least one active radio link. Figure 7.17 is a sequence diagram illustrating the soft handover procedure (see also Chapter 5). In the *hard handover* case, used for inter-frequency handover or when the network does not support soft handover, the radio link between the UE and the UTRAN is completely released at a given moment. Hard handover is executed by means of RRC procedures defined for radio resources reconfiguration. More details on soft and hard handover can be found in Chapter 11.

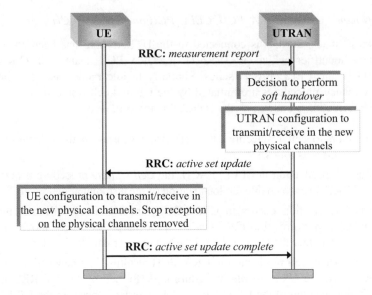

Figure 7.17. *RRC peer-to-peer exchanges during soft-handover procedure*

Inter-RAT mobility

For a UE supporting other radio access technologies (RATs) than UTRA, RRC also assures state transitions to or from those RATs. Figure 7.18 shows state transitions between UTRA and GSM. We notice from this figure that between UTRA and GSM connected modes direct transitions are possible, whereas between UTRA connected mode and GPRS transfer mode transitions are always done via idle modes. This is due to the fact that for GPRS, mobility within the radio access network is based on cell reselection (no handover). In some cases the cell reselection could take up to several seconds, that is why GPRS is not very appropriate for real-time services.

Cell reselection and handover to or from another RAT are in principle based on measurements made on neighboring cells belonging to this other RAT (see Chapter 11). However, UMTS standard enables inter-system blind handover which is executed without prior measurements on the target system. It should be also noted that inter-RAT cell reselection could be ordered by the network using *Inter-RAT cell change order from the UTRAN* procedure (UTRAN → GPRS) and *Inter-RAT cell change order to the UTRAN* procedure (GPRS → UTRAN).

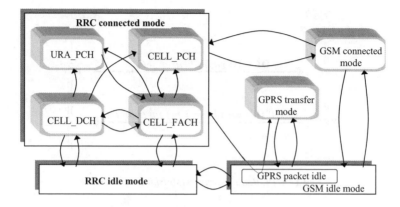

Figure 7.18. *Transition between UTRA and GSM/GPRS radio states*

7.8.7. *Radio bearer management*

Radio resources are rare and CDMA access technology is very sensitive to interferences. However, one of the main requirements for UMTS system specification is the capability to provide various bearer services, dynamically configurable during a call. That is why the system has been designed to give to the network flexible management of resources allocated to UEs.

Radio bearer establishment

On the user plane, a radio bearer is composed of a stack of resources through the different protocol layers (physical and transport channels, RLC and PDCP entities). A radio bearer could also be seen as one section of a RAB, which is a communication path between the UE and the core network (see Chapter 3). As illustrated in Figure 7.19, the radio bearer establishment is initiated by RRC in the RNC by sending *RADIO BEARER SETUP* message to its peer entity in the UE. This message conveys instructions to RRC in the UE for the configuration of lower layers (PDCP, RLC, MAC and physical layer) to support the QoS for the requested service. At the physical layer, it consists of configuring the type of transport channel (DCH or RACH/FACH), the TFCS, the uplink scrambling code, the frequency carrier, etc. The UE will send back to the RNC the *RADIO BEARER SETUP COMPLETE* message to acknowledge the radio bearer establishment request.

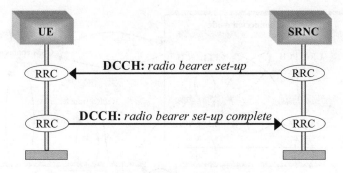

Figure 7.19. *Radio bearer establishment procedure*

Signaling radio bearers

In the control plane, RRC messages are transferred on specific radio bearers called SRB for *Signaling Radio Bearer*. The SRBs have a predefined configuration summarized in Table 7.5. In addition to RRC signaling messages, SRBs are used to convey NAS signaling and SMS messages.

Signaling radio bearer	Logical channel	RLC transfer mode	Type of message transferred on the SRB
SRB0	CCCH	UL: TM, DL: UM	RRC messages (URA/CELL update)
SRB1	DCCH	UM	RRC messages
SRB2	DCCH	AM	RRC messages (except *Direct transfer*)
SRB3	DCCH	AM	NAS messages with high priority
SRB4	DCCH	AM	NAS messages with low priority (e.g. SMS). This SRB is optional

Table 7.5. *Signaling radio bearer configuration*

Example of radio bearers multiplexing in downlink

Let us consider the example illustrated in Figure 7.20 where a user is downloading data while simultaneously receiving signaling messages from the UTRAN and the CN. In order to support the related traffic, the UTRAN has established one RAB for user data and four SRBs. The UE/UTRAN signaling messages are supported by SRB1 and SRB2 whereas UE/CN signaling messages are supported by SRB3 and SRB4. For user data, a RAB configured to support interactive traffic is established.

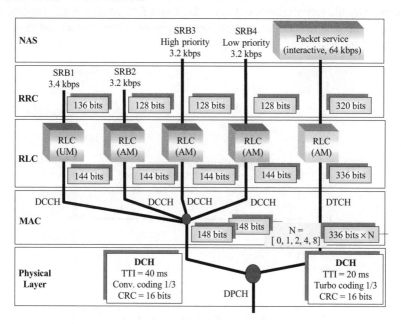

Figure 7.20. *Example of the realization of data and signaling radio bearers*

Reconfiguration and release of radio resources

Several RRC procedures have been defined for a flexible management of radio resources in the UTRAN. Indeed, a radio bearer involved in a communication between the UE and the network could be released by using the RRC *RADIO BEARER RELEASE* message. It could be also modified by means of *RADIO BEARER RECONFIGURATION*, *TRANSPORT CHANNEL RECONFIGURATION* and *PHYSICAL CHANNEL RECONFIGURATION* messages. Due to the dependency between the type of resources at the different protocol layers (e.g. a common transport channel can only be mapped on a common physical channel), reconfiguration of a given layer could lead to the reconfiguration of the other layers. That is why the names given to the above messages are misleading, since each of these messages could be used to reconfigure all the radio layers. For example, *TRANSPORT CHANNEL RECONFIGURATION* procedure could also cover reconfiguration of the physical channel, RLC and PDCP entities.

Configuration or release of radio resources is always requested by the UTRAN and the UE shall send in response either a completion message, in the case of a successful operation, or an error message in case of failure. Figure 7.21 is a generic example of radio resources reconfiguration process. When a physical layer is concerned by the reconfiguration, the UTRAN will first configure the physical layer in order to be able to send and receive on the new physical channel and then it will send to UE the RRC complete message.

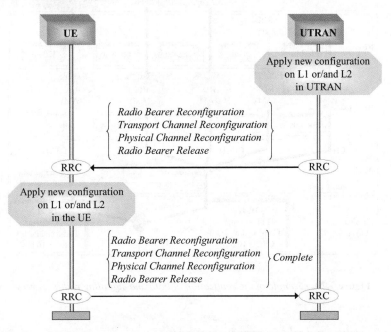

Figure 7.21. *Radio resource reconfiguration procedure*

7.8.8. *Measurement control*

Measurement management is a fundamental function as several other functions (mobility management within the UTRAN, radio bearer reconfiguration, etc.) could rely on measurement results. UMTS standard has specified several types of measurement that the UE shall execute and whose results should be reported to the UTRAN. These measurement types are detailed in Chapter 11. We can distinguish for example:

– received signal power measured on current and neighboring cells. The results of these measurements are used in the UE for cell selection and reselection, and in the UTRAN for handover handling;

– traffic volume measurement on uplink logical channels. The results of these measurements enable the UTRAN to detect the necessity to reconfigure radio resources allocated to the UE. For example, when a UE transmitting on a RACH indicates to the UTRAN (measurement report) that the upper threshold of transmit buffer is reached, the UTRAN could perform a radio bearer reconfiguration procedure by replacing the RACH with a DCH;

– quality measurement based on BLER and used for QoS monitoring and handling.

7.8.9. *Ciphering and integrity*

Ciphering and integrity protection are essential for the safety of exchange over radio interface. The former assures confidentiality in message exchange and the latter, which is a new feature compared to GSM, enables the addressee of the signaling message to authenticate the originator. Ciphering is executed by the RLC layer in case of acknowledged or unacknowledged RLC transfer mode, and by the MAC layer for RLC transparent transfer mode. As for integrity protection, it is executed by the RRC layer. Both ciphering and integrity protection procedures are activated when requested by the core network (see Chapter 8).

The *security mode control* procedure shown in Figure 7.22 makes it possible to start or to modify ciphering or integrity protection. It is also used for stopping ciphering. Upon receipt of the security configuration command from a given CN service domain (CS or PS), the UTRAN sends to UE the RRC message SECURITY MODE COMMAND, containing among other parameters the integrity algorithm (UIA), the ciphering algorithm (UEA) and the parameter FRESH. If the values of configuration parameters received by UE are not acceptable because, for example, they are not compatible with its capabilities, the UE then sends to the UTRAN as a response the RRC message SECURITY MODE FAILURE. If the received security configuration message is acceptable, the UE then takes into account the parameters and responds to the UTRAN with a RRC message SECURITY MODE COMPLETE.

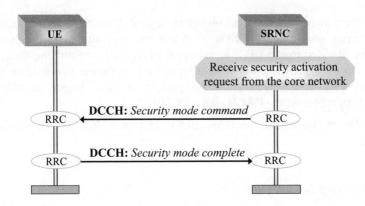

Figure 7.22. *Security mode control procedure*

Ciphering and integrity processes

The ciphering process (see Figure 7.23a) applies separately to each active radio bearer. It is based on the *f8* algorithm receiving as input parameters the ciphering key CK (see Chapter 8), the dynamic parameter COUNT-C that is incremented for every new message, the bearer identity BEARER, the transmission direction DIRECTION (uplink, downlink) and the size of the ciphering block LENGTH. The ciphering block at the output of *f8* is applied in bit-by-bit basis to the message to be transmitted. The reverse operation is performed at the receiving side to *decipher* the received message.

Regarding the integrity protection mechanism (see Figure 7.23b), it is separately applied to each active SRB. It is based on the *f9* algorithm receiving as input parameters the integrity key IK (see Chapter 8), the integrity sequence number COUNT-I formed with the local RRC hyper frame number and the RRC message sequence number, the random number FRESH generated by the RNC, the transmission direction bit DIRECTION and the signaling message to be transmitted including the SRB identity. The algorithm *f9* produces as output an integrity message MAC-I (*Message Authentication Code*) which is simply appended to the signaling message. At the receiving side, the expected integrity message XMAC-I (*eXpected MAC*) is generated and compared to the received MAC-I, and the received RRC signaling message will only be accepted if XMAC-I is equal to MAC-I.

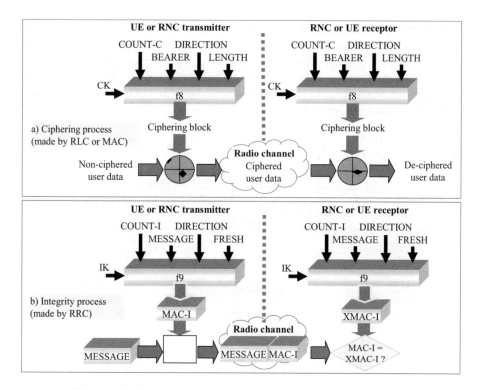

Figure 7.23. *Illustration of ciphering and integrity protection processes*

7.8.10. *Outer loop power control*

Outer loop power control is used by RRC to set the target value for the inner loop power control executed by the physical layer. For each downlink physical channel used for inner loop power control, the UE sets a target value of the *Signal-to-Interference Ratio* (SIR). This SIR value is set to a value which makes it possible to achieve the error ratio requested by the network for the related transport channel. For a transport channel of type DCH, the error ratio BLER (*Block Error Rate*) on transport blocks is generally used. More details about the outer loop power control procedure are given in Chapter 10.

7.8.11. *Protocol layers termination in the UTRAN*

Protocol termination will depend on the type of active transport and logical channel, as depicted in Figure 7.24. The physical layer is always terminated in Node B, except in the case where macrodiversity is used and for which the termination node is RNC. A dedicated channel can only be established on an existing RRC connection and, as such, the involved layers RRC, RLC, MAC, PDCP and BMC are located in the UE and the SRNC. Logical channels exist regardless of the existence of an RRC connection and are therefore handled by the CRNC. BCH represents an exception for which RRC is distributed between Node B and CRNC, and the other protocols terminate in Node B.

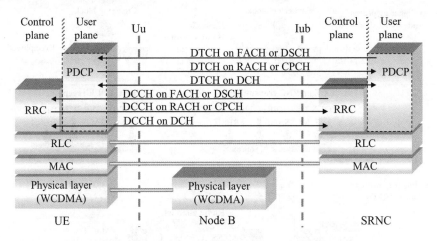

a) Location of radio protocol in UTRAN: dedicated channels

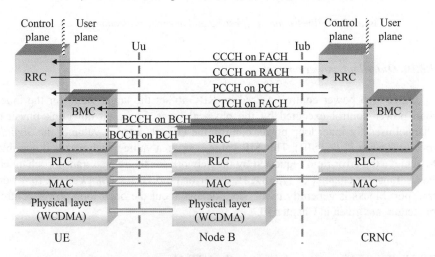

b) Location of radio protocol in UTRAN: common channels

Figure 7.24. *Location of radio protocols in the UTRAN. Dashed lines mean "if used"*

Chapter 8

Call and Mobility Management

8.1. Introduction

In a wired telephone network system, the terminal is permanently attached to the network from a fixed point. To make a call, the user has to simply pick up the handset and dial the number of the remote user at any time. The case of a mobile network is more complex for the following reasons:

– the user equipment is neither permanently attached to the network nor always localized by the network;

– the user equipment is a *mobile* equipment and therefore, in order to be reachable at any time, it must keep its localization up-to-date;

– a geographic area could be covered by several different mobile operator networks and, in this case, a user equipment shall perform the selection of an authorized network. The selected network shall in turn perform admission control on mobile equipments attempting attachment to it.

We can easily see from the above statements that call establishment on a mobile network is not a direct procedure as on a wired network. Indeed, to make the call establishment possible, the following conditions must be fulfilled:

– a successful selection of a mobile network;

– a successful attachment to the selected network;

– updating of the user equipment location.

These functions are the responsibility of *Non-Access Stratum* (NAS) protocols as shown in Figure 8.1.

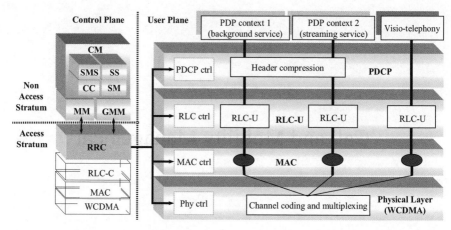

Figure 8.1. *Example showing the location of NAS protocol layers in the UE control plane*

Signaling between the UE and the network is the responsibility of three protocol sub-layers which constitute the protocol layer 3 of the system:

– RRC (*Radio Resource Control*) sub-layer (already described in Chapter 7) is in charge of secure signaling exchange in the radio interface;

– MM (*Mobility Management*) is in charge of all the functions related to the mobility of the UE with regard to the core network (PLMN selection, network attachment, location update, etc.). It is composed of the protocol entities MM and GMM (*GPRS Mobility Management*) respectively responsible for the mobility management of the CS and the PS network domains;

– CM (*Connection Management*) is composed of four types of entities:

  - CC (*Connection Control*) for the handling of connections in the CS domain,

  - SM (*Session Management*) for the handling of sessions in the PS domain,

  - SS (*Supplementary Service*) used for the activation of supplementary services,

  - SMS (*Short Message Service*) for the sending and receiving of text messages.

8.2. PLMN selection

The PLMN (*Public Land Mobile Network*) is a mobile network composed of an access network and a core network. Each PLMN in the world is uniquely identified by a PLMN identifier composed of two fields:

– the MCC (*Mobile Country Code*) indicating the country where the network is located. For example, to a network located in France will be attributed an MCC equal to 208. North America represents an exception with the attribution of several MCCs to the same country. Attribution of MCCs is under the unique responsibility of ITU. Due to this centralization, the uniqueness of the MCC attributed to a network is guaranteed;

– the MNC (*Mobile Network Code*) which is used to differentiate networks within a same country. MNCs attribution is under the responsibility of the regulation authority of the network country.

Equivalent PLMNs

With UMTS, the concept of "equivalent PLMNs" was introduced. PLMNs are equivalent to each other with regard to PLMN selection/reselection, cell selection/reselection and handover procedures. An operator owning both GSM and UMTS network in a same country could use a different PLMN code for each of them. If the two networks are located in different countries, it will be compulsory to use different PLMN codes as well. Due to the list of equivalent PLMNs sent by the serving network, the UE can select a cell belonging to an equivalent network regardless of the radio access technology. A UMTS network shared by different operators and the optimization of radio coverage at country borders are other benefits of the introduction of the "equivalent PLMN" concept.

Regarding the definition of a PLMN, a given mobile will have to differentiate:

– the HPLMN (*Home PLMN*) for which the user has subscribed a network service provision. The operator of the HPLMN is the owner of the USIM in the user equipment;

– the VPLMN (*Visited PLMN*) which is a PLMN the UE has accepted in a roaming situation and which is different from the HPLMN;

– the RPLMN (*Registered PLMN*) which is the last PLMN on which the UE has registered successfully. It could be either the HPLMN or a VPLMN.

PLMN selection can be either manual or automatic and follows rules defined in 3GPP specification [TS 23.122]. PLMNs other than the HPMLN could be ranked in different PLMN selector lists with, in the case of a dual RAT mobile, the possibility to associate one or several radio access technologies (UTRA, GSM) to each entry of these lists. Here we have the different PLMN selector lists:

– the user controlled selector list with its associated radio access technologies;

– the operator controlled selector list with its associated radio access technologies;

– the PLMN selector list without associated radio access technologies.

When the UE is switched on or after recovery from out-of-network coverage, the UE will first select the RPLMN by using all supported radio access technologies. If the RPLMN is found, it then tries to re-attach itself to it. If the RPLMN has not been found or the registration has failed, the UE will then execute the PLMN selection algorithm according to the configured mode (manual or automatic), as described in the following sections.

8.2.1. *Automatic PLMN selection mode*

Automatic PLMN selection is performed in the following priority order of available and allowed PLMNs:

– the HPLMN;

– PLMNs of the user controlled list, if it exists, in the priority order;

– PLMNs of the operator controlled list, if it exists, in the priority order;

– the selection list without associated access technologies, if user controlled and operator controlled selector lists do not exist;

– list of other available PLMN/RATs with high quality received signal in a random order;

– list of other PLMN/RATs in order of decreasing quality of received signal.

The standard does not specify the priority between RATs when several radio access technologies are associated to a PLMN. The PLMN selection procedure is completed as soon as a PLMN is found and its registration succeeded, or when the PLMN search has failed on all existing lists. For the last two lists, the UE performs its search in all the supported RATs before proceeding to the selection. For this purpose, quality comparison criteria are specified between the different radio access technologies.

8.2.2. *Manual PLMN selection mode*

In manual selection mode, upon being switched on, the UE searches for supported RATs on all available PLMNs and presents to the user the list regardless of whether PLMNs are *forbidden* or not. The list is presented in the same order as the one for automatic selection mode. One or several RATs with or without priority order could be associated to each PLMN of the list. The user chooses one PLMN from which list presented by the mobile that will then attempt to register on it.

8.2.3. PLMN reselection

PLMN reselection consists of an attempt to select a PLMN with a higher priority than the current serving one. Only possible in idle mode, PLMN reselection is either requested by the user or automatically triggered under the control of a periodic timer in a roaming situation (i.e. when the current serving PLMN is a VPLMN).

PLMN reselection requested by the user

As for the selection, PLMN reselection requested by the user could be executed either in automatic mode or in manual mode. Manual mode and automatic PLMN reselection procedures are quite similar to the equivalent procedure for PLMN selection. However, for automatic reselection in case of failure of the reselection attempt, the UE will try to return to the PLMN which was selected before the reselection was triggered.

PLMN reselection in a roaming situation

When the UE is roaming, it can periodically attempt to reselect its HPLMN or a PLMN (from the selector lists) with higher priority than the VPLMN. The reselection will be, however, performed with the following restrictions:

– the reselection is only possible in idle mode;

– when an *equivalent PLMN* list is provided by the VPLMN, reselection of the HPLMN can only be attempted if it is not included in the *equivalent PLMN* list, and a PLMN from the user or the operator list is selected only if its priority is higher than the priority of all the equivalent PLMNs;

– only the PLMNs with the highest priority among the VPLMNs in that country are candidates for reselection.

The periodicity of reselection attempts is set according to a value of a parameter that could be stored in the USIM. When present, this parameter could take values in the range of 6 minutes to 8 hours in 6 minute steps, or a value meaning that no periodic reselection shall be made. If the parameter is not present in the USIM, a default value of 60 minutes is taken as reselection period.

8.2.4. Forbidden PLMNs

An attempt to register to a VPLMN could be rejected due to *PLMN not allowed*. In this case, the related PLMN will no longer be used for automatic PLMN reselection. Indeed, the PLMN becomes a *forbidden PLMN* (FPLMN) and its identity is stored in the *forbidden PLMNs* list used for this purpose in the USIM. A PLMN will be removed from the forbidden PLMN list in case of successful registration following its manual selection. The list of *forbidden PLMNs* in the USIM will not be lost when the UE is switched off or the USIM is extracted. An optional extension of this list may be stored in the terminal's own memory.

In addition to the list of *forbidden PLMNs* valid for both CS and PS services, the UE handles another list named *forbidden PLMNs for GPRS service* which is only valid for PS services. A VPLMN is added to the list when it has been rejected due to *GPRS services not allowed in this PLMN* during a registration attempt. This list, which is stored in the terminal's memory, is deleted in case of terminal switch off or removal of the USIM. A PLMN is removed from the list of *forbidden PLMNs for GPRS service* after a successful registration on that PLMN following its manual selection.

8.3. Principle of mobility management in UMTS

The handling of the UE location is a key function for a mobile communication system. It is based on the location areas each including a number of cells uniquely identified. When a UE is not in RRC connected mode, in which its location is known at the cell level, it provides to the network the identity of its current location area in order to be reachable if needed without an exhaustive (network wide) paging. Mobility management procedures include also security aspects on the link between the network and the UE. All terminal location related procedures are triggered by protocols MM and GMM for, respectively, the CS and PS service domains. Location updates can only be performed in certain states of protocols MM and GMM. These states are illustrated by Figure 8.2 and described in Table 8.1.

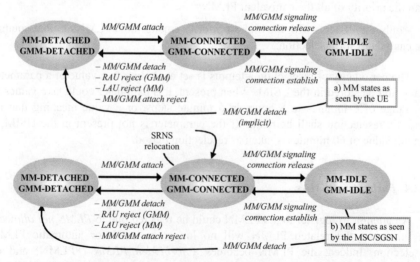

Figure 8.2. *CS and PS MM service states. In the CS domain, the UE performs MM attach and MM detach if signaled by the network only*

	MM/DETACHED	MM/GMM IDLE	MM/GMM CONNECTED
State description	The UE is detached from the CS/PS domain and its location is unknown by the MSC/SGSN	The UE is attached to the CS/PS domain and its location is known by the MSC/SGSN at the LA/RA. No signaling connection is established between the UE and the CS/PS domain	The UE is attached to the CS/PS domain. A signaling connection is established between the UE and the CS/PS domain. Data and signaling could be transferred between the UE and the network
Location update **normal**	No	Yes	No, CS domain Yes, PS domain
periodic	No	Yes	No

Table 8.1. *Description of MM states for the CS and the PS domains*

8.3.1. *Location areas*

Two location area types are handled by the core network (see Figure 8.3):

– LA (*Location Area*) for the CS domain. It consists of a set of cells under the control of an RNC and managed by a unique 3G-MSC/VLR;

– RA (*Routing Area*) for the PS domain. It consists of a sub-set of cells within an LA that are under the control of an RNC and managed by a unique 3G-SGSN.

For unambiguous identification of each location area, the identification codes LAI (*Location Area Identification*) and RAI (*Routing Area Identification*) include the MCC and MNC enabling a universal identification of these location areas (see Figures 8.4a and 8.4b). The MCC identifies the network country and the MNC identifies a network in a given country. An LAC uniquely identifies a location area of the CS domain within a network, while an RAC identifies a routing area of the PS domain within an LAC. LAs and RAs are respectively managed in the core network by VLRs and SGSNs.

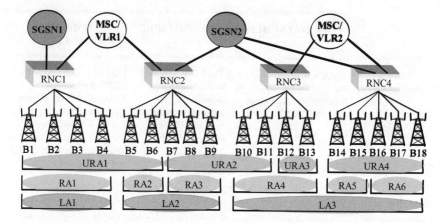

Figure 8.3. *Relations between location areas handled by the core network and the UTRAN*

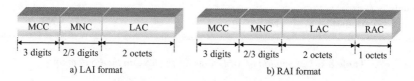

Figure 8.4. *Formats for LAI and RAI*

Table 8.2 below summarizes the division of location/routing area management between the CN and the UTRAN. The equivalent division in the GSM/GPRS network is also indicated for comparison purpose. We can see that the UTRAN is in charge of the mobility at cell level for UMTS, whereas in the case of GPRS networks this mobility is handled by the SGSN. In addition, UMTS introduces the concept of URA (*UTRAN Registration Area*) which is also managed by the UTRAN (see Chapters 6 and 7). There is no direct relationship between URA and the other registration areas LA and RA handled by the CN.

	MSC/VLR (CS)		SGSN (PS)		UTRAN (CS and PS)
	GSM	UMTS	GPRS	UMTS	UMTS
Cell	No	No	Yes	No	Yes
URA	-	No	-	No	Yes
RA	-	No	Yes	Yes	No
LA	Yes	Yes	No	No	No

Table 8.2. *Comparison of location areas in GSM, GPRS and UMTS networks*

8.3.2. *Service states in the core network and the UTRAN*

The management of the UE mobility is shared between the core network and the UTRAN. Finding the exact correspondence between service states in the core network and the UTRAN is not straightforward, as the UE could be in only one RRC mode regardless of its registration in both CS and PS network domains. For instance, the mobile could be in MM-IDLE mode and operate in RRC connected mode if PS service state is GMM-CONNECTED. In general the following can be stated:

– in RRC connected mode, at least one of the core network domains is in connected state (MM-CONNECTED or GMM-CONNECTED) and the UE mobility is controlled by the UTRAN. When the UE is in one of the RRC connected states CELL_DCH, CELL_FACH, CELL_PCH or URA_PCH, it does not take into account system information broadcast in the cell which gives identities of current RA and LA. This means that the terminal will perform *location updating* and *routing area updating* procedures – further described in this chapter – only if it is explicitly signaled by the SRNC (e.g. after SRNS relocation);

– in RRC idle mode, the UE is in MM-IDLE and GMM-IDLE states and its mobility is controlled by the core network. The UE reads system information broadcast in the current cell and performs *location updating* or *routing area updating* procedures respectively;

We notice that in the MM-DETACHED and GMM-DETACHED states, the UE is also in RRC idle mode, however, in these service states, neither the UTRAN nor the CN are involved in handling its mobility.

8.4. Network access control

During initial network registration, when the UE is switched on or in subsequent registration updates, security procedures may be performed to ensure confidentiality, authenticity and integrity of the messages exchanged between the UE and the network.

8.4.1. *Allocation of temporary identities*

Each subscriber of a mobile network has a permanent identifier called IMSI (see Chapter 3). For the purpose of identity confidentiality (protection against identification or localization of a user by an intruder), the network will not frequently transmit the permanent identifier on the radio interface. This is why, at the initial registration, the VLR (CS domain) or the SGSN (PS domain) allocates to the user a temporary identifier which is dynamically chosen and is only valid within a given location area. The temporary identifier is called TMSI (*Temporary Mobile Subscriber Identity*) for the CS domain and P-TMSI (*Packet Temporary Mobile Subscriber Identity*) for the PS domain. The allocation of a temporary identity is performed within the registration (initial or update) procedures or in a dedicated procedure. Figure 8.5 and Table 8.3 illustrate the temporary identity allocation procedure (see also [TS 24.008]).

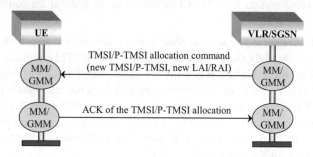

Figure 8.5. *Temporary identity allocation procedure*

Type of procedure	Command message	Ack message
IMSI attach or *Location updating* in the CS domain	LOCATION UPDATING ACCEPT	TMSI REALLOCATION COMPLETE
GPRS attach (PS domain) or *Combined GPRS attach* (PS and CS domains)	ATTACH ACCEPT	ATTACH COMPLETE
RA updating (PS domain) or *Combined RA updating* (PS and CS domains)	ROUTING AREA UPDATE ACCEPT	ROUTING AREA UPDATE COMPLETE
Dedicated procedures *IMSI* (CS domain) or *P-TMSI reallocation* (PS domain)	TMSI/P-TMSI REALLOCATION COMMAND	TMSI/P-TMSI REALLOCATION COMPLETE

Table 8.3. *Types of message used in UE temporary identity allocation procedures*

8.4.2. *UE identification procedure*

The network can identify the UE either from its IMSI or its *International Mobile Equipment Identity* (IMEI). The IMSI is requested by the network when it can no longer (e.g. after a database failure) identify the mobile from a received temporary identity (see Figure 8.6). This is a weakness in confidentiality and it should be followed by a temporary identity allocation procedure. Regarding the IMEI, it is requested by the network with possibly the software version IMEISV (*IMEI Software Version*) to check whether the UE is in a list of forbidden equipment. This list can contain identifiers of stolen or non-compliant mobile equipments.

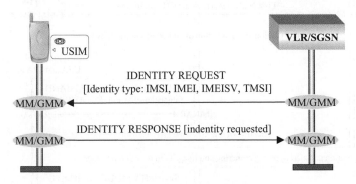

Figure 8.6. *Mobile identification procedure*

8.4.3. *Ciphering and integrity protection activation*

Ciphering is used for the confidentiality of data and signaling conveyed on the radio interface. As for integrity protection, which is new in comparison with GSM, it enables authentication of signaling messages exchanged on the radio access network and represents a key feature in UMTS security system. Both ciphering and integrity protection take place in the *access stratum* but their activation is under the responsibility of the core network for each of the service domains (see also Chapter 7). The activation is initiated by the VLR/SGSN upon receipt of the first NAS level signaling message from the UE (*ATTACH REQUEST, LOCATION UPDATING REQUEST, ROUTING AREA UPDATE REQUEST, CM SERVICE REQUEST* or *PAGING RESPONSE*). This first signaling message contains the UE identity and a parameter KSI (*Key Set Identifier*) which enable the network to identify the ciphering and integrity protection keys stored in the UE without performing the authentication procedure. The ciphering and integrity protection activation (security mode control) procedure is illustrated in Figure 8.7.

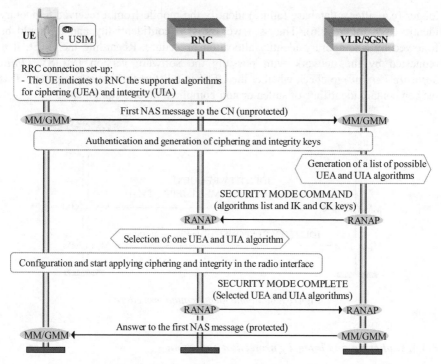

Figure 8.7. *Security mode control procedure*

8.4.4. *Authentication*

The authentication procedure, on the one hand, enables the network and the mobile to authenticate mutually and, on the other hand, enables the network to provide to the mobile parameters which enable the latter to calculate a new UMTS ciphering key or a new UMTS integrity key. It is worth noting this difference with GSM system in which mutual authentication is not supported, as the GSM mobile station never authenticates the network.

The principle of mutual authentication is based on a mutual supply by the network and the UE of the proof that they hold a secret, while not divulging it. The secret consists of the key K stored both in the USIM module of the UE and the *Authentication Centre* (AuC) of the CN. The authentication procedure is secured since the key K, which is a static parameter, is never exchanged between the UE and the CN. Only dynamic parameters generated from the key K and a random number RAND (therefore not predictable) are exchanged. Figure 8.8 shows the mechanism used to generate authentication parameters within the AuC. "⊕" and "∥" represent respectively "exclusive or" and "concatenation" operations.

The AuC first generates a sequence number SQN and a random number RAND[1]. Then, the parameters SQN, RAND, K and AMF (*Authentication Management Field*) are given as inputs to functions *f1* to *f5* (present in the USIM and the AuC) to generate elements of the authentication vector. AMF enables the signaling of supplementary parameters used for a flexible handling of the authentication process, for instance, the indication of the algorithm and key used to generate an authentication vector when several algorithms and keys are possible.

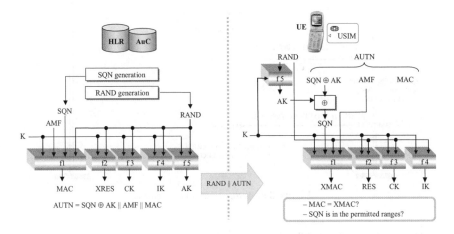

Figure 8.8. *Generation of authentication vector*

In addition to the ciphering and integrity keys CK and IK, the following parameters are output of *fi* functions:

– AK (*Anonymity Key*) is used to conceal the sequence number *SQN* when the latter could enable a passive attack (identification and localization) of the UE;

– XRES (*eXpected RESponse*) which is the response expected from the UE for its authentication;

– MAC (*Message Authentication Code*) which is the authentication code of the message to send to the mobile.

The AuC builds the authentication token AUTN composed of three fields containing respectively the values of (SQN ⊕ AK), AMF and MAC. The authentication vector is a concatenation of parameters RAND, XRES, CK, IK and AUTN. Figure 8.9 illustrates the authentication procedure.

---

1 SQN generation techniques are described in Appendix C of [TS 33.102].

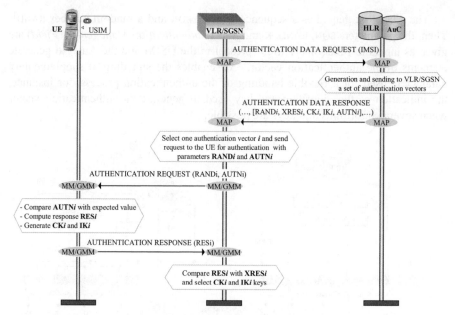

Figure 8.9. *Authentication procedure*

The authentication is performed as follows:

– upon detection of the necessity to authenticate the UE, the VLR/SGSN requests the relevant authentication parameters to the AuC;

– the AuC generates a set of ordered authentication vectors and sends it to the VLR/SGSN;

– the VLR/SGSN stores the received vectors and selects one of them (for example the one corresponding to index value i), then it sends to the UE the parameters RANDi and AUTHi = (SQN $\oplus$ AK) || AMF || MAC of the vector i. The other parameters XRESi, CKi and IKi are not sent to the UE. XRESi will be used to check the validity of the response sent by the UE, while CKi and IKi will be respectively used as ciphering and integrity key if authentication with vector i has been performed successfully;

– from the authentication request, the USIM, which holds at the UE side the secret for producing authentication parameters and ciphering and integrity keys, generates parameters XMACi (*eXpected* MAC), RESi (*RESponse*), CKi and IKi from parameters RANDi and AUTNi, by using the mechanism illustrated in Figure 8.8;

– the UE then compares the parameters MAC received from the network and XMAC locally generated to authenticate the network, checks the validity of the sequence number SQN and, if everything is correct, sends to the network the response to the authentication request with the generated parameter RES;

– with the response from the UE, the VLR/SGSN compares the received parameter RES and XRES to authenticate the UE; if RES is equal to XRES, then the authentication procedure is terminated and CK*i* and IK*i* could be used on both sides for ciphering and integrity protection following the mechanisms described in Chapter 7.

8.5. Network registration

The registration, which is only performed if a valid USIM is active in the UE, consists of the attachment of the subscriber (the USIM) to the network in order to access the provided services. For the CS domain, the network uses an information element *CS DOMAIN SPECIFIC NAS SYSTEM INFORMATION* in the message SIB1 to inform the UE on whether the attach procedure (location update of type *IMSI attach*) is required or not. Once a PLMN and a suitable cell of this PLMN are selected (see Chapter 11), and provided that the network has indicated the requirement to perform the attach procedure, the UE attempts to perform a registration to the network in order to access the provided services. The registration is done for each network service domain (CS and PS) and enables the network to know the initial location of the UE. Also, this secures the link between the UE and the network: mutual authentication, identification of the UE by the network, and allocation of a temporary identifier to the UE by the network.

The registration is performed using NAS signaling procedures *IMSI attach* for the CS domain and *GPRS attach* for the PS domain.

8.5.1. *IMSI attach procedure*

The *IMSI attach* procedure is initiated by the MM protocol layer in the UE. It starts by sending the message *LOCATION UPDATING REQUEST* containing, among others, the following parameters:

– the type of location updating set to the value *IMSI attach*;

– the *Location Area Identifier* (LAI) previously stored in the USIM;

– the UE identifier that could be either the IMSI or a valid TMSI;

– the parameter *follow on request* used to ask the network not to release the signaling connection at the end of the *IMSI attach* procedure when other signaling or traffic messages have to follow.

Figure 8.10 is an example of initial registration in the CS domain. Upon reception of the *LOCATION UPDATING REQUEST* message with a type of location updating set to *IMSI attach*, the VLR executes the authentication procedure (see Figure 8.9) and activates ciphering and integrity protection (see Figure 8.7). The VLR may then optionally proceed with the identification of the UE by either its IMEI or its IMEISV. When this first step is successfully performed, the VLR will indicate to the HLR the new location of the UE with the IMSI and the VLR identifier. The HLR proceeds then with the transfer of subscriber data to the VLR and acknowledges the location updating request. The VLR then allocates a temporary identifier to the UE and sends to it a *LOCATION UPDATING ACCEPT* message that could convey, in addition to the TMSI and the LAI, a *follow on proceed* parameter used to indicate to the UE that the signaling connection is maintained as requested and the *equivalent PLMN* parameter containing a list of equivalent PLMNs. Then, in order to acknowledge the allocation of a temporary identifier by the VLR, the mobile sends to it a *TMSI REALLOCATION COMPLETE* message that completes the procedure.

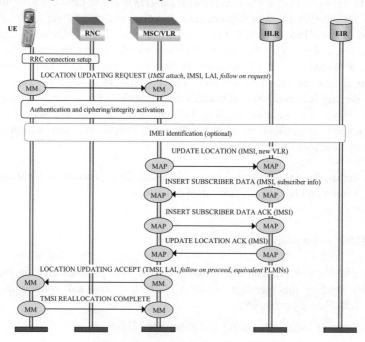

Figure 8.10. *IMSI attach procedure*

8.5.2. *GPRS attach procedure*

This procedure is used either to register only in the PS domain (*GPRS attach* procedure for GPRS services) or to register both in the PS and CS domains (*combined GPRS attach* for GPRS and non-GPRS services). The latter case is only possible when the optional interface Gs between the SGSN and the MSC/VLR is implemented in the CN. The procedure is initiated by the GMM protocol layer in the UE by sending to the network a message *ATTACH REQUEST* with the following parameters:

– the type of requested attachment that could take values *GPRS attach*, *GPRS attach while IMSI attached* or *combined GPRS/IMSI attach* for, respectively, a simple registration in the packet domain, a registration in the packet domain while the UE has already registered in the CS domain or a combined registration in both the CS and the PS domains;

– the UE identifier that could be the IMSI or a valid P-TMSI identifier;

– the signature (if it exists) associated with the P-TMSI in case the P-TMSI is used as identifier;

– the location area identifier associated to the P-TMSI;

– the *TMSI status* used in the case of a combined registration to indicate that a valid TMSI is not available in the UE;

– the parameter *follow on request* used to ask the network not to release the signaling connection at the end of the *GPRS attach* procedure when other signaling or traffic messages have to follow.

Upon receipt of the message *ATTACH REQUEST*, the SGSN can execute the identification and authentication procedures and activate ciphering and integrity protection, depending on the values of the received parameters.

Figure 8.11 illustrates a generic case of combined GPRS attach taking into account a change of SGSN and MSC/VLR since the last detach from the network. When the new SGSN receives the attach request with a P-TMSI as UE identifier, it will identify the old SGSN from the received RAI and send to it a message *SEND IDENTIFICATION* to request the permanent identifier of the UE and one or several authentication vectors. If the old SGSN is not able to identify the mobile because the P-TMSI does not exist in its database or the P-TMSI validity control using the P-TMSI signature has failed, it will indicate to the new SGSN that the subscriber is unknown in its database. In case of a negative acknowledgement, the new SGSN will attempt to identify the UE by its IMSI, then it will execute the authentication procedure and activate ciphering and integrity protection. After the link is securitized, the new SGSN continues with the registration procedure by updating at the HLR the UE location for the PS domain. The parameters "SGSN number" and "SGSN address" in the message *UPDATE GPRS LOCATION* represent respectively the telephone number (ISDN) and the SGSN's IP address. The HLR cancels the UE location in the old SGSN, transfers the UE subscription data to the new SGSN and acknowledges the registration request for the PS domain. The new SGSN carries on with the registration in the CS domain by sending a message *LOCATION UPDATE REQUEST* to the new VLR which is identified thanks to its LAI. The new VLR then performs the registration of the UE in the CS domain. This consists of asking the HLR to update the location of the UE, cancelling the location information in the old VLR, transferring the UE subscription data to the new VLR and acknowledging the location updating request received from the HLR, and then sending an acceptance message of the registration request in CS domain with allocation of a TMSI.

After the successful registration in both CS and PS domains, the new SGSN sends to the UE the registration acceptance message conveying the TMSI, the P-TMSI and possibly the P-TMSI signature. The procedure ends with the sending by the UE of a message *ATTACH COMPLETE* to the new SGSN that in turn sends as response a message *TMSI REALLOCATION COMPLETE*. These two last messages enable the UE to notify the new CN nodes with the new temporary identifiers.

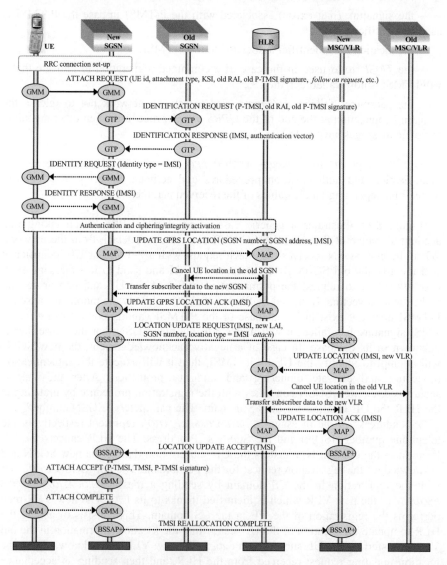

Figure 8.11. *Combined GPRS/IMSI attach procedure*

8.6. UE location updating procedures

After registration, including the initial location updating, the mobile must signal each change of its location to the network in order to be always reachable while moving. This is done using the procedures *location updating* and *routing area updating* for, respectively, CS and PS domains.

8.6.1. *Location updating procedure*

We can distinguish two types of location updating in the CS domain: normal and periodic. These two procedures, as well as the *IMSI attach* procedure, are all initiated by the UE using the message *LOCATION UPDATING REQUEST* – an information element *location updating type* being used to differentiate them.

Normal location updating is performed each time the UE detects a change of LA in the system information broadcast on the BCH of the current cell, or when it is informed by the network in response to a MM connection establishment request that it is unknown in the VLR controlling the current location area.

Periodic location updating is used by the UE to periodically signal to the network its presence in a LA. The period is set by the value of a timer provided by the network in the information element *CS DOMAIN SPECIFIC NAS SYSTEM INFORMATION* transmitted in some RRC messages. The timer could take values in the range of 1 to 25.5 hours by steps of 6 minutes. Value 0 is also used to indicate that periodic location updating shall not be used.

Figure 8.12 shows an example of inter-MSC/VLR location area updating where the old and the new location areas are controlled by different VLRs. The location updating and *IMSI attach* procedures are similarly performed, the information element *location updating type* in the message *LOCATION UPDATING REQUEST* making it possible to distinguish them. In the case of location updating, the information element takes the values *normal location updating* or *periodic location updating*. Compared to the example of *IMSI attach* procedure shown in Figure 8.10, the location updating procedure adds the cancelling of location in the old MSC/VLR.

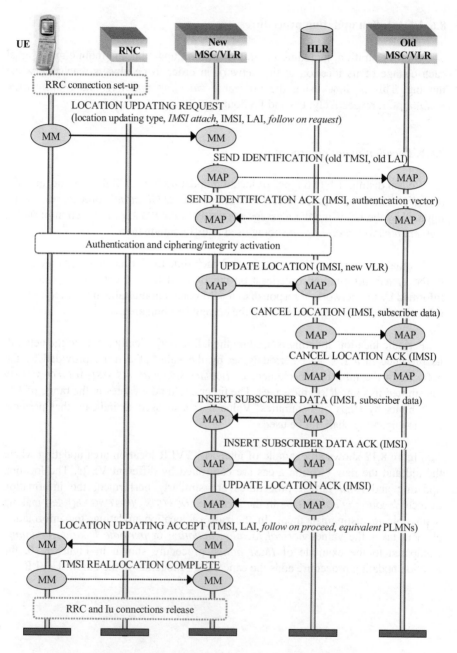

Figure 8.12. *Inter-MSC/VLR location updating procedure*

8.6.2. *Routing area updating procedure*

The routing area updating procedure is used for updating the UE location in the PS domain (*RA updating*) as well as for updating the UE location in both CS and PS domains (*combined RA and LA updating*). We can distinguish different variants for this procedure:

– *normal routing area updating* initiated by the UE when it detects a change of RA from system information broadcast in the current cell, or to re-establish a signaling connection in the PS domain when the RRC connection is released due to "*directed signaling connection re-establishment*";

– *periodic routing area update* used by the UE to regularly signal to the network its presence in an RA. The updating period is set by the value of the timer T3312 provided by the network in the messages *ATTACH ACCEPT* or *ROUTING AREA UPDATE ACCEPT*;

– *combined RA/LA updating* used when the UE has to perform an LA update while it is currently registered in both the CS and PS domains;

– *combined RA/LA updating with IMSI attach* used when the UE, which is only attached to the PS domain, has to perform simultaneously an *RA updating* and a registration procedure in the CS domain (*IMSI attach*).

Figure 8.13 is an example of *routing area updating* procedure that illustrates different cases including inter-SGSN and inter-MSC/VLR location updating, as well as location updating while a signaling connection is active. The latter case is allowed only after *SRNS relocation* (see the following section).

The procedure is initiated by the UE by sending a message *ROUTING AREA UPDATE REQUEST* to the new SGSN via the new SRNC. The information element *Update type* in the message indicates the requested type of routing area updating. If the old RA indicated in the message is not controlled by the new SGSN, the latter will request the *SGSN context* to the old SGSN (message *SGSN CONTEXT REQUEST*). In case a signaling connection is active, meaning that the routing area updating results from SRNS relocation, the old SGSN will request the *SRNS context* to the old SRNC. The SRNS context contains for each PDP context the packet sequence numbers GTP-SND (sequence number of the next GTP PDU to send to the UE), GTP-SNU (sequence number of the next GTP PDU to send to the GGSN) and PDCP-SNU (sequence number of the next PDU PDCP expected from the UE) used for data transfer synchronization after an SRNS relocation. These sequence numbers are included in the message *SGSN CONTEXT RESPONSE* sent to the new SGSN in response to the SGSN context request. The new SGSN can proceed to the UE authentication in case of a failed context request (e.g. if the response to the context request indicates an invalid P-TMSI signature), then it indicates to the old SGSN that it has received PDP context related information and is now ready to take over the message transfer. Upon receipt of the acknowledgement, the old SGSN will route any message addressed to the UE towards the new SGSN.

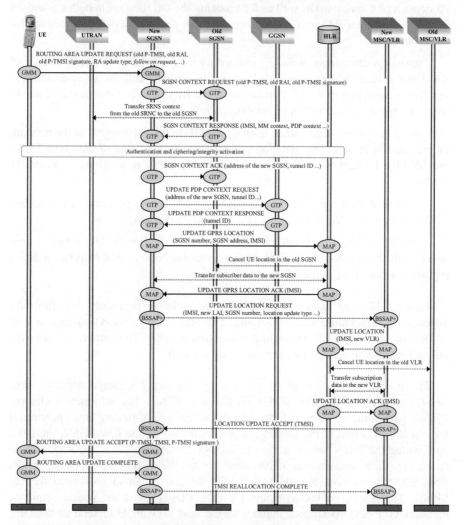

Figure 8.13. *Routing area updating procedure including inter-MSC/VLR location updating*

The new SGSN then updates the PDP context at the GGSN and sends a location updating request to the HLR. The latter then cancels the location information, transfers the UE subscription data to the new SGSN and acknowledges the location updating request.

Location updating in the CS domain can be of two types with regard to the value of information element updating type in the first message of the procedure:

– *IMSI attach* type if the *updating type* is set to *combined RA/LA updating with IMSI attach*;

– *normal location updating* type (intra-VLR or inter-VLR) if the *updating type* is set to *combined RA/LA updating*, the new RA being inside the new LA.

The procedure ends with the sending of a location updating acceptation message that could convey new temporary identifiers (P-TMSI and/or TMSI) and, in this case, the selection of these new identifiers is notified to the new SGSN and MSC/VLR using messages ROUTING AREA UPDATE COMPLETE and TMSI REALLOCATION COMPLETE.

8.6.3. *SRNS relocation*

The SRNS relocation procedure also known as *streamlining* is new compared to GSM. It consists of changing the serving RNC (SRNC) of a connected UE, i.e. the RNC supporting the Iu connection with the network. The decision to perform the procedure is always taken by the SRNC in the following situations:

– while a communication is active, the UE moves away from its SRNC and there is in the path between the UE and the SRNC a drift RNC (DRNC) used as a relay for messages exchange. In order to lighten the traffic on the Iur interface between the DRNC and SRNC, the latter could use the SRNS relocation procedure to change the traffic path by transferring to the old RNC the role of serving RNC;

– following a cell reselection, the UE connected to the PS domain and using common radio resources, initiates RRC *cell update* or *URA update* procedures (see Chapter 7). The target cell is under the control of another RNC (target RNC) that relays the message towards the source RNC by using the protocol RNSAP (message UPLINK SIGNALING TRANSFER INDICATION). If the Iur interface between the source and the target RNCs does not support user traffic (only inter-RNC signaling), the SRNC source will initiate the SRNS relocation procedure to transfer the role of serving RNC to the target RNC;

– based on the measurement results received from the UE, the SRNC which knows the topology of the network, decides to perform a *hard handover* towards a cell controlled by another RNC.

Figures 8.14 and 8.15 illustrate respectively the SRNS relocation with and without the use of the Iur interface. In the latter case, the SRNS relocation procedure is accompanied by a *hard handover*, leading to a resources reconfiguration process.

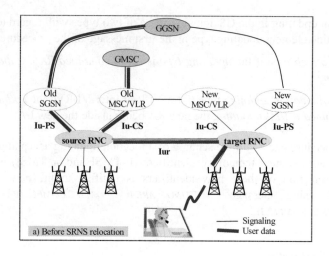

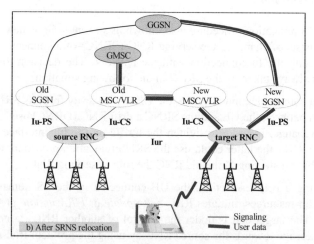

Figure 8.14. *SRNS relocation procedure with the Iur interface supporting user data traffic*

The source and target RNCs can be interconnected to the same MSC/VLR (*intra-MSC/VLR SRNS relocation*) or different MSC/VLR (*inter-MSC/VLR SRNS relocation*) for the CS domain. This statement is also valid for the PS domain where they are called *intra-SGSN SRNS relocation* and *inter-SGSN SRNS relocation*.

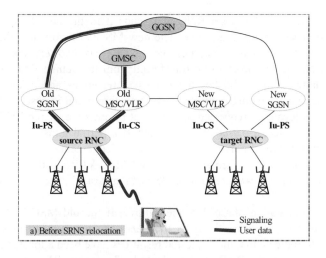

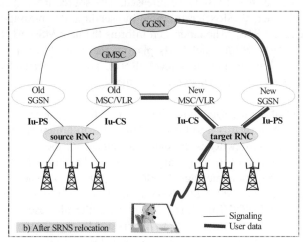

Figure 8.15. *SRNS relocation without Iur interface*

When the UE is connected to both CS and PS domains, the SRNS relocation shall be performed for both domains and the coordination is assured by the target RNC. The procedure is only considered successful if it is successful for the two domains.

Figures 8.16 and 8.17 are respectively examples of inter-MSC/VLR and inter-SGSN SRNS relocation procedure with use of an Iur interface. When the SRNS source takes the decision to perform an SRNS relocation, it informs its MSC/VLR (called old MSC/VLR in Figure 8.16). The latter prepares a *RELOCATION REQUEST* message destined for the target RNC and sends it to the new MSC/VLR in the *PREPARE HANDOVER REQUEST* message. The new MSC/VLR forwards the relocation request to the target RNC. The message *RELOCATION REQUEST* contains all the information on the resources to be transferred from the source RNC to the target RNC. The latter then establishes the requested resources and acknowledges the request. If several RABs are requested to be transferred and in case the target RNC is not able to allocate all requested resources, it informs the source RNC that could either accept or reject a partial RAB transfer. If a partial transfer is accepted, the non-transferred RABs are released. The acknowledgement message *RELOCATION REQUEST ACKNOWLEDGE* is then sent to the old MSC/VLR via the new MSC/VLR that forwards it in a *PREPARE HANDOVER RESPONSE* message.

After the preparation phase described above, if the old MSC/VLR decides to carry on with the procedure, it sends the *RELOCATION COMMAND* message to the source RNC to indicate that the requested resources have been allocated in the target RNC and, if needed, the RABs to be released. The source RNC then initiates the execution of the actual SRNS relocation by sending the message *RELOCATION COMMIT* to the target RNC. The latter then informs the new MSC/VLR by using the message *RELOCATION DETECT* and starts playing the role of serving RNC for the related UE to which it can allocate a new U-RNTI. This is indicated by the UTRAN in the *UTRAN MOBILITY INFORMATION/CONFIRM* messages.

Upon receipt of the message indicating the detection of the relocation execution, the new MSC/VLR forwards it to the old MSC/VLR in a message *PROCESS ACCESS SIGNALING REQUEST* and switches the data exchange with the mobile on the target RNC. After allocation of a temporary identifier, the target RNC sends to its MSC/VLR the completion message *RELOCATION COMPLETE*. The new MSC/VLR then uses the container *SEND END SIGNAL REQUEST* to forward the completion message to the old MSC/VLR. This latter releases the old resources and sends a response *SEND END SIGNAL RESPONSE* to the new MSC/VLR which will end the procedure by releasing resources which were allocated to itself.

The inter-SGSN SRNS relocation procedure (see Figure 8.17) is quite similar to the example illustrated by Figure 8.16 with only the following differences:

– the use of GTP container messages instead of MAP container messages;

– the update of the PDP context at the GGSN with the new SGSN;

– after receiving the with message *RELOCATION COMMIT*, the source RNC stops sending packets to the UE and proceeds with the transfer of buffered packets to the target RNC;

– if requested by the new RNC in the RNTI re-allocation exchanges, the UE will perform *RA updating* procedure.

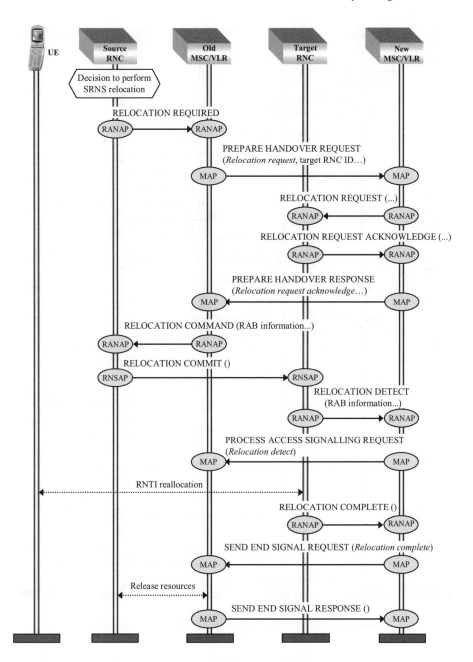

Figure 8.16. *Example of inter-MSC/VLR SRNS relocation*

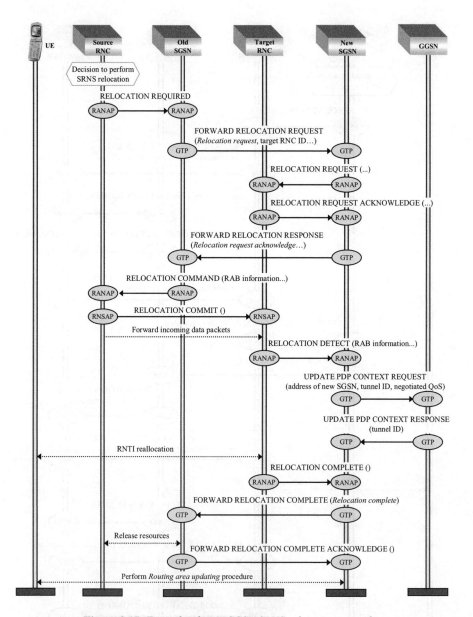

Figure 8.17. *Example of inter-SGSN SRNS relocation procedure*

8.6.4. *Detach procedures*

Two different procedures, *IMSI detach* and *GPRS detach*, are used by the MM/GMM protocol layer for detaching the mobile from, respectively, the CS and PS domains of the core network.

IMSI detach procedure

This procedure can be initiated by the UE or by the network. It is used by the network to detach from the CS domain at switch off or when the USIM is extracted while the UE is on. The procedure consists of sending by the UE an *IMSI DETACH INDICATION* message to the MSC/VLR. Whether the procedure needs to be performed or not is indicated to the UE in the CS domain specific to the system information broadcast in SIB1.

GPRS detach procedure

This procedure can be initiated by the UE as well as by the network. It is initiated by the UE for the same reason as for the *IMSI detach* procedure (switch off, USIM extraction), and by the network when, for instance, the context of the UE has been lost following a network failure. It could be used for a simple detach from only the PS domain, a combined detach from CS and PS domains or also for a simple detach from the CS domain. The procedure is initiated by sending to the UE the message *DETACH REQUEST* and, if the detach cause is not a UE switching off, the request is acknowledged by the message *DETACH ACCEPT*.

8.7. Call establishment

Most of the services provided by a communication network are usually accessed after a connection set-up between the communication end points. This connection is maintained during and released at the end of the exchanges between the end points. Such a function is assured in the UE by the protocol layer CM. Other functions such as radio interface control (reliable and secure messages exchange, cell change, etc.) and mobility management (PLMN selection, location update, etc.) are respectively assured by RRC and MM/GMM protocol layers, and are hidden to the CM layer.

8.7.1. *Circuit call*

Circuit calls could be initiated either by the UE (mobile originated call) or the network (mobile terminated call) for different types of service (voice-telephony, video telephony, data, etc.). Figure 8.18 is an example of mobile originated call.

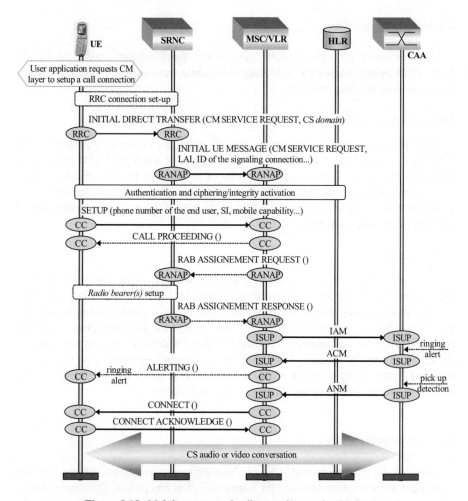

Figure 8.18. *Mobile-originated call procedure in the CS domain*

On a request from an application, the CM protocol sub-layer initiates a call establishment procedure by a service request submitted to the MM sub-layer. The service request (message CM *SERVICE REQUEST*) is sent to the RNC in an RRC *INITIAL DIRECT TRANSFER* message with indication of the CS domain as addressed core network service domain. The RNC then establishes a signaling connection (Iu connection) with the MSC/VLR and forwards to it the CM *SERVICE REQUEST* message. This message contains the requested type of service (normal call establishment, emergency call establishment, SMS sending, supplementary service activation), the mobile identity, etc.

After authentication, ciphering and integrity protection activation, the UE sends to the MSC the message *CC SETUP* which contains all the information necessary for the call establishment. The MSC/VLR could optionally send to the UE a *CALL PROCEEDING* message to inform that the call establishment request has been accepted and is being processed. If required for the requested service, radio resources (RAB) are then established. The MSC/VLR may also send to the UE an *ALERTING* message when the alerting of the distant user is started. Upon receipt of the latter message, the calling UE locally generates a call ringing tone that could be also generated by the network if the codec had been already activated. When the called UE accepts the call, it is indicated (message *CONNECT*) by the MSC/VLR to the calling UE. The latter then activates the voice codec, sends an acknowledgement (*CONNECT ACKNOWLEDGE*) to the network and stops the local ringing tone if it is ongoing.

A mobile terminated call procedure is initiated by the network using a paging message. Upon reception of this message, the RRC layer in the UE informs the MM layer and establishes a RRC connection if not yet done. Then it uses the container *INITIAL DIRECT TRANSFER* to convey the message MM paging response towards the SRNC with indication of the addressed core network service domain (CS domain). The SRNC uses the container RANAP *INITIAL UE MESSAGE* to forward the message paging response to the MSC/VLR. Then the MSC/VLR performs security procedures and sends to the UE the *SETUP* message. At this point, the CC layer in the UE can confirm to the network that the incoming call request has been received (message call confirmed) and send an alerting message as soon as the corresponding local signal has been generated. The procedure is completed by the sending of a *CONNECT* message by the UE and the acknowledgement by the network.

8.7.2. *Packet call*

A call (session) establishment in the PS domain consists of a *PDP context* activation for user data exchange. It could be initiated either by the UE or by the network. A service such as SMS sending does not require PDP context activation, only a service connection must be established before the message transfer. Figure 8.19 is an example of PDP context activation initiated by the network.

At the beginning of the procedure illustrated by Figure 8.19, the UE is attached to the PS domain but an RRC connection is not active (the UE is in the RRC idle state). The incoming data call is initiated when the GGSN receives from the external packet data network a PDU consisting of a request to establish a *PDP context*. If no *PDP context* has been established with the addressed UE, the GGSN requests to the HLR the information needed for routing the call request (message *MAP SEND ROUTING INFORMATION FOR GPRS*). If the UE is reachable, the HLR provides the GGSN with the address of the SGSN controlling the RA where the UE is located.

The GGSN then notifies the SGSN of the *PDP context* establishment request and the SGSN, after some verifications (IMSI known, UE currently attached, etc.), accepts the request and sends a paging request to the serving RNC of the UE. As there is no active RRC connection, the RNC sends to the mobile a "paging type 1" (if there was an active RRC connection the RNC would send a "paging type 2").

Upon reception of the paging message, the UE establishes the RRC connection and sends to the network a SERVICE REQUEST with the cause (*service type*) set to "paging response". This message is sent in a container RRC INITIAL DIRECT TRANSFER to the RNC and forwarded by the latter to the SGSN in a container RANAP INITIAL UE MESSAGE, the routing decision being taken in accordance with the value of the information element *CN domain identity* set to *PS domain*.

After the authentication and ciphering and integrity protection activation is performed, the SGSN sends to the UE via the RNC a message indicating that a PDP context activation is required. The UE then sends a PDP context activation request to the SGSN that in turn asks the GGSN to create the PDP context, establishes necessary RABs and sends to the UE an accept message acknowledging the PDP context activation request. From this point onwards the traffic channel is considered open and user data exchange can start.

In the case of a session initiated by the UE, it is started by sending a SERVICE REQUEST message with a *service type* set to "signaling". Upon acceptation of the service request by the network, the UE starts the actual PDP context activation and the remaining part of the procedure is similar to the case of an incoming call. An active *PDP context* could be modified by the UE, the SGSN or the GGSN. The modification is about parameter values that have been negotiated at the context activation, such as QoS. A PDP context modification procedure is for instance initiated by the SGSN when subscription data such as subscribed QoS have been modified by the HLR. On a radio link failure or upon detection of a lasting inactivity of the link, the RNC may release established RABs or the Iu connection which have as a consequence the local modification of PDP contexts both in the UE and in the network for streaming and conversational traffic types, in order to suspend packet data exchange. On radio link re-establishment, the UE re-activates the PDP contexts by means of the PDP context modification procedure. This modification procedure is also initiated by the network to provide to the UE the PDP address when external PDN address allocation is performed using DHCPv4 or Mobile IPv4. In this case the PDP context with the PDP address is set to 0.0.0.0 (the received ACTIVATE PDP CONTEXT ACCEPT message contains a PDP address equal to 0.0.0.0). Then the external PDN address is allocated to the TE (by using DHCP or Mobile IP between the TE and the external PDN via a DHCP relay in GGSN in the case of DHCP or a mobile IP foreign agent in GGSN in case of the MIP), and updated in the MT with a network initiated PDP context modification procedure (the GGSN sends an update PDP context request to the SGSN which then sends to the MT a modify PDP context request containing the allocated external PDN address).

Similarly to the activation and modification, the *PDP context deactivation* could be initiated by the UE, the SGSN or the GGSN. The *PDP Context Deactivation* procedure is based on the messages *DELETE PDP CONTEXT REQUEST* and *DELETE PDP CONTEXT RESPONSE* between SGSN and GGSN, and the messages *DEACTIVATE PDP CONTEXT REQUEST* and *DEACTIVATE PDP CONTEXT ACCEPT* between the UE and the SGSN. The procedure ends with radio resources release.

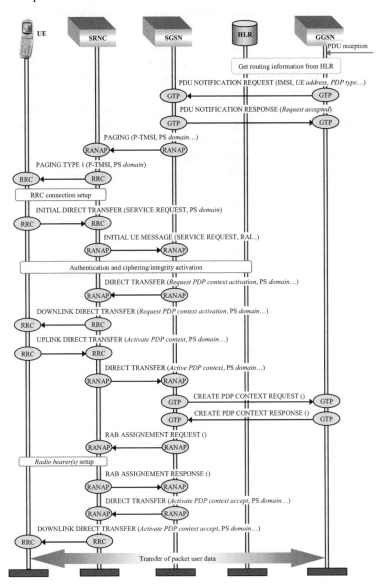

Figure 8.19. *Example of PDP context activation initiated by the network*

8.8. Intersystem change and handover between GSM and UMTS networks

3GPP specifications enable dual mode terminals of Type 2 (see Chapters 3 and 4) to roam through GSM and UMTS networks both for CS and PS services.

8.8.1. *Intersystem handover from UMTS to GSM during a CS connection*

Figure 8.20 illustrates the signaling exchange for a UMTS to GSM handover. The two networks could belong to the same operator or different operators and share or not the same MSC. In the chosen example the handover is of type inter-MSC.

Based on measurement results reported by the UE, the SRNC notifies the 3G-MSC of the necessity to continue the communication on a more suitable (better quality) GSM cell than the current UMTS cell (message RANAP *RELOCATION REQUIRED*). Through the "E" interface, the 3G-MSC sends to its peer 2G-MSC the handover request by using the MAP message *PREPARE HANDOVER RESPONSE*. The 2G-MSC then forwards the request to the GSM BSC by using the message BSSMAP *HANDOVER REQUEST*. The BSC contacts the target GSM BTS for the allocation of required radio resources and sends an acknowledgement message BSSMAP *HANDOVER REQUEST ACK* to the 2G-MSC. This message is forwarded to 3G-MSC in a container MAP *PREPARE HANDOVER RESPONSE*. Upon receipt of the acknowledgement, the 3G-MSC uses ISUP signaling to negotiate with its peer 2G-MSC the establishment of a CS connection.

When the connection is established, the 3G-MSC orders the UE via the SRNC (message RRC *HANDOVER FROM UTRAN COMMAND*) to change from its current UTRA cell to the new GSM cell. Once the UE has successfully completed the radio access in the GSM cell, this is indicated to the 2G-MSC by the BSC using a *HANDOVER DETECT* message. The *HANDOVER COMPLETE* message is used by the UE to indicate to the BSC and the 2G-MSC that the handover has been successfully performed. This indication is then forwarded to the 3G-MSC (message MAP *SEND END SIGNALING REQUEST*) to inform it that the SRNC can now release resources on the Iu interface. The CS communication is then resumed with the new data path "remote user ↔ 3G-MSC ↔ 2G-MSC ↔ BSC ↔ UE", whereas the old data path (before the handover) was "remote user ↔ 3G-MSC ↔ RNC ↔ UE". It should be noted that the 3G-MSC keeps the call control role and will indicate the end of call to 2G-MSC by using ISUP signaling and the message MAP *SEND END SIGNAL RESPONSE*.

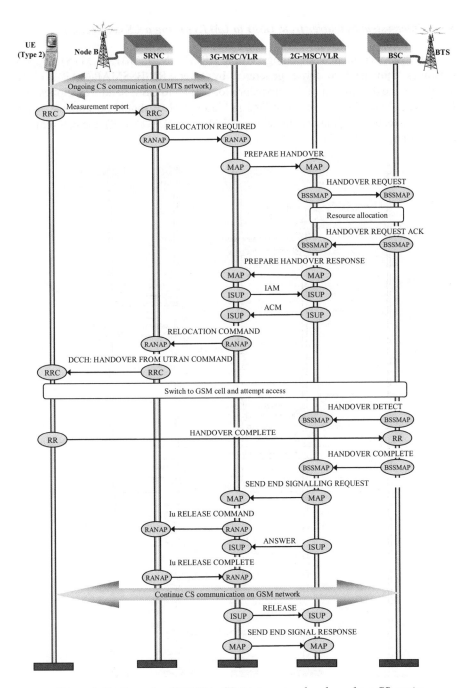

Figure 8.20. *Example of UMTS to GSM intersystem handover for a CS service*

8.8.2. Intersystem handover from GSM to UMTS during a CS connection

For handover related signaling, layers RRC/RANAP/MAP (UMTS) support messages equivalent to those generated by layers RR/BSSMAP/MAP (GSM). Therefore, the description of the procedure for UMTS to GSM handover applies also for GSM to UMTS handover illustrated in Figure 8.21. It should be similarly noted that after execution of the cell switching (GSM cell to UMTS cell), the call remains controlled by the 2G-MSC.

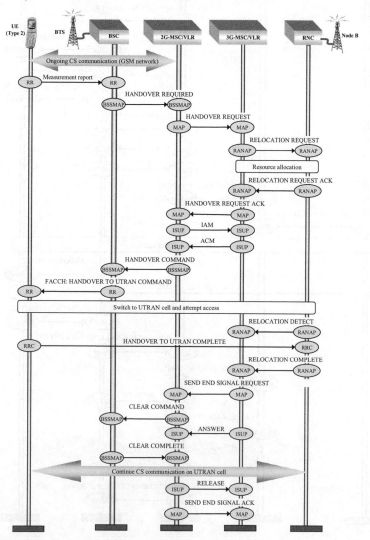

Figure 8.21. *Example of GSM to UMTS intersystem handover for a CS service*

8.8.3. *Intersystem change from UMTS to GPRS during a PS session*

The Move from UMTS to GPRS network consists of a cell reselection process from UMTS to GPRS. The procedures to be performed will depend on the service state of the mobile before the cell change:

– case of UE in GMM-IDLE state. If the RAs of UMTS and GPRS networks are different, the GPRS procedure *RA updating* will be performed. If UMTS and GPRS networks share the same RA, a variant of the GPRS *RA updating* procedure, named *selective RA update*, will be performed instead [TS 23.060];

– case of UE in GMM-CONNECTED state. The UE will perform the GPRS *RA updating* procedure regardless of whether the same RA is used or different RAs are used for the UMTS and the GPRS networks. A combined RA/LA updating could also be performed.

Figure 8.22a shows the case where the two GSM and UMTS systems share the same SGSN. In this example, it is assumed that the UE is in GMM-CONNECTED and a packet transfer is ongoing when the SRNC has decided to perform a cell change from UMTS to GPRS. As soon as the cell change decision is taken, the UE stops data transfer. After the establishment of an LLC connection (specific to GPRS system) between the UE and the SGSN, the UE sends to the SGSN the GMM message *ROUTING AREA UPDATE REQUEST* via the GSM network. In order to resume the user data transfer that had been suspended during the cell change, the SGSN requests the SRNC to transfer to it parameters GTP-SND, GTP-SNU and PDCP-SND (RANAP messages *SRNS CONTEXT REQUEST* and *SRNS CONTEXT RESPONSE*).

At this stage, the SRNC stops the packet transfer to the UE and starts buffering new packets received from the GGSN. Before pursuing the procedure, the SGSN could check the UE identity and subscription information. If any problem is detected, the SGSN will reject the RA updating request. Otherwise, the SGSN will request the SRNC (RANAP message *DATA FORWARD COMMAND*) to transfer to it the buffered downlink packets that have not yet been delivered to the UE. After a while, the Iu connection is released and the SGSN updates the *PDP context*. Then, if necessary, the SGSN will allocate a new P-TMSI to the UE indicated in the message GMM *ROUTING AREA UPDATE ACCEPT* and in this case the UE completes the procedure by sending to the SGSN the message GMM *ROUTING AREA UPDATE COMPLETE*.

8.8.4. *Intersystem change from GPRS to UMTS during a PS session*

Rules described in the previous section for a change from UMTS to GPRS apply also for a change from GPRS to UMTS. It should simply be noted that a state GPRS-STANDBY is equivalent to the GMM-IDLE state and that the GPRS-READY state is equivalent to the GMM-CONNECTED state. Figure 8.22a illustrates the message exchange between the UE and the network when the BSC has decided a change to a UMTS cell from a GPRS cell. We assume in this example that just before the cell change decision the UE was in the GPRS-READY state. This example applies to the case where the UE was transmitting packets before the system change. Upon reception of the intersystem cell change order, the UE must release the LLC connection and perform the *RA updating* process in the UMTS system.

The main difference between the procedures illustrated by Figures 8.22a and 8.22b is that in the latter case there is no packet retransmission from BSC to SGSN, as data are instead buffered by the SGSN. Given the difference of radio access technologies, resource re-allocation mechanisms for the resumption of uplink and downlink data transfer are also different. In the case of the system change of type inter-SGSN, the new SGSN shall, upon receipt of the message *ROUTING AREA UPDATE REQUEST*, request to the old SGSN to send to it the characteristics of active PDP contexts by using the message GTP *SGSN CONTEXT REQUEST* [TS 23.060].

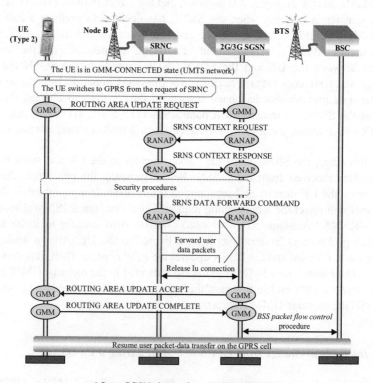

a) Intra-SGSN change from UMTS to GPRS

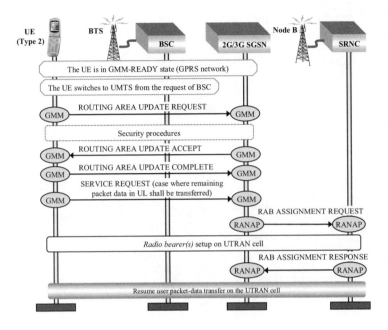

b) Intra-SGSN change from GPRS to UMTS

Figure 8.22. *Examples of intersystem change during a packet switched session*

Chapter 9

UTRA/FDD Transmission Chain

9.1. Introduction

Besides the use of CDMA, other key radio parameters differentiate the physical layer of UTRA/FDD from that of GSM. In contrast with GSM/GPRS, the UTRA/FDD physical layer was designed to support different applications with different *Quality of Service* (QoS) profiles onto one connection. For this purpose, a very flexible channel-multiplexing scheme was specified, including powerful channel coding and interleaving algorithms.

In order to transfer the binary information throughout the propagation channel with the specified QoS, the UTRA physical layer introduced the concept of transport channels to support sharing physical resources (channels) between multiple data flows. They correspond to the parallel active services (e.g. data, video, voice) or signaling messages, each having specific QoS constraints such as delay, bit rate and bit error rate. Both the transport and the physical channels are then defined by the operations applied to them, which are based on the required QoS. Such operations are illustrated in Figure 9.1 and they are studied in the following sections.

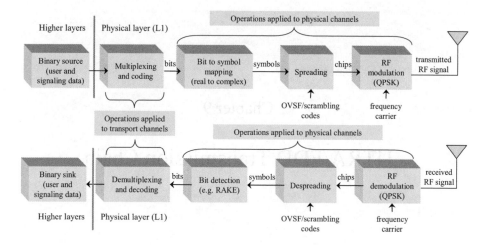

Figure 9.1. *Simplified transmission chain in the UTRA/FDD physical layer*

9.2. Operations applied to transport channels

9.2.1. *Multiplexing and channel coding in the uplink*

High data rate services or a combination of lower rate transport channels may be multiplexed into one or several physical channels. This flexibility enables numerous transport channels (services) of varying data rates to be efficiently allocated to physical channels within the same RRC connection. Figure 9.2 illustrates the different operations that define a *Dedicated Transport Channel* (DCH). Similar operations apply to common transport channels. The single output data stream from the coding and multiplexing operations is called *Coded Composite Transport Channel* (CCTrCH). In general, there can be more than one CCTrCH. The bits of one CCTrCH may be conveyed by one *Dedicated Physical Channel* (DPCH) or by several DPCHs based on the multicode transmission approach.

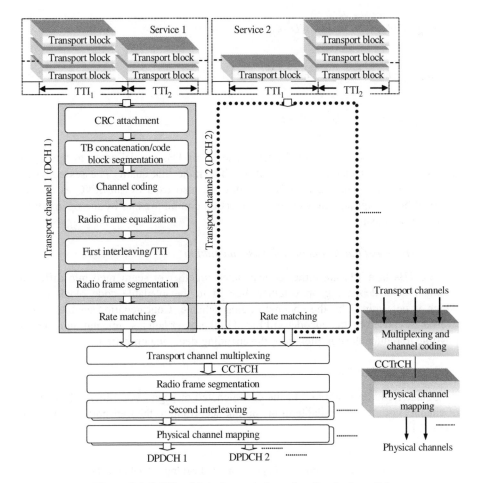

Figure 9.2. *DCH multiplexing and channel coding in the uplink*

9.2.1.1. *CRC attachment*

CRC (*Cyclic Redundancy Check*) is used to add error detection information to each *transport block* (TB). The CRC length is independently determined for each TrCH by the higher layers and can be 0, 8, 12, 16 or 24 bits. The more bits the CRC contains, the lower the probability of having undetected errors in the receiver. However, the user data rate is also lower because of this overhead. The size also defines which of the cyclic generator polynomials will be used:

$$g_{CRC\_24}(D) = D^{24} + D^{23} + D^6 + D^5 + D + 1$$

$$g_{CRC\_16}(D) = D^{16} + D^{12} + D^5 + 1$$

$$g_{CRC\_12}(D) = D^{12} + D^{11} + D^3 + D^2 + D + 1$$

$$g_{CRC\_8}(D) = D^8 + D^7 + D^4 + D^3 + D + 1$$

Regardless of the result of the CRC check, all TBs are delivered to the upper layers with the associated error indications. The radio link quality is assessed based on such error estimates at L2 (RLC). They are also used by the RNC as quality information for uplink macro-diversity selection/combining and by the open power control mechanism.

9.2.1.2. Transport block concatenation/segmentation

The TBs in a TTI are either concatenated together or segmented into different coding blocks depending on whether they fit in the available code block size – which is determined by the channel coding scheme. Concatenation enables lower overheads caused by the encoder tail bits. On the other hand, segmentation makes it possible to reduce the complexity of the encoding/decoding operations.

9.2.1.3. Channel coding

Channel coding is applied to the concatenated or segmented blocks. Two types of channel coding are available in the uplink according to the transport channel (see Table 9.1).

Transport channel	Type of channel coding	Coding rate
RACH	convolutional coding	1/2
DCH, CPCH	convolutional coding	1/2, 1/3
	turbo coding	1/3

Table 9.1. *Channel coding schemes used in the uplink*

9.2.1.4. Convolutional coding

Convolutional coding is used with relative low data rate services requiring a BER in the order of 10^{-3} (e.g. conversational services). The constraint length of this encoder is 9 and the coding rate can either be 1/2 or 1/3. The resulting convolutional encoders are depicted in Figure 9.3 [TS 25.212].

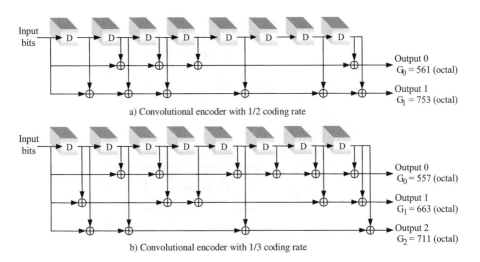

Figure 9.3. *Convolutional encoder in UTRA/FDD. "D" denotes a unitary delay*

For every bit arriving at the input of the encoder, two bits are obtained at the output from the rate 1/2 in the order {*output 0, output 1*}. Similarly, the output from the rate 1/3 is in the order {*output 0, output 1, output 2*}. Before encoding, 8 tail bits with binary value 0 are added to the end of the code block. In Figure 9.3, the shift registers are initialized to "0" when starting the encoding process. The choice of the coding rate depends on the QoS defined for the active service(s). Such a choice is a trade-off between performance, decoding complexity and increase of the overhead [PRO 95].

9.2.1.5. *Turbo coding*

In contrast to convolutional coding, turbo coding is applied for higher data rates achieving BERs as low as 10^{-5} [BER 93]. Whether turbo coding is used or not depends on the UE radio capability parameters. The structure of the original turbo encoder is very simple: the parallel concatenation of two convolutional encoders with an interleaver in-between (see Figure 9.4). The coding rate is always 1/3. On reception, the turbo decoder consists of the corresponding decoders with an internal interleaver and employs iterative decoding with passing decoding information from one iteration to another. The iterative decoding is based on the MAP (*Maximal A Posteriori Probability*) algorithm [BAH 74]. However, this can be replaced by other simpler algorithms like the log-MAP one and the SOVA (*Soft Output Viterbi*

Algorithm) [HAG 89]. They can be considered as approximations of the MAP approach.

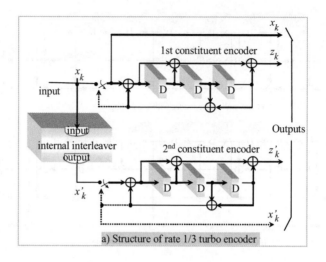

a) Structure of rate 1/3 turbo encoder

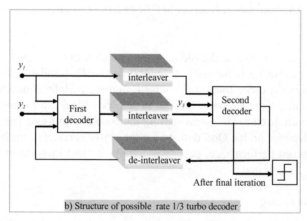

b) Structure of possible rate 1/3 turbo decoder

Figure 9.4. *UTRA/FDD 1/3 turbo encoder and example of associated decoder*

9.2.1.6. *Radio frame equalization*

The purpose of radio frame equalization is to divide the data block arrived after channel coding evenly between 10 ms radio frames by adding padding bits. Note that radio frame size equalization task is only applied to the UL – in the DL, the rate matching function already delivers blocks of equal size per frame.

9.2.1.7. *First interleaving*

In the entire physical layer of the coding chain there are in total two interleavers, plus the internal one of the turbo encoder. The first interleaver is used when the interleaving period is greater than 10 ms, i.e. when the TTI is 20, 40 or 80 ms. This is performed separately for all transport channels (whereas the second interleaving happens after the transport channel multiplexing on the CCTrCH). The first interleaver is a block interleaver with inter-column permutations. The permutation pattern is fixed and identical for the applications in UL and DL. Figure 9.5 is an example showing the principle of the first interleaver operation.

TTI	Number of columns	Order of column permutation
10ms	1	{0}
20 ms	2	{0,1}
40 ms	4	{0,2,1,3}
80 ms	8	{0,4,2,6,1,5,3,7}

a) *Order of column permutation in the first interleaver*

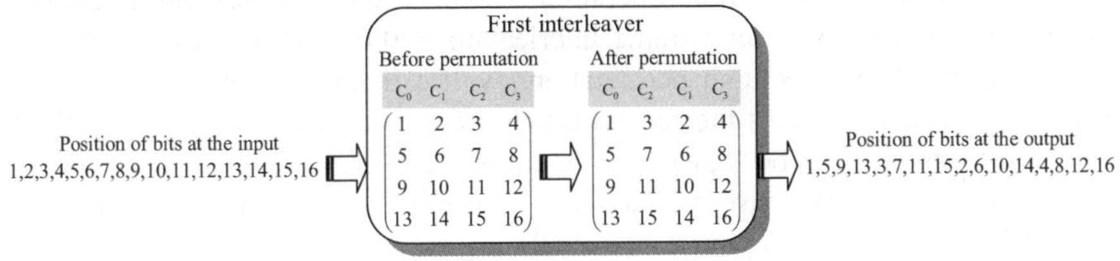

b) *First interleaver operation: example with TTI = 40 ms*

Figure 9.5. *a) First interleaver principle and b) example with TTI = 40 ms*

9.2.1.8. *Frame segmentation*

If the first interleaving operation is used, the frame segmentation will distribute the data coming from the first interleaver over 2, 4 or 8 consecutive frames in line with the interleaving length.

9.2.1.9. *Rate matching*

Rate matching performs bit puncturing or repetition on each TrCH separately but in a coordinated way between the simultaneously transmitted TrCH. The idea is to match the number of bits to be transmitted with the number of bits available in the frame. The amount of puncturing or repetition for each service is a function of the service combination and their associated QoS. However, puncturing may deteriorate the radio link quality and therefore repetition is preferred. In fact, puncturing is used to avoid multicode transmission or when facing the limitations of the UE transmitter or Node B receiver.

RRC provides a semi-static parameter, the *Rate Matching* (RM) attribute which is used to calculate the rate matching amount when multiplexing several TrCHs for the same frame. The RM parameter together with the TFCI are enough for the receiver to calculate the current rate matching parameters and perform the inverse operation [TS 25.212].

9.2.1.10. *Transport channel multiplexing*

The different transport channels are multiplexed together by the transport channel multiplexing operation. Indeed, each TrCH delivers one radio frame every 10 ms to the TrCH multiplexing. These frames are then concatenated to form a single binary stream: the CCTrCH.

9.2.1.11. *Physical channel segmentation and second interleaving*

The bits composing the CCTrCH are distributed within the different physical channels associated to that CCTrCH. When only one physical channel is used, this physical channel segmentation operation is not used. The second interleaver, sometimes referred to as intra-frame interleaving (10 ms radio frame interleaver), consists of block inter-column permutations with a fixed number of columns (see Table 9.2). At the output of the second interleaver, the amount of bits is exactly the number of bits that can be transmitted within the physical channel for the selected spreading factor. In the case of multicode transmission, the second interleaver is separately applied for each physical channel.

Number of columns	Order for column permutations
30	{0, 20, 10, 5, 15, 25, 3, 13, 23, 8, 18, 28, 1, 11, 21, 6, 16, 26, 4, 14, 24, 19, 9, 29, 12, 2, 7, 22, 27, 17}

Table 9.2. *Inter-column permutation for second interleaver*

9.2.1.12. *Example of channel coding and multiplexing in the uplink*

Figure 9.6 shows an example of the coding for UL DPDCH and DPCCH logical channels following the parameters of Table 9.3. Other examples are available in [TS 25.101, TS 25.994]. In this example, the DTCH carries a 12.2 kbps voice channel and the DCCH carries a 2.4 kbps signaling channel. Each of these channels is convolutionally coded and interleaved. The DTCH uses 20 ms (TTI) frames and the segmentation operation splits it into two parts to fit into 10 ms radio frames. On the other hand, the DCCH uses 40 ms (TTI) frames and is split into four parts to fit into the 10 ms radio frames. Rate matching is applied in a way that an amount equivalent to 22% of the bits is repeated in each channel. These are then multiplexed to create a CCTrCH and, after a second interleaving, this is mapped onto a DPDCH running at 60 kbps. The operations applied on the physical channel after this point are discussed in the following sections.

Parameters	DTCH/DCH (12.2 kbps)	DCCH/DCH (2.4 kbps)
Transport block size	244	100
Transport block set size	244	100
TTI	20 ms	40 ms
CRC size	16	12
Channel coding type, coding rate	convolutional coding, 1/3	
Rate matching	22% (repetition)	

Table 9.3. *Uplink multiplexing and coding parameters (12.2 kbps voice service example)*

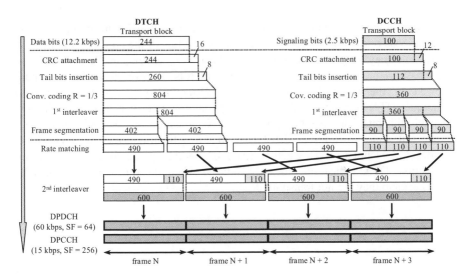

Figure 9.6. *Example of multiplexing and coding in the uplink (adapted from [TS 25.101])*

9.2.2. *Multiplexing and channel coding in the downlink*

Figure 9.7 shows the multiplexing channel coding in the downlink. Most of the operations applied in the uplink are identically used in the downlink: CRC attachment, channel coding, interleaving, TrCH multiplexing, physical channel segmentation and mapping. The channel coding schemes applied to the downlink TrCHs are depicted in Table 9.4.

Transport channel	Channel coding	Coding rate
BCH, PCH	convolutional coding	1/2
DCH, DSCH, FACH	convolutional coding	1/2, 1/3
	turbo coding	1/3

Table 9.4. *Transport channel coding in the downlink*

In the uplink, a dynamic rate matching scheme is used for matching the total instantaneous rate of the CCTrCH to the channel bit rate of the DPDCH. In this way, the bit rate may vary in a frame-by-frame basis. Conversely, in the downlink the transmissions are interrupted if the number of bits is lower than the maximum allowed by the DPDCH according to the *Discontinuous Transmission* (DTX) technique. DTX is hence used to fill up the radio frame of a dedicated physical channel with bits. The insertion point of DTX indication bits depends on whether fixed or flexible positions of the TrCHs in the radio frame are used. The RNC decides for each CCTrCH whether fixed or flexible positions are used during connection. DTX indication bits only indicate when the transmission should be turned off and they are not transmitted.

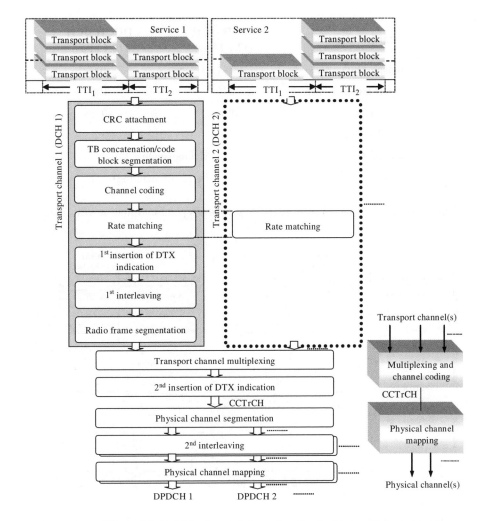

Figure 9.7. *Multiplexing and channel coding in the downlink*

9.2.2.1. *Example of channel coding and multiplexing in the downlink*

An example of channel coding and multiplexing is shown in Figure 9.8 following the parameters depicted in Table 9.5. The example considers two logical channels DTCH and DCCH, each mapped into one DCH with data rates of 12.2 kbps and 2.5 kbps, respectively. These two DCHs are multiplexed into one DPDCH.

Parameters	DTCH/DCH (12.2 kbps)	DCCH/DCH (2.4 kbps)
Transport block size	244	100
Transport block set size	244	100
TTI	20 ms	40 ms
CRC size	16	12
Channel coding type, rate	channel coding 1/3	
Rate matching	14.7% (puncturing)	

Table 9.5. *Downlink multiplexing and coding parameters (12.2 kbps voice service example)*

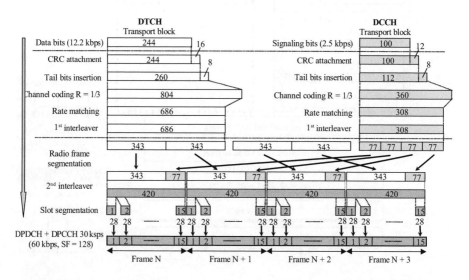

Figure 9.8. *Example of multiplexing and coding in the downlink (adapted from [TS 25.101])*

9.3. Operations applied to physical channels

9.3.1. *Characteristics of physical channels in UTRA/FDD*

The radio parameters that characterize a physical channel in UTRA/FDD are given and compared with GSM in Table 9.6. These parameters are:

– frequency carrier;

– time scheduling: time information of the duration of a physical channel;

– relative phase in uplink between I and Q branches (0 or $\pi/2$);

– a channelization code;

– a scrambling code.

	UTRA/FDD	GSM
Multiple access	CDMA/FDMA	TDMA/FDMA
Duplexing	FDD	FDD
Chip rate (Mcps)	3.84	...
Carrier bandwidth	5 MHz	200 kHz
Channel coding (coding rate)	turbo coding (1/3), convolutional coding (1/2 or 1/3)	convolutional coding (1/2)
Frame length	10 ms (15 slots)	4.615 ms (8 slots)
Data modulation	BPSK (UL), QPSK (DL)	GMSK
Power ctr. frequency	1,500 Hz	2 Hz

Table 9.6. *Key radio characteristics of UTRA/FDD compared with GSM*

The time unit in UTRA/FDD is based on the chip duration equal to 1/3.84 MHz $\approx 0.26\,\mu s$. The duration of a physical channel is also measured by the following parameters:

– a *radio frame* comprising 38,400 chips (10 ms);

– a time *slot* whose length is 2,560 chips (≈ 0.667 ms). There are 15 time *slots* in a frame.

Another time metric is the cell *System Frame Number* (SFN) which is controlled by the UTRAN [TS 25.331] and is used in a cell for time reference ranging from 0 to 4,095 frames. Broadcast system information is scheduled by SFN which is also used for paging groups.

9.3.2. *Channelization codes*

Node B transmits unique channels to many UEs and each UE receiver should be able to distinguish its own channels from all the channels transmitted by Node B. This function is provided by the channelization codes in the downlink. In the uplink, they make it possible to separate the symbols of the dedicated physical data channel

(DPDCH) from those of the dedicated physical control channel (DPCCH) belonging to the same user.

9.3.2.1. *Generation of channelization codes*

Channelization codes are also known as *Orthogonal Variable Spreading Factor* (OVSF) codes. These codes are orthogonal and their length is known as *Spreading Factor* (SF). They are hence uniquely defined by $C_{ch,SF,k}$, where SF is the spreading factor of the code and k is the code number such that $0 \leq k \leq SF - 1$. Channelization codes are part of the spreading operation where each data symbol is transformed into a number of chips. The number of chips per data symbol equals the SF.

OVSF codes are originated from the code tree illustrated in Figure 9.9. Each level of the code tree defines the channelization codes $C_{ch,SF,k}$ leading to a given symbol rate. Although the orthogonal property is maintained across different symbol rates, the selection of one OVSF code will prevent the usage of the "sub-tree", i.e. a code that is on an underlying branch with higher SF. Similarly, a smaller SF code on the path to the root of the tree cannot be used.

For instance, if the code $C_{ch,4,0}$ is allocated to one user, the codes $C_{ch,8,0}$, $C_{ch,8,1}$, $C_{ch,16,0}$, $C_{ch,16,1}$, $C_{ch,16,2}$, $C_{ch,16,3}$... are "blocked" and cannot be allocated to other users. This is also the case for code $C_{ch,2,0}$ in the root path of $C_{ch,4,0}$.

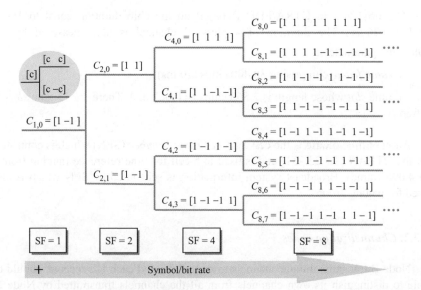

Figure 9.9. *Tree-structure that allows generation of "OVSF codes" (channelization codes)*

In the downlink the OVSF codes are limited radio resources in a cell and, as such, their allocation is managed by the RNC. On the contrary, in the uplink this problem does not exist and each UE autonomously manages the code tree. The orthogonal property of OVSF codes is only preserved when the channels are perfectly synchronized at the symbol level. For this reason, they cannot be used to distinguish different users in the uplink. The orthogonality is also lost due to multi-path and this is the main motivation for using scrambling codes.

9.3.3. Scrambling codes

The loss of cross-correlation of OVSF codes is compensated by the additional scrambling operation. Indeed, scrambling codes have good correlation properties and are always used on top of the channelization codes without impacting on the transmission bandwidth. Scrambling codes are used to separate different cells in the downlink and different UEs in the uplink. Like OVSF codes, scrambling codes are also applied at fixed rate of 3.84 Mcps. Note that scrambling codes enable OVSF code reuse among UE and Node B within the same geographic location.

9.3.3.1. *Uplink scrambling codes*

Each UE output signal is scrambled with a unique complex-valued scrambling code that enables the Node B receiver to discern one UE from another. Complex scrambling is used to distribute the power evenly between the two orthogonal I/Q branches by continuously rotating the constellation. The RNC is in charge of assigning the scrambling codes in uplink to the different UEs by considering two alternatives.

Long scrambling codes

They span over one 10 ms frame length (38,400 chips with 3.84 Mcps) and are used if in Node B a Rake receiver is used. The ith element of the n complex-valued long scrambling sequence C_n is defined as:

$$C_{long,n}(i) = c_{long,1,n}(i)\left(1 + j(-1)^i c_{long,2,n}\left(2\lfloor i/2 \rfloor\right)\right) \tag{9.1}$$

where $i = 0...2^{25} - 2$, $\lfloor x \rfloor$ is the rounding to the nearest lower integer of x and $c_{long,1,n}$ and $c_{long,2,n}$ are generated from polynomials $X^{25} + X^3 + 1$ and $X^{25} + X^3 + X^2 + X + 1$ as depicted in Figure 9.10.

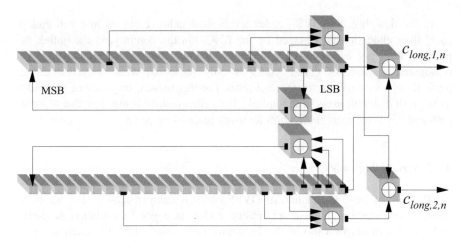

Figure 9.10. *Long-scrambling code generator in the uplink [TS 25.213]*

Short scrambling codes

They span over one symbol and are 256 ms in length. Short codes are used for simplifying the use of advanced receiving technologies in Node B such as multi-user detection. Short scrambling code *n* is generated from:

$$C_n = c_{short,1,n}(i \bmod 256)\left(1 + j(-1)^i c_{short,2,n}(2\lfloor (i \bmod 256)/2 \rfloor)\right) \qquad [9.2]$$

where $i = 0, 1, 2$, etc. and the sequences $c_{short,1,n}$ and $c_{short,2,n}$ are defined from a sequence of the periodically extended S(2) codes [KUM 96]. Such sequences are generated from sequences $a(i)$, $b(i)$, $d(i)$ as shown in Figure 9.11. These three sequences are associated to generator polynomials $x^8 + x^5 + 3x^3 + x^2 + 2x + 1$, $x^8 + x^7 + x^5 + x + 1$ and $x^8 + x^7 + x^5 + x^4 + 1$, respectively.

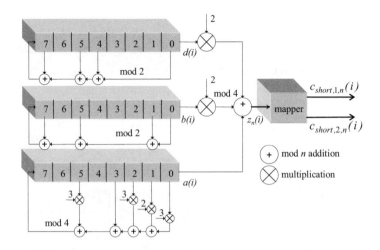

Figure 9.11. *Short scrambling code generator in the uplink [TS 25.213]*

9.3.3.2. *Downlink scrambling codes*

All data channels forming the output signal of each Node B are multiplied by a unique scrambling code. Only long scrambling codes are used in the downlink. Though there are $2^{18} - 1$ scrambling codes denoted $S_{dl,n}$, only 8,192 of those are used in practice in order to speed up the cell search procedure, i.e. $n = 0, 1,..., 8,191$. These codes are divided into 512 sets, each set has one primary scrambling code with code number $n = 16 \times i$ ($i = 0...511$) and 15 secondary scrambling codes with code number $n = 16 \times i + k$ ($k = 0...511$).

The set of primary scrambling codes is further divided into 64 groups of 8 primary scrambling codes. The jth scrambling code group comprises the primary scrambling codes obtained as $16 \times 8 \times j + 16 \times m$, where $j = 0...63$ and $m = 0...7$. Finally, each primary scrambling code n is unambiguously associated with a left (denoted $n + 8,192$) and a right alternative scrambling code (denoted $n + 16,384$) – they can be used during the downlink compressed mode.

In the downlink, the combination of two real-valued sequences $c_{n,1}$ and $c_{n,2}$ makes it possible to generate the complex-valued scrambling code $S_{dl,n}(i) = c_{n,1}(i) + jc_{n,2}(i)$. The generator polynomials for the m-sequences $c_{n,1}$ and $c_{n,2}$ are respectively $X^{18} + X^7 + 1$ and $X^{18} + X^{10} + X^7 + X^5 + 1$. Figure 9.12 illustrates this operation.

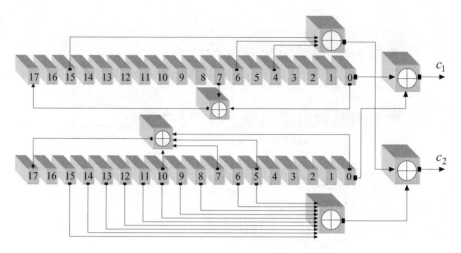

Figure 9.12. *Long scrambling code generator in downlink [TS 25.213]*

9.3.4. *UTRA/WCDMA transmitter*

The radio characteristics of the UE and Node B are specified in [TS 25.101] and [TS 25.104], respectively. The combination of OVSF and scrambling codes provide the baseband spread signal. As shown in Figure 9.13, the resulting symbols (chips) from the spreading operation shall be converted into the analog domain with the help of a *Digital-to-Analog Converter* (DAC). Special attention should be paid to the adjacent channel interference, since power leakage from adjacent channels contributes to the noise floor of the channel and hence to the system capacity. Channel interference is determined by the response of the baseband filtering (RRC filter) and the anti-alias filter which follows the DAC. The result of the baseband filter operation is a complex-valued chip sequence that is QPSK modulated after passing through analog reconstruction filters. In this case QPSK modulation refers to the up conversion operation of modulating the RF carrier with the I/Q baseband signal. The modulated signal is finally amplified with a *Power Amplifier* (PA). The non-linear effects of the PA represent the main source of adjacent channel leakage: it directly affects the co-existing systems on adjacent channels.

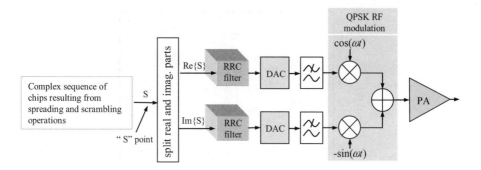

Figure 9.13. *QPSK modulation process for all physical channels*

Transmit pulse shape filter

The impulse response of the baseband filter is based on *Root Raised Cosine Filtering* (RRC) to eliminate intersymbol interference at the symbol sampling point and to constrain the transmitted energy within the channel bandwidth. The filtering is generally distributed between the transmitter and receiver (matched filtering) and implemented as an RRC filter in both the transmitter and receiver. Such a filter uses a roll-off factor defined as $\alpha = 0.22$ [PRO 95] giving an impulse response obtained from [TS 25.101, TS 25.104]:

$$p(t) = \frac{\sin\left(\pi\frac{t}{T_c}(1-\alpha)\right) + 4\alpha\frac{t}{T_c}\cos\left(\pi\frac{t}{T_c}(1+\alpha)\right)}{\pi\frac{t}{T_c}\left(1-\left(4\alpha\frac{t}{T_c}\right)^2\right)} \qquad [9.3]$$

where $T_c = 1/3.84$ MHz ≈ 0.26042 µs is the time chip. Figure 9.14 illustrates the time and frequency response of the RRC filter.

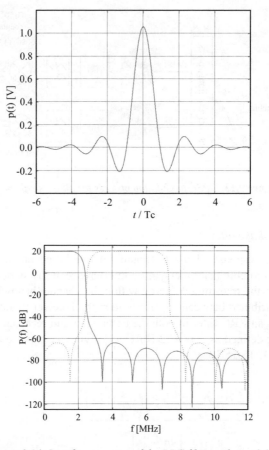

Figure 9.14. *Impulse response of the RRC filter with* α = 0.22

Frequency spectrum allocation

The frequency spectrum allocation for UMTS is not uniform, which means that different regions with different spectrum allocation for UMTS have to cohabit. In Europe (ITU Region 1), which is one of those major regions, the UTRA/FDD mode spectrum allocation consists of one frequency band at 1,920-1,980 MHz (uplink) and one at 2,110-2,170 MHz (downlink). The frequency separation is therefore 190 MHz for the fixed frequency duplex mode and between 134.8 MHz and 245.8 MHz for the variable frequency duplex mode. As shown in Figure 9.15a, the required channel spacing is 5 MHz and the channel grid (raster) is 200 kHz, which is the same as in GSM – this supports the reconfigurability of the channel selection for UMTS and GSM operating mode. Each carrier frequency f_0 can hence be qualified by the *UTRA Absolute Radio Frequency Channel Number* (UARFCN) N_t that satisfies:

$$N_t = f_0, 0 \text{ MHz} \leq f_0 \leq 3276.6 \text{ MHz}$$

ACLR

The transmit spectrum of the UE must not interfere with either adjacent UTRA frequency channels or nearby narrowband systems when transmitting on channels at the top of the spectrum. The adjacent channel performance within the UTRA band is determined by the *Adjacent Channel Leakage Power Ratio* (ACLR), which is defined as the ratio of the transmitted power to the power measured after a receiver filter in the adjacent channel. The ACLR sets the limits on the transmitter saturation response of the PA which causes amplitude compression and phase variations of the output signal and therefore influences the output spectrum. Figure 9.15b illustrates minimum ACLR levels $ACLR_1$ and $ACLR_2$ corresponding to the power level integrated respectively over the first and second adjacent carriers at 5 MHz and 10 MHz. For the UE, these ACLR values correspond to power class 3 and power class 4 devices.

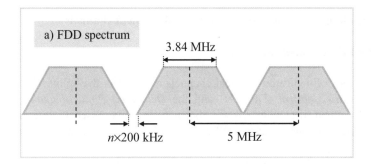

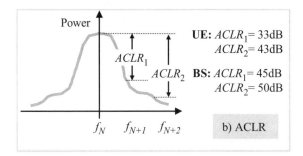

Figure 9.15. *a) UTRA FDD spectrum and b) ACLR for first and second adjacent carriers*

Error measurement

The modulation accuracy for the UMTS system is specified in terms of *Error Vector Magnitude* (EVM) [TS 25.101]. The EVM is the magnitude of the phasor difference according to the time between an ideal reference signal and the measured transmitted signal after it has been compensated in timing, amplitude, frequency, phase and DC offset. The 3GPP specification for the maximum tolerated EVM is 17.5% rms which is measured at the chip rate before despreading of the waveform.

Frequency accuracy

The necessary frequency accuracy for the carrier signal transmitted by the UE shall be in the range of ± 0.1 ppm compared with the carrier frequency received by Node B. Similarly, the frequency accuracy for the carrier signal transmitted by Node B shall be in the range of ± 0.05 ppm compared with the carrier frequency received by the UE. Any frequency error which exceeds this value will affect the EVM. For instance, if the emitted frequency carrier is 2 GHz, the UE shall ensure that variations around 2 GHz are less than ± 200 Hz. Due to errors in Doppler the UE/Node B carrier signal varies, so that the signal received by Node B/UE has to be averaged over a sufficient period of time to meet the frequency accuracy constraints.

9.4. Spreading and modulation of dedicated physical channels

9.4.1. *Uplink dedicated channels*

Two types of dedicated physical channels are used in the uplink: the *Dedicated Physical Data Channel* (DPDCH) and the *Dedicated Physical Control Channel* (DPCCH). When these channels are allocated to a given user, there is always one DPCCH and zero, one or several DPDCHs within the same L1 radio connection.

The DPDCH carries user plane (digitized voice, video, etc.) data as well as layer 3 signaling (UE measurements, active set update, etc.) data, whereas DPCCH conveys physical layer control information and TFCI control data generated by MAC (TFCI bits) as depicted in Figure 9.16.

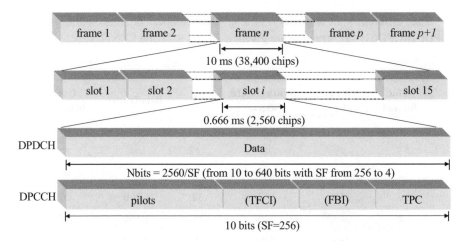

Figure 9.16. *Uplink DPDCH/DPCCH frame and slot structures*

The control information within a DPCCH slot contains:

– *Pilot symbols.* Known by the Node B receiver, the pilot symbols are used for channel estimation and coherent detection/averaging. They may also be used to measure the *Signal to Interference Ratio* (SIR) involved in the detection of TPC symbols;

– *TFCI.* The *Transport-Format Combination Indicator* symbols tell the Node B receiver the slot format and data rate in a given frame (CCTrCH parameters information). Note however that including TFCI symbols is optional: they can be omitted, for instance, in a fixed-rate service. When used, they are encoded based on the (32,10) sub-code of the second order Reed-Muller code [TS 25.212]. Errors in TFCI will cause the whole frame to be destroyed;

– *FBI.* The *FeedBack Information* between the UE and the UTRAN is used in the L1 closed loop mode Tx diversity. They are also involved in macrodiversity situations when power control is based on *Site Selection Diversity Transmission* (SSDT) approach;

– *TPC.* The *Transmit Power Control* symbols are commands sent to Node B to make it increase or reduce the power of the DPCH in the downlink. They are estimated in a *slot*-by-*slot* basis (1,500 Hz rate).

The different slot formats defining the number of bits within the UL DPDCH/DPCCH slot structure are given in Tables 9.7a and 9.7b [TS 25.211].

Slot format	Bit rate (kbps)	Symbol rate (ksps)	SF	Bits/frame	Data bits/slot
0	15	15	256	150	10
1	30	30	128	300	20
2	60	60	64	600	40
3	120	120	32	1,200	80
4	240	240	16	2,400	160
5	480	480	8	4,800	320
6	960	960	4	9,600	640

a) *Slot formats of DPDCH in the uplink*

Slot format	Bit rate (kbps)	Symbol rate (ksps)	SF	Bits /frame	Bits /slot	Pilots /slot	TPC /slot	TFCI /slot	FBI /slot
0						6	2	2	0
1						8	2	0	0
2	15	15	256	150	10	5	2	2	1
3						7	2	0	1
4						6	2	0	2
5						5	1	2	2

b) *Slot formats of DPCCH in the uplink*

Table 9.7. *Slot formats for uplink a) DPDCH and b) DPCCH*

9.4.1.1. *Uplink DPDCH/DPCCH coding and modulation*

The payload user and signaling data in DPDCH are transmitted on the "I" path of the QPSK modulator, whereas the pilot, TFCI, FBI and TPC bits of DPCCH are transmitted on the "Q" path (see Figure 9.17). With regard to the DPDCH, the RRC layer in the UE estimates the minimum allowed SF from the set of TFC indicated to the UE by the admission control functionality in the UTRAN. The OVSF code is then selected such that $C_{SF,SF/4}$. Conversely, the OVSF code and the SF are fixed in the DPCCH such that $C_{256,0.}$ Each channel is separately spread. The amplitude of the resulting chips can be individually adjusted by the gain parameters β_c and β_d associated to the DPCCH and DPDCH, respectively. Upper layers indicate to the physical layer the ratio β_c/β_d or this is calculated as described in [TS.25 213]. The ratio is quantized into 4 bit words.

Before modulation, the composite spread signal is scrambled with the complex sequence $C_n(i)$, $i = 0...38,399$, obtained from a long (expression [9.1]) or short (expression [9.2]) code. In contrast to the downlink, in the uplink the scrambling operation results from a special complex function that limits the signal transitions across the I/Q plane and the 0° phase shift transitions. This function is commonly known as *Hybrid Phase Shift Keying* (HPSK) although this term is not used in 3GPP specifications. HPSK eliminates the zero-crossing for every second chip, hence reducing the probability to 1:8 (rather than 1:4). This improves the peak-to-average (PAR) power ratio and therefore the efficiency of the UE power amplifier. This technique assumes pairs of consecutive identical chips achieved by appropriate choice of orthogonal OVSF codes for I and Q branches. This is done by selecting channelization codes from the top half of the code tree.

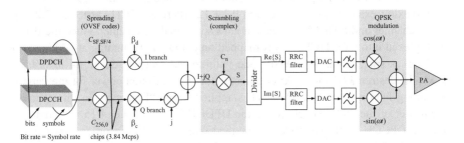

Figure 9.17. *Spreading and modulation for uplink DPDCH/DPCCH physical channels*

9.4.1.2. *Variable rate transmission in uplink*

While the bit rate in the DPCCH is fixed and equals 15 kbps, that in the DPDCH is variable and may change frame-by-frame (10 ms) as function of the service (see Figure 9.18). It is even possible to interrupt transmission in the DPDCH, for instance, during periods of silence in a voice conversation, as studied in Appendix 1. Note that the DPCCH is always transmitted in order to keep active the inner loop power control (TPC bits) and the Tx diversity close loop in case this is used (FBI bits). Moreover, continuous transmission in the uplink regardless of the bit rate reduces audible interference problems. Higher bit rates require more transmission power in the DPDCH as well as in the DPCCH in order to enable accurate channel estimation.

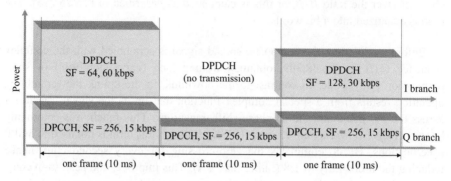

Figure 9.18. *Example of variable rate transmission in the uplink*

9.4.1.3. *Example of coding in the uplink*

The channel coding operation in uplink is illustrated in Figure 9.19 based on the multiplexing example given in section 9.2.1.12 where a DTCH carries a 12.2 kbps voice channel and a DCCH conveys a 2.4 kbps signaling channel. Once the data is multiplexed into a CCTrCH, this is mapped onto a DPDCH running at 60 kbps. The DPDCH is spread with an OVSF code with spread factor equal to 64 in order to reach the desired 3.84 Mcps. After adjusting the transmission power for the variable spreading factor, the spread DPDCH is applied to the "I" branch. In this example, the power of the DPCCH relative to the DPDCH is – 6 dB, i.e. 4/15. The data rate for the DPCCH is always 15 kbps obtained with SF = 256 in order to reach 3.84 Mcps. The DPCCH is then applied to the "Q" branch. The resulting I and Q signals are then filtered and used to modulate the RF carrier.

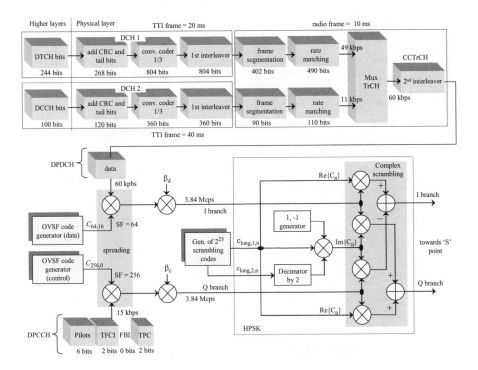

Figure 9.19. *Example of spreading, scrambling and modulation in the uplink*

9.4.1.4. *Multicode transmission in the uplink*

In order to increase the bit rate in the uplink, the data within a CCTrCH can be mapped into multiple DPDCHs transmitted in parallel and using the same spreading factor. This approach is called multicode transmission and depends on the capabilities of the UE and Node B. In uplink, the UE can use only one CCTrCH. Up to six DPDCHs can be used. They all have spreading factors equal to four and may share the same code providing that one is in "I" and the other one in "Q" branches, which makes them orthogonal. The used OVSF codes are shown in Figure 9.20. In this figure we note that only one DPCCH is present within a multicode operation. It is worth noting that a multicode transmission in the uplink increases the PAR and makes the design of the power amplifier more difficult, although few services today require in practice higher bit rates in the uplink.

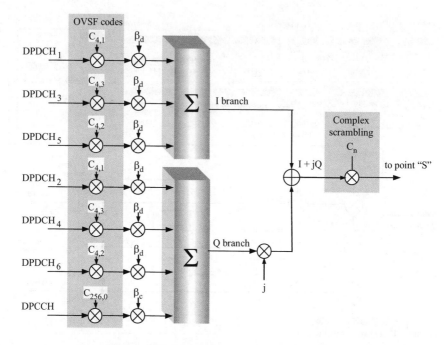

Figure 9.20. *Multicode transmission in the uplink*

9.4.1.5. *Data rates in the uplink*

Table 9.8 shows the channel bit rates achievable in the uplink with a single and multiple DPCCHs. They are a function of the SF and the number of parallel channels involved in the transmission. Voice services typically require SF = 128, while SF = 16 may be needed for video telephony and SF = 32 for real-time packet services (e.g. streaming). We are more interested, however, in "practical" user data rates that can be observed on top of MAC/RLC layers. By assuming 1/2 coding rate and by neglecting MAC/RLC headers, coder tail bits and CRC bits it is possible to accommodate 2 Mbps or an even higher data rate in a multicode transmission.

Slot format	SF	No. of bits in DPDCH/slot	Gross bit rate (kbps)	Practical bit rate (kbps)
0	256	10	15	7.5
1	128	20	30	15
2	64	40	60	30
3	32	80	120	60
4	16	160	240	120
5	8	320	480	240
6	4	640	960	480
Multicode with 6 codes and SF = 4		640 per code	5,760	2,880

Table 9.8. *Achievable user data rates in the uplink*

9.4.2. *Downlink dedicated channel*

In contrast to the uplink, there is only one type of downlink dedicated physical channel referred to as DPCH (*Dedicated Physical Channel*). This channel conveys the information resulting from the DCH – information originally generated at L2/L3 or above, including user or signaling data. The DPCH can be seen as a time multiplex of two channels: one carrying actual data (DPDCH) and the other control information generated at L1 (DPCCH) as shown in Figure 9.21. The data bits are accommodated into two fields within the DPCH named *data 1* and *data 2*. With the exception of FBI bits, the same type of control information as in the uplink DPCCH is carried out: known pilot bits, TPC commands and optional TFCI. It is the UTRAN that determines if a TFCI should be transmitted, hence making mandatory for all UEs to support the use of TFCI in the downlink or to apply blind estimation as described in [TS 25.212]. In the absence of TFCI bits, DTX bits can be accommodated instead. Table 9.9 gives the possible slot structures (formats) for the DPCH in the downlink – this is specified by upper layers within a radio channel transmission.

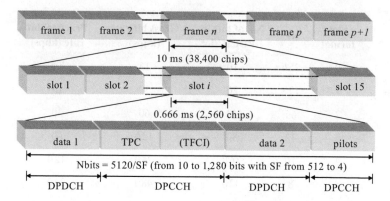

Figure 9.21. *Slot structure of the DPCH in the downlink*

Slot format	Bit rate (kbps)	Symbol rate (ksps)	SF	Bits/slot DPDCH		Bits/slot DPCCH		
				Data 1	Data 2	TPC	TFCI	Pilots
0	15	7.5	512	0	4	2	0	4
1	15	7.5	512	0	2	2	2	4
2	30	15	512	2	14	2	0	2
3	30	15	256	2	12	2	2	2
4	30	15	256	2	12	2	0	4
5	30	15	256	2	10	2	2	4
6	30	15	256	2	8	2	0	8
7	30	15	256	2	6	2	2	8
8	60	30	128	6	28	2	0	4
9	60	30	128	6	26	2	2	4
10	60	30	128	6	24	2	0	8
11	60	30	128	6	22	2	2	8
12	120	60	64	12	48	4	8*	8
13	240	120	32	28	112	4	8*	8
14	480	240	16	56	232	8	8*	16
15	960	480	8	120	488	8	8*	16
16	1,920	960	4	248	1,000	8	8*	16

Table 9.9. *Downlink DPCH slot formats. If TFCI bits denoted "*" are not transmitted, these fields can be used to insert DTX bits*

9.4.2.1. *Downlink DPCH coding and modulation*

Figure 9.22 depicts a simplified transmitter in Node B with coding and modulation operations applied to the DPCH. The bits within the DPCH slot structure are mapped into I/Q paths: even bits go to the "I" branch, while odd bits are accommodated in the "Q" branch so that a 2 bit symbol is composed. The resource management functionality in the RNC selects the channel code to be used and this is communicated to the UE during the RRC connection. The same channelization code is then used to spread I and Q branches. The amplitude of the resulting complex signal is weighted by the *Gi* factor. The downlink DPCHs complex signals of all the UEs active in the cell are added and then scrambled with the scrambling code distinguishing that cell or sector. Channel filtering and QPSK RF modulation is finally applied as illustrated in the "S" point in Figure 9.13.

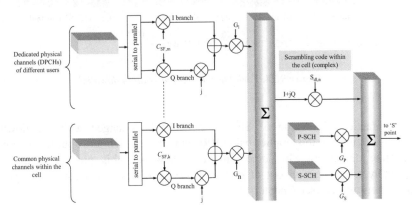

Figure 9.22. *Simplified scheme of a Node B emitter illustrating DPCH operations*

It should be noted that the amplitude of the data DPDCH symbols is estimated with respect to the amplitude of the control DPCCH symbols. This feature provides increased protection to TFCI, TPC and pilot symbols and is defined in dB by the network parameters PO1, PO2 and PO3, respectively (see Figure 9.23).

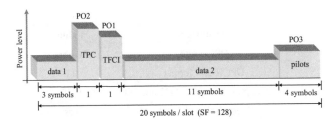

Figure 9.23. *Example of variable power levels of control fields in respect to data fields*

9.4.2.2. *Variable rate transmission in the downlink*

In contrast to the uplink where the value of the SF can change from one frame to another, in the downlink this parameter is fixed during an active communication (see Figure 9.24) and the data rate is varied by rate matching operation. Thus, when the DPDCH data rate decreases, rather than attempting to modify the value of SF, transmission is maintained at the same rate and the positions of *data 1* and *data 2* can be used to insert DTX bits.

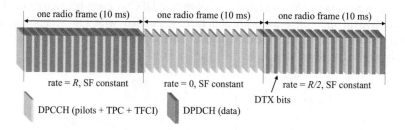

Figure 9.24. *Variable rate transmission in downlink*

In practice, the value of SF can actually be changed within a downlink communication but this implies using RRC signaling which turns out to be complex and time demanding. In some situations, however, this process is necessary especially when running out of downlink OVSF codes and during a compressed mode transmission (see Chapter 11).

9.4.2.3. *Example of coding in the downlink*

Figure 9.25 shows an example of a downlink chain based on the example given in section 9.2.2.1. This concerns a DTCH carrying a 12.2 kbps voice service within 20 ms frames and a DCCH conveying a 2.4 kbps data stream on a 40 ms frame. They use two DCHs that are multiplexed together to form the CCTrCH. The CCTrCH is interleaved and mapped onto a DPDCH with the channel parameters of Table 9.10.

Parameters	Value
DPCH rate	30 kbps
DPCH slot format	11
Relative power for PO1, PO2 and PO3	0 dB
Transmission of TFCI	yes

Table 9.10. *Example of downlink DPCH parameters for a voice service at 12.2 kbps*

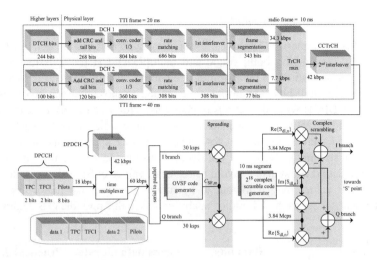

Figure 9.25. *Example of spreading, scrambling and modulation in downlink*

9.4.2.4. *Multicode transmission in the downlink*

Similarly to the uplink, in the downlink higher bit rates can be obtained with a multicode transmission scheme. However, in the downlink more than one CCTrCH may be employed for one UE: each CCTrCH can have a different spreading factor. Furthermore, in contrast to the uplink, the maximum bit rate is achieved with three parallel codes using the same SF value. Indeed, in downlink the same OVSF code tree is shared by all users within the cell and the minimum value for SF is four: three of these codes can be allocated to the same UE, whereas the fourth is reserved for the downlink common channels. As observed in Figure 9.26, only one set of control information (pilot, TPC and TFCI bits) is associated to the parallel data payload of the DPCHs involved in a multicode transmission.

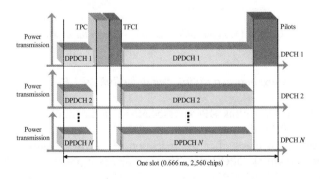

Figure 9.26. *Slot format of a DPCH within a multicode transmission*

9.4.2.5. *Data rates in the downlink*

Table 9.11 summarizes "practical" user data rates that can be achieved in the downlink by considering both, single and multiple DPCH transmission schemes. Practical rates are estimated by assuming 1/2 coding rate and neglecting MAC/RLC headers and CRC/tail bits. The bit rates are similar to those that can be obtained in the uplink. A DPCH with SF = 256 is typically used for a voice service using a half-rate codec and SF =128 can, for instance, use a more efficient codec such as full-rate. Lower SF values are needed for higher data rate applications such as video telephony or streaming. Note that multicode transmission can be used not only to increase the data rate but also to avoid using small values of SF: using parallel codes with high SF values may work out to be more robust against radio propagation degradations – as far as the UE radio capability makes this possible [TS 25.306].

Slot format	SF	Data bits data (1 + 2)/slot	Gross data bit rate approx. (kbps)	Practical data bit rate (kbps)
0	512	4	6	3.0
3	256	14	21	10.5
8	128	34	51	25.5
12	64	60	90	45.0
13	32	140	210	105.0
14	16	288	432	216.0
15	8	608	912	456.0
16	4	1,248	1,872	936.0
Multicode with 3 codes and SF = 4		1,248 per code	5,616	2,808.0

Table 9.11. *Achievable user data rates in the downlink*

9.4.3. *Time difference between uplink and downlink DPCHs*

There is always an uplink DPCH associated to a downlink DPCH. However, as shown in Figure 9.27, the transmission of a DPCH slot in the uplink takes place at approximately 1,024 chips after the reception of the first significant path of the corresponding downlink DPCH slot. This fact makes it possible to reduce delays in the estimation of TPC and FBI commands (if Tx Diversity is activated) [TS25.214].

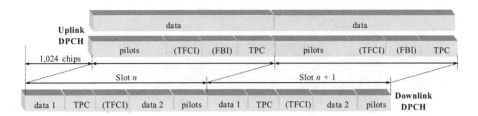

Figure 9.27. *Time difference between uplink and downlink DPCHs*

9.5. Spreading and modulation of common physical channels

9.5.1. *The Physical Random Access Channel (PRACH)*

The *Physical Random Access Channel* (PRACH) carries the RACH transport channel used by the UE to request RRC connection establishment or to send small messages of user data. The slot structure of the PRACH is shown in Figure 9.28. Each slot consists of two parts, a data part on which the RACH transport channel is mapped and a control part that transports layer 1 control information – data and control parts are transmitted in parallel. The data part is spread with OVSF codes using SF = 256 (15 kbps), 128 (30 kbps), 64 (60 kbps) or 32 (120 kbps). Note that in UL symbols/s = bits/s. The control part is in fixed rate (15 kbps) and comprises 8 known pilot bits and 2 TFCI bits that correspond to SF = 256.

PRACH transmissions begin with a short preamble pattern that alerts Node B of the forthcoming RACH access message. The PRACH preamble part consists of 256 repetitions of a signature of length 16 chips ($16 \times 256 = 4,096$). They are based on the Hadamard code set of length 16 and as such, there is a maximum of 16 available signatures [TS 25.213]. The length of the message part may be one (10 ms) or two (20 ms) radio frames according to the value of the TTI in the current RACH as defined by higher layers. A message part of 20 ms is well suited for cells of big size for which a low coding rate can be used and hence, a better coverage can be obtained. Due to its low capacity, fast power control does not apply to PRACH, which is a fact that limits still further its usage for transporting user data with high QoS requirements.

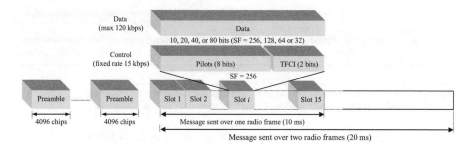

Figure 9.28. *PRACH slot and frame structures*

9.5.2. *The Physical Common Packet Channel (PCPCH)*

The *Physical Common Packet Channel* (PCPCH) carries the CPCH and is used for uplink user packet data and signaling transmission. The PCPCH is a more efficient way to send packet data in the uplink when compared to the RACH, since a packet transmission may span more than two frames. This requires, however, Node B to control the PCPCH power transmission.

The PCPCH access transmission consists of one or several *Access Preambles* (AP) whose length is 4,096 chips, one *Collision Detection Preamble* (CDP) whose length is 4,096 chips, a DPCCH *Power Control Preamble* (PCP) which is either 0 or 8 slots in length and a message of variable length $N \times 10$ ms frames (see Figure 9.29). The length of the message and the *PCPCH PCP* part are higher layer parameters.

The *PCPCH AP* uses the same RACH preamble signature sequences. The number of sequences used could be less than the ones used in the RACH preamble. The scrambling code could either be a different code segment of the Gold code used to form the scrambling code of the RACH preambles or it could be the same scrambling code in case the signature set is shared. The *PCPCH CDP* part uses a scrambling code chosen to be a different code segment of the Gold code used to form the scrambling code for the RACH and CPCH preambles.

The frame and slot structure of the PCPCH message is equivalent to that of an uplink DPCH. Indeed, each slot consists of two parts, a data part that carries layer information and a control part that carries L1 control information. Both parts are transmitted in parallel (see Figure 9.29). The control part is transmitted at 15 kbps fixed rate (SF = 256), whereas the rate in the data part depends on the value of SF chosen to be in the set $\{256, 128, 64, 32, 16, 8, 4\}$ resulting in channel rates of 15 kbps, 30 kbps, 60 kbps, 120 kbps, 240 kbps, 480 kbps or 960 kbps, respectively. A long or short code sequence may be used to scramble the PCPCH message part.

The PCPCH is associated to the control part of a downlink DPCH (DPCCH) – this is needed in order to provide inner loop power control. When a PCPCH is used, higher layer information in the downlink is transmitted to the UE via a FACH/S-CCPCH.

NOTE.– CPCH/PCPCH channels have not been used in real networks so far. This feature will likely be removed from *Release 5* onwards, with the aim to simplify UE requirements and enable a smooth evolution with less complexity [3GPP RP-050144].

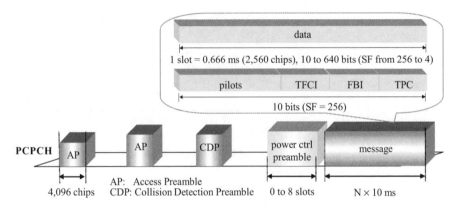

Figure 9.29. *PCPCH access transmission structure*

9.5.3. *The Physical Downlink Shared Channel (PDSCH)*

The *Physical Downlink Shared Channel* conveys DSCH information and is used to dish out packet data to a number of UEs. Indeed, on a radio frame basis, the UTRAN may allocate different PDSCHs under the same PDSCH root channelization code to different UEs based on code multiplexing. Similarly, multiple parallel PDSCHs, with the same SF may be allocated to a single UE following a multicode-like transmission scheme. The value of SF can, however, change from one frame to another. For PDSCH, the allowed SFs may vary from 256 to 4, which gives channel bit rates of 32, 64, 28, 256, 384 and 512 kbps.

As shown in Figure 9.30, a PDSCH frame is associated in the downlink with one DPCH, although they do not necessarily have the same SF and are not necessarily frame aligned. From the figure, we can infer that a PDSCH frame starts somewhere between 3 and 18 slots from the end of a DPCH frame. Note that the PDSCH does not carry L1 control information: this is transmitted on the DPCCH part of the associated DPCH and includes power control commands and TFCI symbols. The TFCI field of the associated DPCH is used to indicate to the UE that there are data to decode on the DSCH and provides it with the instantaneous transport format as well as the channelization code of the PDSCH. The associated DPCH may also be used to transmit user data – a voice service on the DPCH may be active in parallel to the reception of a packet data service on the PDSCH.

NOTE.– DSCH feature is optional and not used in real networks. Moreover, the introduction of HSDPA diminishes the benefits of this channel. Hence, this feature will likely be removed from *Release 5* onwards, with the aim to simplify UE requirements and enable smooth evolution with less complexity [3GPP RP-050144].

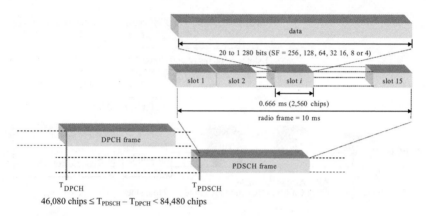

Figure 9.30. *PDSCH slot structure and time difference with its associated DPCH*

9.5.4. *The Synchronization Channel (SCH)*

The *Synchronization Channel* (SCH) consists of two sub-channels, the *Primary Synchronization Channel* (P-SCH) and the *Secondary Synchronization Channel* (S-SCH) as depicted in Figure 9.31. These channels are composed of two codes known as *Primary Synchronization Code* (PSC) and *Secondary Synchronization Code* (SSC). The PSC is composed of a fixed 256 chip code and is broadcast by all Node Bs every slot. The code is constructed from a generalized hierarchical Golay sequence [TS 25.213]. This is used by the UE during initial acquisition to determine if a Node B is present and establish the slot boundary timing within the cell covered by that Node B.

The SSC consists of repeatedly transmitting a (length 15) sequence of modulated codes whose length is 256 chips. The SSC is denoted $c_s^{i,k}$ where $k = 0, 1\ldots14$ and $i = 0, 1,\ldots,63$. When a UE decodes 15 consecutive SSC transmissions, it can determine the Node B frame boundary timing, as well as derive information that will aid in the identification of the Node B scrambling code (see Chapter 10).

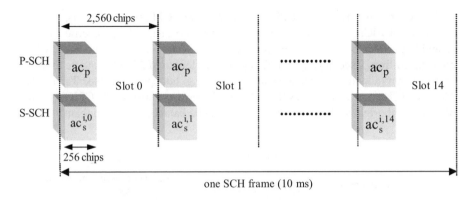

Figure 9.31. *SCH frame and slot structure. In the figure, "a" is a gain factor*

9.5.5. *The Common Pilot Channel (CPICH)*

The *Common Pilot Channel* (CPICH) carries a pre-defined bit/symbol sequence at fixed rate (15 ksps with SF = 256) as depicted in Figure 9.32. This channel is a continuous loop broadcast of the scrambling code identifying Node B. Moreover, the CPICH is used by the UE as a coherent reference for precise measurement of the Node B time reference as well as to determine the signal strength of the surrounding Node Bs before and during cell site handover and cell selection/reselection procedures. Since no additional spreading is applied to the CPICH, and given its low rate, it is quite easy for the UE to acquire a lock to this reference before any other channels can be received.

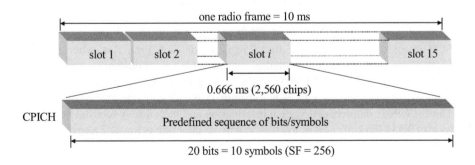

Figure 9.32. *CPICH slot structure*

There are two types of CPICH: the *Primary* CPICH (P-CPICH), and the *Secondary* CPICH (S-CPICH). The P-CPICH consists of a specific primary scrambling code to be used for the whole cell/sector, whereas the S-CPICH is constructed from a secondary scrambling code of length of 256 chips to be used in a narrow beam (adaptive antennae/beam steering) e.g. hot spot areas (see Table 9.12). The P-CPICH somehow defines the cell boundaries: by adjusting the CPICH powers of two neighboring Node Bs, the traffic load can be balanced between them.

	P-CPICH	S-CPICH
Channelization code	Fixed $C_{256,0}$	A code among those generated with SF = 256
Scrambling code	A primary scrambling code	A primary or secondary scrambling code
Channels per cell	Only one	Zero, one or several
Coverage	Broadcast over the entire cell	Broadcast over the entire cell or just a part of it

Table 9.12. *Differences between the P-CPICH and the S-CPICH*

9.5.6. *The Primary Common Control Physical Channel (P-CCPCH)*

The *Primary Common Control Physical Channel* (P-CCPCH) is the physical channel carrying the BCH transport channel in the downlink. The SF and the channelization code of the P-CCPCH are fixed to $C_{256,1}$. Neither TPC commands, nor TFCI and pilot bits are transmitted on this channel.

As shown in Figure 9.33, the P-CCPCH is time multiplexed with the SCH. The SCH is transmitted during the first 256 chips of each time slot, whereas the P-CCPCH is off. This fact makes it possible to constantly transmit the P-SCH/S-SCH and the P-CCPCH at the same power. During the remaining 2,304 chips of each slot the P-CCPCH is transmitted, which contains 18 bits of the BCH. The P-CCPCH delivers cell system information and therefore it needs to be decoded by all UEs in the cell. This is the reason why its data rate is low (27 kbps) and its Tx power is high.

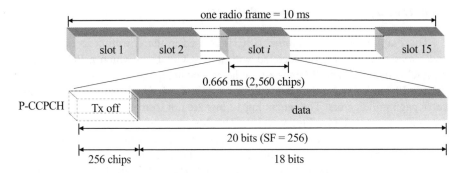

Figure 9.33. *P-CCPCH slot structure*

9.5.7. *The Secondary Common Control Physical Channel (S-CCPCH)*

The *Secondary Common Control Physical Channel* (S-CCPCH) carries FACH and PCH which can either share the same physical channel (same frame) or both have their own channel. The slot structure of the S-CCPCH is shown in Figure 9.34. The SF = 256, 128, 64, 32, 16, 8 or 4 yield channel symbol rates varying from 15 to 960 ksps. In contrast to P-CCPCH, the S-CCPCH may carry user data on FACH and a slow power control scheme is applied to enhance the quality of service. Variable rates are also supported with the help of the TFCI field.

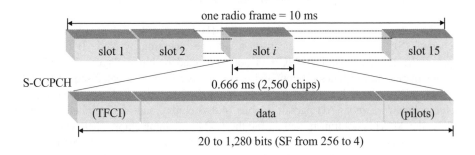

Figure 9.34. *S-CCPCH slot structure*

9.5.8. *The Paging Indicator Channel (PICH)*

The *Paging Indicator Channel* (PICH) is used in the downlink to carry *Paging Indicators* and is always associated with an S-CCPCH to which a PCH transport channel is mapped. The PICH is a pure layer 1 channel and the structure of its radio frame (10 ms length) comprises 300 bits $b_0...b_{299}$, of which only 288 are used to carry paging indicators. The remaining 12 bits are not formally part of the PICH and are not transmitted (see Figure 9.35). The low rate of the PICH obtained with SF = 256 and the high power with which it is transmitted make this channel easy to be detected by all the UEs within the cell. The paging procedure showing the interaction of the PICH and the S-CCPCH (PCH) is described in more depth in Chapter 11.

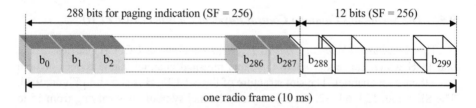

Figure 9.35. *PICH frame structure*

9.5.9. *The Acquisition Indicator Channel (AICH)*

The *Acquisition Indicator Channel* (AICH) is a fixed rate downlink physical channel (SF = 256) used to indicate in a cell the reception by Node B of PRACH preambles (signatures). The AICH is not connected to higher layers. The channel spans over 2 radio frames (20 ms) and is constructed from a repeated sequence of 15 consecutive access slots (AS) whose length is 5,120 chips. Each AS comprises an *Acquisition-Indicator* (AI) part of 32 bits (4,096 chips), and as illustrated in Figure 9.36, the AICH is not transmitted during the 1,024 remaining chips. There can be up to 16 signatures acknowledged on AICH at the same time. If the AICH information is not present, the terminal considers the cell barred and proceeds to a cell re-selection. The role of the AICH in the random access process is described in Chapter 10.

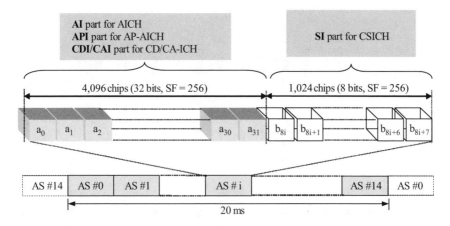

Figure 9.36. *Generic frame structure valid for AICH, AP-AICH, CD/CA-ICH and CSICH*

9.5.10. *Other downlink physical channels associated with the PCPCH*

The *Access Preamble Acquisition Indicator Channel* (AP-AICH), the *Collision Detection/Channel Assignment Indicator Channel* (CD/CA-ICH) and the *CPCH Status Indicator Channel* (CSICH) are downlink physical channels involved in the PCPCH access transmission process. All these channels are fixed rate (30 kbps) and use SF = 256. The structure of the AP-AICH and the CD/CA-ICH is illustrated in Figure 9.36. This is identical to that of the AICH and may or may not use the same channelization codes. The main difference is the *Access Preamble Indicator* (API) part in the case of the AP-AICH, and the *Collision Detection Indicator/Channel Assignment Indicator* (CDI/CAI) part in the case of the CD/CA-ICH. The size of these two parts is of 32 bits and their length is of 4,096 chips. The part of the slot with no transmission (1,024 chips) is reserved for possible use by CSICH.

The CSICH is always associated with an AP-AICH and uses the same channelization and scrambling codes. As illustrated in Figure 9.36, the CSICH comprises two parts: one part whose duration is 4,096 chips with no transmission that is not formally part of the CSICH and a *Status Indicator* (SI) part consisting of 8 bits denoted by b_{8i}... b_{8i+7}, where i is the access slot number. The part of the slot with no transmission is reserved for use by AICH, AP-AICH or CD/CA-ICH.

Table 9.13 summarizes the main characteristics of the UTRA/FDD physical channels.

Physical channel	Associated transport channel	SF	Channel rate (kbps)	Channelization code	Scrambling code	Power control	Macrodiversity	TX diversity
DPDCH (UL)	DCH (UL)	4-256	960-15	variable, $C_{SF,SF/4}$	long or short	yes (CL)	yes	-
DPCCH (UL)	none	256	15	fixed, $C_{256,0}$	long or short	yes (OL/CL)	yes	-
PRACH (UL)	RACH	32-256	120-15	variable (OVSF)	long	yes (OL)	no	-
PCPCH (UL)	CPCH	4-256	960-15	variable (OVSF)	long or short	yes (OL/CL)	no	-
DPCH (DL)	DCH (DL)	4-512	1,920-30	variable (OVSF)	prim. or sec.	yes (CL)	yes	yes (OL/CL)
PDSCH (DL)	DSCH	4-256	1,920-30	variable (OVSF)	prim. or sec.	yes (CL)	on asso. DPCH	yes (OL/CL)
P-CPICH (DL)	none	256	30	fixed, $C_{256,0}$	primary	no	no	yes (OL)
P-CCPCH (DL)	BCH	256	27	fixed, $C_{256,1}$	primary	no	no	yes (OL)
S-CCPCH (DL)	FACH and/or PCH	4-256	1,920-30	variable (OVSF)	prim. or sec.	slow (FACH)	no	yes (OL)
SCH (DL)	none		specific code with 256 chips per *slot*			no	no	yes (TSTD)
PICH, AICH, AP-AICH, CD/CA-ICH, CSICH (DL)	none	256	fixed, ≈ 30	variable $C_{256,n}$	primary	no	no	yes (OL)

Table 9.13. *Physical channels characteristics in UTRA/FDD. OL: open loop, CL: closed loop*

Chapter 10

UTRA/FDD Physical Layer Procedures

10.1. Introduction

This chapter gives examples of the most common Layer 1 procedures, enabling a better understanding of the practical usage of some of the physical channels described in Chapter 9.

10.2. The UE receptor

Key Layer 1 procedures in the UE concern cell synchronization, power-control (TPC) and feedback (FBI) command generation, and a number of measurements used in cell-search and handover procedures (see Chapter 11). The performance of these procedures is highly dependent on the performance of the UE receiver architecture. Such architecture is not, however, specified in the standards with the purpose of enabling mobile manufactures to differentiate from each other in terms of performance and cost.

Figure 10.1 shows an example of a UE receiver. The front end of the receiver processes the RF signal and provides an analog baseband signal that is converted to the digital domain with an *Analog-to-Digital Converter* (ADC). At the output of the RRC filter, the Rake receiver attempts to separate multipath components and combine them based, for instance, on the MRC approach described in Chapter 5. The delay positions at which significant energy arrives are estimated by a path searcher algorithm using pilot symbols within the physical channels. These are also used by the Rake fingers to track the fast changing phase and amplitude values caused by the propagation channel, so that these effects are mitigated. The receiver operates in conjunction with the *Automatic Gain Control* (AGC) and the *Automatic Frequency Control* (AFC). The AGC improves the dynamic range of the receiver by maintaining a constants signal level at the input of the ADC. The AFC loop adjusts the crystal frequency to match the Node B frequency.

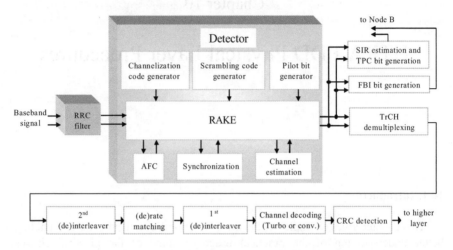

Figure 10.1. *Possible UE receiver architecture*

The UE receiver also performs the following L1 operations in the downlink.

After the UE is powered-on:

– the UE gets time/frequency synchronization with a given Node B from the SCH. Based on operations applied to the P-SCH and the S-SCH, the UE obtains slot and frame synchronization, respectively. The S-SCH also gives information about the scrambling code group used in the cell. This is finally exploited by the UE so that the cell scrambling code is determined from correlations applied to the CPICH;

– with the scrambling code estimated, the UE is able to detect the BCH cell system information conveyed by the P-CCPCH. With this information, the RACH access process can start so that the UE can make a request to the UTRAN for establishing an RRC connection. The UTRAN acknowledges this UE request on the FACH carried by the S-CCPCH. A DPCH can then be allocated to the UE to perform network registration and other NAS procedures.

Once registered to the network, the UE is put in an idle radio state where the PICH is periodically monitored. If a paging channel indication is detected, the UE shall demodulate the PCH message carried by the S-CCPCH.

During an active communication on the DPCH:

– the UE measures the SIR from the pilot bits/symbols in the control part of the downlink DPCH. This is then used to generate TPC commands for Node B to adjust the power of the downlink DPCH on a slot-by-slot basis;

– if closed loop Tx diversity is used, the UE generates FBI bits once every 2 or 4 slots and sends it to Node B in order to adjust the antenna phase and/or the amplitude depending on the Tx diversity mode;

– the UE decodes TFCI bits to find out which transport channels were active in the given DPDCH frame as well as their bit rate and channel decoding parameters. This operation is applied on a frame-by-frame basis (every 10 ms);

– with the estimated TFCI bits, the UE decodes the DCH message and determines how often data must be delivered to higher layers according to the TTI value: 10 ms, 20 ms, 40 ms or 80 ms.

10.3. Synchronization procedure

The synchronization procedure, also called cell search or acquisition procedure, is the first L1 task that the UE performs after power-on. This is also used to monitor/measure neighboring cells once the UE established first communication with the network. The purpose is twofold: obtain common channel slot/frame synchronization and determine the downlink scrambling code. This procedure is carried out in four steps (not mandatory in the standard) and involves three channels to facilitate the search: S-SCH, P-SCH and the CPICH. We can infer from Figure 10.2 that once the UE is time-aligned with the SCH slot/frame, it is automatically synchronized with the P-CCPCH slot/frame which is the time reference for all the other downlink channels within the cell.

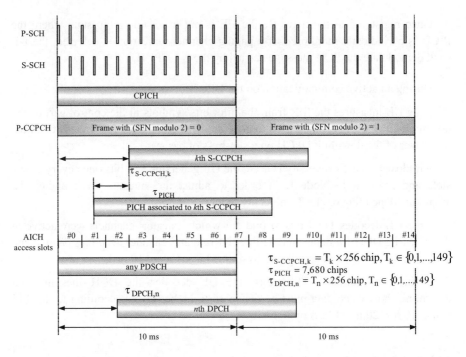

Figure 10.2. *Downlink physical channel time reference [TS 25.211]*

10.3.1. *First step: slot synchronization*

During the first step the UE uses the P-SCH to acquire slot synchronization within a cell. This is done from correlations performed on the received signal and the P-SCH sequence which is the same for all cells within the UTRAN. The correlation windows are 256 chips wide, i.e. the size of the P-SCH sequence. The outputs of the correlation operations are used to compose a "correlation profile". The correlation window is shifted on a chip-by-chip basis or a fraction of it in case the received signal is oversampled. Figure 10.3 shows an example of a profile involving 2,560 correlation points. The issue now is to select the strongest correlation *peak* which shall (potentially) indicate the beginning of a time slot. Due to low operating SNR, the correlation estimates have to be non-coherently accumulated over several slots in order to get reliable decision statistics.

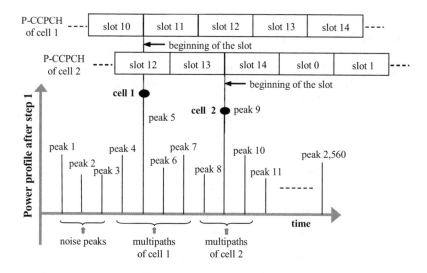

Figure 10.3. *UE correlation profile resulting from the slot synchronization step*

10.3.2. *Second step: frame synchronization and code-group identification*

After the first step, the start of a given slot is estimated but its position on the frame structure is unknown. This is indeed the role of the second step in which the UE uses the S-SCH to find frame synchronization and identify the scrambling code-group of the cell found in the first step. Unique cyclic S-SCH code sequences are assigned to each of the 64 groups. Each sequence comprises 15 codes of 256 chips in size corresponding to the number of slots within one frame (see Table 10.1). The 15 codes are chosen from a set of 16.

Group	Slot number within the code sequence associated to each group														
	0	**1**	**2**	**3**	**4**	**5**	**6**	**7**	**8**	**9**	**10**	**11**	**12**	**13**	**14**
0	$c_s^{1,0}$	$c_s^{1,1}$	$c_s^{2,2}$	$c_s^{8,3}$	$c_s^{9,4}$	$c_s^{10,5}$	$c_s^{15,6}$	$c_s^{8,7}$	$c_s^{10,8}$	$c_s^{16,9}$	$c_s^{2,10}$	$c_s^{7,11}$	$c_s^{15,12}$	$c_s^{7,13}$	$c_s^{16,14}$
1	$c_s^{1,0}$	$c_s^{1,1}$	$c_s^{5,2}$	$c_s^{16,3}$	$c_s^{7,4}$	$c_s^{3,5}$	$c_s^{14,6}$	$c_s^{16,7}$	$c_s^{3,8}$	$c_s^{10,9}$	$c_s^{5,10}$	$c_s^{12,11}$	$c_s^{14,12}$	$c_s^{12,13}$	$c_s^{10,14}$
i	...	...	...	...	...	...	...	...	...	...	...	...	...	...	...
62	$c_s^{9,0}$	$c_s^{11,1}$	$c_s^{12,2}$	$c_s^{15,3}$	$c_s^{12,4}$	$c_s^{9,5}$	$c_s^{13,6}$	$c_s^{13,7}$	$c_s^{11,8}$	$c_s^{14,9}$	$c_s^{10,10}$	$c_s^{16,11}$	$c_s^{15,12}$	$c_s^{14,13}$	$c_s^{16,14}$
63	$c_s^{9,0}$	$c_s^{12,1}$	$c_s^{10,2}$	$c_s^{15,3}$	$c_s^{13,4}$	$c_s^{14,5}$	$c_s^{9,6}$	$c_s^{14,7}$	$c_s^{15,8}$	$c_s^{11,9}$	$c_s^{11,10}$	$c_s^{13,11}$	$c_s^{12,12}$	$c_s^{16,13}$	$c_s^{10,14}$

Table 10.1. *Code sequences associated to the groups composing the S-SCH*

By correlating the received data with all 16 S-SCH possible codes, a matrix of 15×16 hypothesis denoted by $p_{i,k}$, where $i \in \{1, 2 \ldots 16\}$ and $k = 0,1\ldots14$ is formed (see Table 10.2). The correlation points $p_{i,k}$ are added following the 64 possible patterns in Table 10.1, thus giving a total number of 960 hypothesis (64×15). The hypothesis with the largest metric forms the most probable candidate. As P-SCH is always transmitted along with the S-SCH, the P-SCH correlation can serve as phase reference in order to cope with phase rotations caused by the frequency error and by fading.

Code number	Slot number				
	0	**1**	**k**	**13**	**14**
code 1	$p_{1,0}$	$p_{1,1}$	$p_{1,k}$	$p_{1,13}$	$p_{1,14}$
code 2	$p_{2,0}$	$p_{2,1}$	$p_{2,k}$	$p_{2,13}$	$p_{2,14}$
code i	$p_{i,0}$	$p_{i,1}$	$p_{i,k}$	$p_{i,13}$	$p_{i,14}$
code 15	$p_{15,0}$	$p_{15,1}$	$p_{15,k}$	$p_{15,13}$	$p_{15,14}$
code 16	$p_{16,0}$	$p_{16,1}$	$p_{16,k}$	$p_{16,13}$	$p_{16,14}$

Table 10.2. *Correlation matrix with hypothesis* $p_{i,k}$

10.3.3. *Third step: primary scrambling code identification*

At this step, the UE is synchronized at the slot/frame level and knows the scrambling code group of the found cell. The P-CCPCH and/or the CPICH symbols are used to search for 1 out of 8 possible scrambling codes which correspond to the code group of interest. In general, the UE performs symbol-by-symbol correlations over the CPICH with all 8 codes within the code group identified in the second step by keeping in mind that the channelization code of the CPICH is known and equals $C_{256,0}$.

10.3.4. *Fourth step: system frame synchronization*

The UE will not acquire full synchronization with the network before decoding the SFN (*Cell System Frame Number*). This fact makes it possible to read broadcast cell-system information within the BCH necessary for RACH/CPCH access and PICH monitoring procedures. Therefore, after the primary scrambling code is identified, the P-CCPCH can be detected and the system and cell specific BCH information can be read. Figure 10.4 shows the time relation between the P-CCPCH and the other physical channels involved in the synchronization process.

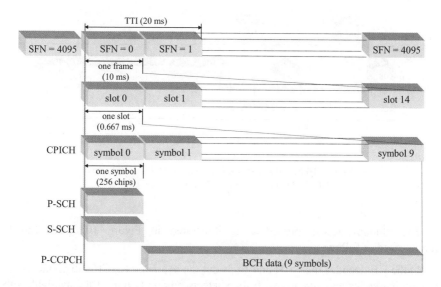

Figure 10.4. *Physical channels involved in the synchronization process*

10.4. Random access transmission with the RACH

Transmitting user or signaling data via the RACH involves a random access procedure based on the slotted ALOHA approach with fast acquisition indication [RAP 96]. The process is commonly used to request the establishment of an RRC connection, to answer to a paging or to send a short message (SMS).

The random access operation starts at the beginning of a number of well-defined time intervals, denoted access slots. Higher layers indicate to the UE what access slots are available by considering that there are 15 access slots per two frames and they are spaced 5,120 chips apart (see Figure 10.5). The total set of uplink access channels is divided into 12 sub-sets called RACH sub-channels.

The UE shall decode cell broadcast system information on the BCH before starting the transmission on the RACH. The information concerns: the preamble scrambling code, the message length (either 10 or 20 ms), the AICH transmission timing and channelization code, the set of available signatures and the set of available RACH sub-channels for each *Access Service Class* (ASC). The ASC parameter defines different priorities of RACH usage when the RRC connection is set up. There are eight different ASCs varying from 0 (highest priority used in case of emergency call) to 7 (lowest priority). The ASC is always associated with a persistence level (1-8) and a persistence scaling factor.

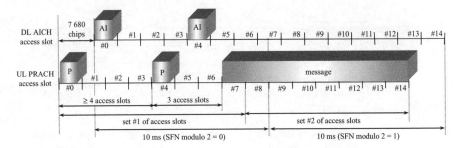

Figure 10.5. *Time relation between PRACH (uplink) and AICH (downlink) access slots as seen by the UE*

The random access procedure is illustrated in Figure 10.6 and can be summarized as follows:

1) the UE decodes the BCH and finds out the available RACH sub-channels and scrambling codes and signatures. It then selects randomly one of the available sub-channels and signatures;

2) the UE determines the maximum number of preamble retransmissions. Also, the downlink power is measured and the initial preamble power level is set with a proper margin due to open loop inaccuracy;

3) the first preamble is transmitted with the selected signature s and at the set power;

4) Node B according to the signature s detected on the PRACH preamble derives the symbols of the AI part and replies by repeating the preamble using the AICH;

5) the UE decodes the AI part in the AICH to see whether Node B has detected the RACH preamble. If the AICH is not detected, the preamble is resent with 1dB higher transmit power. If AICH is detected, a 10 or 20 ms long message part is transmitted with the same power as the last preamble. After a number of unsuccessful attempts, an error message is transmitted to the upper layers.

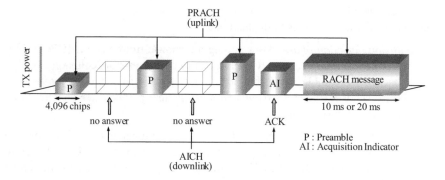

Figure 10.6. *RACH access procedure*

10.5. Random access transmission with the CPCH

Although not used in real networks, the CPCH access process is given here for the sake of consistency with 3GPP specifications. The CPCH transmission is a special case of random-access transmission based on DSMA-CD (*Digital Sense Multiple Access with Collision Detection*) with fast acquisition indication, collision resolution, pre-data power control, closed loop power control of the message and access preamble ramp-up identical to the RACH. CPCH is designed for transfer of signaling as well as user packet data in non-real-time applications (SMTP, HTTP, FTP) – the size of the message part is not limited to 2 frames as in the RACH.

The CPCH access procedure requires the UE to read system information messages containing:

– *Access Preamble* (AP) and *Collision Detection* (CD) scrambling codes and signature sets;

– AP-AICH and CD-AICH preamble channelization codes;

– CPCH scrambling code and DPCCH DL channelization code.

RRC also informs L1 about parameters involved in the CPCH open loop power control algorithm. Before starting the CPCH access procedure, the UE checks CPCH *Status Indication Channel* (CSICH) if there is a free channel (code and signature). In contrast to the RACH procedure, if no sub-channels are free, transmission is not attempted:

– *access phase*. The UE randomly selects a signature and an access slot from the set of CPCH signatures and access slots available in the cell. The AP is then sent with the selected signature, within the selected access slot and at the initial power set by MAC layer. The power of the AP is increased in case the UE does not receive an acknowledgement from Node B on the AP-AICH. When the AP is positively acknowledged, the UE passes to the next phase;

– *collision detection phase*. After getting a response from Node B (CPCH API), the *Collision Detection* preamble (CDP) is sent to detect possible collisions. Node B echoes back the preamble using the *Collision Detection Indication Channel* (CD-ICH). It also sends a *Channel Assignment* message that points a free channel to UE for actual packet transmission;

– *power control adjustment phase*. This phase is optional and may last from 0 to 5 ms as indicated by higher layers. During this period, both the UL PCPCH control part and the associated DL DPCCH are transmitted prior to the start of the uplink PCPCH data part;

– *message transmission phase*. In case the above phases run successfully, the actual message part of the CPCH is transmitted.

As previously mentioned, a DL DPCCH is always associated to the PCPCH (see Figure 10.7). This channel conveys layer 1 information including power control commands and pilot bits for estimating channel propagation conditions.

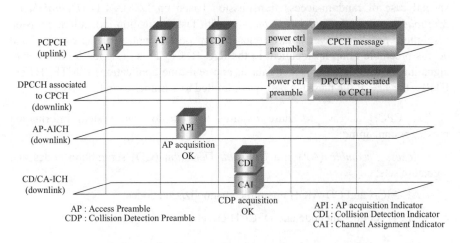

Figure 10.7. *CPCH access procedure*

10.6. Paging decoding procedure

The paging procedure is used by the network to reach the UE when this is in an idle state and no common or dedicated resources are allocated to it. For instance, the network may need to alert the UE of an incoming call in the RRC idle, URA_PCH and

CELL_PCH states (type 1 paging messages) or simply to inform the UE that some system information messages in the serving cell have changed.

Similarly to GSM networks, the *Discontinuous Reception* (DRX) approach is used in order to prolong the UE battery life: the UE listens only to the PCH within its DRX group and the network will only page the UE in that group of paging channels. In contrast to GSM, in UTRA the PCH (S-CCPCH) is not monitored as such but, rather, the associated PICH channel which can easily be demodulated at the Layer 1 level.

All the UEs camping within a given cell are divided into groups and *Paging Indicator* (PI) messages per paging group are periodically transmitted in the PICH when some UE in the group is paged. In each PICH frame are transmitted N_p PIs, where N_p is a cell-specific parameter set by the RNC. Possible values for this parameter are 18 (16 bits are repeated), 36 (8 bits are repeated), 72 (4 bits are repeated) or 144 (only 2 bits are repeated). When DRX is used the UE needs only to monitor the PI in one so called *Paging Occasion* (i.e. one PICH frame) per DRX cycle. The paging occasion is estimated by the UE by involving the IMSI, the SFN, the value of N_p and the length of a DRX cycle [TS 25.304]. The DRX cycle length equals 2^k frames, where k is a cell-based parameter set by RNC.

The UE has to listen to the PI messages in the time interval defined for its paging group. If the PI reception indicates that someone in the group is paged (PI = "1"), then all members of the group have to decode the next PCH frame from the S-CCPCH. The less frequently a PI is sent, the less often must the UE awake from the sleep (idle) mode to listen to the channel, but also the longer the delay in the connection establishment. The time relationship between PICH and S-CCPCH frames is shown in Figure 10.8 where an example is also given with $N_p = 18$ and where the UE reads the 13th paging indicator within a DRX cycle of 8 frames.

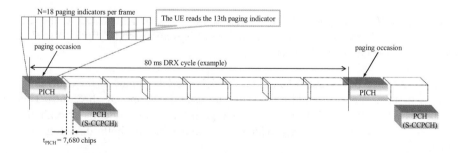

Figure 10.8. *UE paging monitoring process*

10.7. Power control procedures

Power control is used in WCDMA systems for adjusting the transmit power of the UE and Node B to the minimum levels within the required QoS and for creating minimal interference to the other users. Furthermore, power control is needed to increase the capacity of the system, compensate fading on the radio channel, eliminate/reduce near-far effect and reduce battery consumption. The standard defines open and inner loop power control procedures. Special cases of power control during *soft handover* are studied in Chapter 11.

10.7.1. *Open loop power control*

Open loop power control is the capability of the UE to set initial transmit powers for the first RACH preamble and for the DPCCH to a certain value before starting inner loop power control. This ensures that at call initialization the UEs do not use excessive transmissions powers and therefore reduce near-fact effect. The transmit power of the first PRACH preamble (and that of the PCPCH) is estimated from:

$$Preamble\_Initial\_Power = Primary\_CPICH\_power - CPICH\_RSCP$$
$$+ UL\_interference + Constant\_Value$$

where *Primary_CPICH_power*, *UL_interference* (measured by Node B) and *Constant_Value* are broadcast on the BCH [TS 25.331].

Similarly, the first DPCCH power level for the uplink inner loop power control is started as:

$$DPCCH\_Initial\_power = DPCCH\_Power\_offset - CPICH\_RSCP$$

where *DPCCH_Power_offset* is calculated at the RNC and provided to the UE during a radio bearer or physical channel reconfiguration, and *CPICH_RSCP* is measured by the UE on the CPICH.

The open loop power control assumes that UL and DL pathlosses are equivalent, which is not true since different frequencies are used in UL and DL directions. This means that fast fading distortions are not properly compensated, although the pathloss and shadowing effects which are comparable in both directions are mitigated.

10.7.2. *Inner loop and outer loop power control*

10.7.2.1. *Inner loop power control*

The inner loop power control is based on the feedback information from the other end on the radio link. It compensates the fading of the radio channel in UL and DL and keeps the quality of the connection at the desired level. This also reduces the effects of the near-far phenomenon in the uplink and mitigates neighboring cell interference in the downlink (see Chapter 5).

Inner loop power control in the uplink

In the uplink, the inner loop power control is used to adjust the UE transmitted power of the uplink DPCH and PCPCH (message part) upon reception of the transmit power control commands (TPC) in the downlink.

Node B receives the target SIR (SIR_{target}) from the RNC (outer loop power control). The value of the SIR_{target} is then compared with the estimated SIR (SIR_{est}) on the pilot symbol of the uplink DPCCH. If $SIR_{est} > SIR_{target}$, Node B generates a TPC command with value "0". The UE estimates this command and reduces its transmission power. Conversely, if $SIR_{est} < SIR_{target}$, a TPC command with value "1" is generated and sent to the UE in order for this to increase its transmission power (see Figure 10.9). The UE should be able to adjust the output power by the inner loop power control with step sizes of 1 dB, 2 dB or 3 dB. The latter is used when operating in compressed mode.

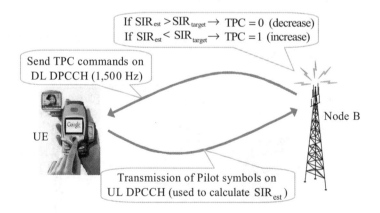

Figure 10.9. *Inner loop power control in the uplink*

The standard defines two algorithms for interpreting the TPC commands sent by Node B [TS 25.214]. The actual algorithm to be used by the UE is communicated by the UTRAN.

ALGORITHM 1.– this algorithm represents the ordinary power control as described above and is applied on a slot-by-slot basis. It is used to compensate for fast fading channels. Two power control steps are possible: 1 dB (well suited for UEs with speed $v < 30$ km/h) and 2 dB (well suited for UEs moving in the range $30 < v < 80$ km/h).

ALGORITHM 2.– the second algorithm makes it possible to use step sizes smaller than 1 dB by turning off uplink power control during certain number of slots. Indeed, the UE does not change its transmission power until it has received 5 consecutive TPC commands. If all five consecutive commands are "1" (increase) the transmit power is increased by 1 dB. If all five TPC commands are "0" (decrease) the transmit power is reduced by 1 dB. Otherwise, the transmit power is not changed. Note that increasing transmission power by 1 dB until the reception of the fifth slot is equivalent to applying a power control step of 0.2 dB per slot. Algorithm 10.2 is well suited for mobiles with very slow (< 3 km/h) or high (> 80 km/h) speed and is used to compensate slow fading rather than rapid fluctuations. In algorithm 2 errors may appear in detecting one or several TPC commands on a five-slot cycle. This makes algorithm 2 to be less robust than algorithm 1 against burst channel errors.

Inner loop power control in the downlink

The inner loop downlink power control sets the power of downlink DPCH transmitted by Node B. A similar algorithm applied to the uplink is used as illustrated in Figure 10.10.

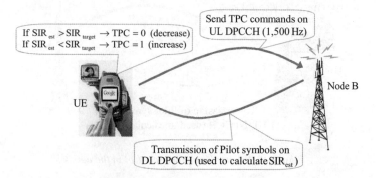

Figure 10.10. *Inner loop power control in the downlink*

Before generating the TPC commands the UE verifies the parameter DPC_MODE as indicated by RRC. Two modes for downlink are specified by the standard. If DPC_MODE = 0, the UE sends a unique TPC command to each slot. If DPC_MODE = 1, the UE repeats the same TPC command over three slots. Accordingly, Node B updates the radio link power on every slot or on every third slot depending on the DPC_MODE. The step size of 1 dB is mandatory and other step sizes are optional [TS 25.104].

Power control on downlink common channels

The transmit power of downlink common channels is determined by the network, though this is not specified in the standards. Table 10.3 gives examples of typical power levels.

Downlink common channel	Power level (typical)	Comment
P-CPICH	30-33 dB	Set during NW planning: 5-10% of maximum Node B Tx power ($\approx$20W)
P-SCH/S-CH	-3 dB	Relative to P-CPICH power
P-CCPCH	-5dB	Relative to P-CPICH power
PICH	-8 dB	Relative to P-CPICH power and the number of paging indicators per frame
AICH/AP-AICH	-8 dB	Power of one acquisition indicator (AI) compared to P-CPICH
S-CCPCH	-5 dB	Relative to P-CPICH and SF = 256. The configuration covers FACH power and PCH power
PDSCH	Inner loop power control	TPC commands sent by the UE on the associated uplink DPCCH

Table 10.3. *Examples of power control settings in downlink common channels*

10.7.2.2. *Outer loop power control*

The goal of the outer loop power control is to produce an adequate SIR_{target} for the inner loop power control in order to maintain the quality requirement on each bearer service. The outer loop power control mechanism makes it possible to compensate the imperfections in the inner loop power control, such as TPC command errors and E_b/N_o requirements in, for instance, fast moving UEs (see Figure 10.11).

When the CRC bits are present, it is possible to estimate a $BLER_{target}$ for each DCH transport channel within the same RRC connection, but only one SIR_{target} estimate is calculated per CCTrCH. An example of the algorithm is as follows:

- if CRC NOK (incorrect transport block): $SIR_{target} = SIR_{target} + \alpha$;

- if CRC OK (correct transport block): $SIR_{target} = SIR_{target} - \beta$.

where $\beta = \alpha/(1 / BLER_{target} - 1)$. For instance, for $\alpha = 1$ dB and $BLER_{target} = 1\%$, $\beta = 0.01$. Note that the estimation of the SIR_{target} is not standardized and may be a function of the FER, BLER, E_b/N_o, etc. The ideal value for the SIR_{target} is such that the desired quality is obtained with minimal transmission power for Node B and the UE. The outer loop power control operates with a frequency higher than the TTI value, i.e. in the range of 1 to 100 Hz.

Since the outer loop does not know that the power control and SIR have saturated, it may keep on increasing the target. A similar phenomenon happens in the lower bound as well. This phenomenon called *integrator windup* can be mitigated by bounding the range of the received SIR with an appropriate filter.

In the uplink, the RNC sets the SIR_{target} for Node B, whereas in the downlink the outer loop power control function is implemented in the UE. The target SIR value is adjusted by the UE by using a proprietary algorithm that provides the same measured quality as the quality target set by the RNC. The value of the downlink outer loop power control quality target (e.g. $BLER_{target}$) is controlled by the RNC for each downlink DCH.

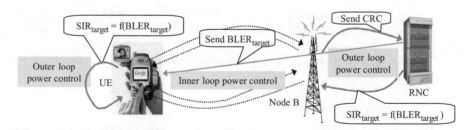

Figure 10.11. *Outer loop power control principle*

10.8. Transmit diversity procedures

In UTRA standard, transmission diversity means that the same signal is transmitted via more than one antenna at Node B (see Chapter 5). This technique makes it possible to create multipath diversity against fading and shadowing, and therefore the capacity in the downlink can be increased. The transmit diversity methods used in UTRA/FDD comprise open loop and closed loop approaches as shown in Table 10.4. Their performance depends on channel delay characteristics and the speed of the UE. For instance, in low speed and small delay spread channels, the gain against fading is high whereas in high-speed and large delay spread scenarios the gain is low.

The idea behind open loop transmit diversity is to transmit delayed copies of the same signal from different antennae without any feedback sent by the UE to Node B. Its implementation is transparent to the UE receiver. Two methods are defined in the standard: *Time Switched Transmit Diversity* (TSTD) and *Space Time block coding Transmit Diversity* (STTD).

Within closed loop transmit diversity, feedback information from the UE to Node B is sent to optimize the transmission from the diversity antenna. Based on this information, Node B can adjust the phase and/or amplitude of the antennae.

In the case where the transmit diversity (open or closed loop) is used on every downlink channel in the cell, the CPICH shall be transmitted from both antennae using the same channelization and scrambling code. In this case, the pre-defined bit sequence of the CPICH is different for Antenna 1 and Antenna 2 [TS 25.211].

Physical channel	Open loop Tx diversity		Closed loop Tx diversity
	TSTD	STTD	
P-CCPCH, S-CCPCH	no	yes	no
DPCH, PDSCH	no	yes	yes
SCH	yes	no	no
PICH, AICH	no	yes	no
CSICH, AP-AICH, CD/CA-ICH	no	yes	no
DL-DPCCH for CPCH	no	yes	yes

Table 10.4. *Tx diversity techniques used in UTRA/FDD*

10.8.1. *Time Switched Transmit Diversity (TSTD)*

The TSTD approach utilizes the time orthogonality between two antennae and is used only on the synchronization channel SCH time multiplexed with the P-CCPCH. The principle relies on switching the transmitted signal in time (slot-by-slot) between the antennae as illustrated in Figure 10.12. We can observe in the figure that the amplitudes of the P-SCH and the S-SCH are the result of a modulation by symbol a which enables the UEs within a cell to know whether the P-CCPCH is

STTD encoded ($a = +1$) or not ($a = -1$). The synchronization procedure is not affected when TSTD is used in the cell.

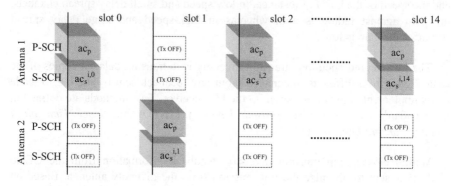

Figure 10.12. *TSTD principle when applied to the SCH*

10.8.2. *Space Time block coding Transmit Diversity (STTD)*

The idea behind the STTD algorithm is to transmit the same signal over two antennae after applying time block encoding as depicted in Figure 10.13. This procedure operates by blocks of 4 symbols denoted by b_0, b_1, b_2, b_3 [TS 25.211]. STTD encoding creates a kind of space-time diversity exploited by the UE Rake receiver. When the UE knows that STTD is used for a given channel, the *Maximum Ratio Combining* (see Chapter 5) takes into consideration the encoding made at the transmission to enhance the detection performance in the UE receiver. The STTD encoding is optional in the UTRAN, whereas the STTD support is mandatory at the UE.

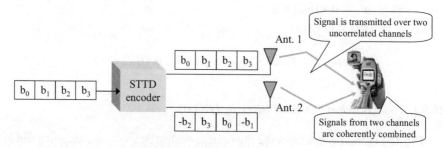

Figure 10.13. *STTD encoding principle*

10.8.3. *Closed loop transmit diversity*

Within closed loop transmit diversity, the UE measures the relative phase (and power) of two CPICH channels transmitted over two different antennae. It calculates then the most suitable (complex) weights w_1 and w_2 so that the received power at the desired UE is maximized and the interference to other UEs is minimized (see Figure 10.14). Closed loop transmit diversity only applies to DPCH and DSCH for which two modes are specified [TS 25.214].

Mode 1

The UE uses the CPICH transmitted from both Node B antennae ($CPICH_1$ and $CPICH_2$) to calculate (only) the phase difference between the two antennae. The UE sends one bit feedback command (FBI) per slot to Node B on the uplink DPCCH. The weight w_2 is an average of two consecutive FBI commands and thus, four different phase values can be applied.

Mode 2

In this mode the phase difference and transmit power are adjusted with 4 bits sent over 4 consecutive slots on the uplink DPCCH. Thus, the UE can generate 16 combinations of phase and amplitude. Three of these four bits are used to adjust the phase (45° resolution) and one bit is used to adjust amplitude (0.8/0.2).

Mode 2 is most suitable for low terminal speeds (slow feedback) and for higher data rates. Despite its higher complexity, Mode 2 is better than Mode 1 roughly until 30 km/h of the UE speed. Note that Mode 1 is better than STTD roughly until 70 km/h of the UE speeds. In general, the diversity gain with closed loop modes is degraded when the UE speed gets higher.

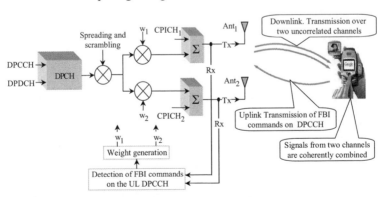

Figure 10.14. *Principle of closed loop TX diversity (valid for Modes 1 and 2)*

Chapter 11

Measurements and Procedures of the UE in RRC Modes

11.1. Introduction

After having been switched on, the UE performs a number of measurements so that the cell offering the best radio quality can be selected. During an active communication, the UE continues monitoring and reporting the received quality of the neighboring cells and if one of them turns out to have better quality than the current one, the communication is transferred to this new cell with a *handover* procedure. Even during idle periods where the UE waits for an indication of an incoming call, the quality of the current and the neighboring cells is periodically evaluated and a *cell reselection* process may take place if a quality criterion is fulfilled in one of the monitored cells. This chapter describes the measurements realized by the physical layer in the different RRC modes as well as the procedures enabled by the results of these measurements such as cell selection/reselection and handover procedures.

11.2. Measurements performed by the physical layer

The measurement of various quantities is one of the key services provided by the physical layer. The measurement process is controlled by the RRC layer in the UE and by NBAP and *frame protocols* in Node B. Note that the standard will not specify the method used to perform these measurements or stipulate that the list of measurements must all be performed. While some of the measurements are critical to the functioning of the network and are mandatory for delivering the basic functionality, others may be used by the network operators to optimize the radio resources. Most of the measurements are performed by the physical layer, although some of them, like traffic volume statistics, are performed by higher layers such as MAC.

When the UE is in CELL_DCH state, the measurement process is initiated by the UTRAN via the RRC *MEASUREMENT CONTROL* message [TS 25.331]. This indicates:

– the type (intrafrequency, interfrequency, intersystem, etc.);

– the measurement ID;

– a command (*set-up*, *modify* or *release*);

– the measurement quantity (pathloss, CPICH RSCP, CPICH Ec/No, etc.);

– the reporting quantity and criteria (periodical, event-triggered) as well as the mode (acknowledged, unacknowledged).

In RRC idle, CELL_PCH, URA_PCH and CELL_FACH states, the measurement control message is broadcast by the UTRAN in SIB 11 and SIB 12 on the BCH.

The measurement reporting is defined by the network and should be based on measurements (i.e. event-triggered) or time (i.e. periodical, like in GSM). The result is sent to the network using the *MEASUREMENT REPORT* message, including the measurement ID and the results (case of CELL_DCH and CELL_FACH states). Event-triggered reporting means that the UE sends a report to the UTRAN only when the reporting criterion is fulfilled. Combining the event triggered and the periodical approach is also possible. From the system quality point of view, the periodic reporting mode can perform better than the event-triggered mode. However, the resulting signaling load becomes considerably higher. In other words, at the same level of signaling load, the event-triggered mode results in a better system quality.

11.2.1. *Measurement model for physical layer*

As shown in Figure 11.1, internal layer 1 filtering is applied to the inputs measured at point "A" in order to average out the impact of multipath fading. Actual filtering is dependent on implementation. What the standard specifies are the performance requirements and the reporting rate at point "B" in the model [TS 25.133].

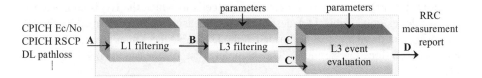

Figure 11.1. *Physical layer model at the UE*

Point "B" represents a measurement reported by the physical layer to the upper layers after L1 filtering. The reporting rate is defined by the standard and its measurement is type specific. In order to give the operator the possibility to have a better control over the accuracy of the measurements, a special higher layer filter model (L3 filtering) is specified. Thus, the measurement reported at point "B" is filtered again following the configuration provided by RRC signaling (UE measurements) or NBAP signaling (Node B measurements). The L3 filter is standardized and this enables the operator to easily increase the accuracy of the measurements but causes an increased delay. At point "C" the measurement is received after processing in the L3 filter. This and other measurements are finally used to evaluate the reporting criteria and, if fulfilled, a measurement report message will be sent via the radio interface (or Iub interface in case of Node B) at point "D".

11.2.2. *Types of UE measurements*

The measurements performed by the UE are categorized as follows:

– *intrafrequency* measurements on DL physical channels at the same frequency used in the current cell or in the cells in the so-called *active set*;

– *interfrequency* measurements on the DL physical channels at frequencies that differ from the frequency of the current cell or the cells in the *active set*;

– *intersystem* measurements on the DL physical channels belonging to another radio access system (e.g. GSM). They are also called inter-RAT (*Radio Access Technology*) measurements. When different RATs are used, the measurement quantity is translated into a second measurement quantity using *mapping functions*. The second measurement quantities enable the comparison of different RATs signal qualities. The nature and the parameters of the mapping functions are broadcast in system information messages;

– *traffic volume* measurements on UL measured by L2;

– *quality* measurements such as the DL transport block error rate (BLER);

– internal measurements such as the UE TX power, the received signal level of a UTRAN carrier and the TX-RX time difference on a dedicated channel.

In UTRA FDD the *active set* contains the cells to which the UE is connected in soft handover state. Table 11.1 describes some measurements performed by the UE by noting that these are also performed at the UTRAN side. Other measurements are exploited on location services. The complete list of measurements is given in [TS 25.215].

Type of measurement	Description
CPICH RSCP (*Received Signal Code Power*)	– Received power on one code measured on the primary CPICH – Can be intrafrequency or interfrequency – Used in cell selection/reselection, handover, DL/UL open loop power control and path loss calculation
P-CCPCH RSCP (*Received Signal Code Power*)	– Received power on one code measured on P-CCPCH (TDD) – Used for monitoring UTRA TDD cells
SIR (*Signal-to-interferenceRatio*)	– Signal-to-interferenceratio typically measured on theDPCCH: (RSCP/interference)× (SF) – Can be intrafrequency or interfrequency – Used in DL/UL open and closed loop power control
RSSI (*Received Signal Strength Indicator*)	– Received signal power strength measured on a frequency carrier – Can be intrafrequency or interfrequency (UTRA carrier RSSI) or intersystem (GSM RSSI carrier) – Used to evaluate interfrequency or intersystem *handover*
CPICH E_c/N_o	– CPICH Energy per chip (Ec) divided by the power density in the band (E_c/N_o = RSCP/RSSI) – Can be intrafrequency or interfrequency – Used to evaluate *handover* and cell selection/reselection
BLER (*Block Error Rate*)	– Estimation of the transport channel block error rate based on evaluating the CRC on each transport block – Intrafrequency – Used to evaluate the DL radio link quality

Table 11.1. *Examples of measurements performed by the UE*

11.3. Cell selection process

After a UE is switched on, it attempts to make contact with a UMTS network (PLMN). The UE then looks for a *suitable* cell of the chosen PLMN and enters the idle mode by *camping* on that cell. Once camped on the chosen cell, the *cell selection process* is completed. The UE finally registers its presence in the location/routing area (LA/RA) of the chosen cell, if necessary, by means of a location updating or IMSI attach procedure (see Chapter 8).

The purpose of camping on a cell is threefold:

– it enables the UE to receive cell broadcast system information;

– it enables the UE to access the subscribed services by initially accessing the network on the RACH of the cell on which it is camped;

– it enables the UE to receive a call since the network knows the cells of the LA/RA in which the UE is camped. The network can then send pages for the UE on the BCH of all the cells in the LA/RA.

The UE can camp on a *suitable cell* to obtain normal service whereas, when camped on an *acceptable cell*, the UE can receive limited service only. The requirements to be fulfilled by an *acceptable* and *suitable* cell are discussed later on.

11.3.1. *PLMN search and selection*

The UE tries to select as a priority a cell belonging to the network of the operator (PLMN) where the user subscribed a service contract. If not found, another PLMN from a priority list (see [TS 23.122]) is selected and a cell belonging to that PLMN is searched. The PLMN selection process is controlled by the *Non-Access Stratum* (NAS), whereas the cell search process is performed by the layers in the *Access Stratum* (AS) as illustrated in Figure 11.2. In the case of a dual-mode UE (Type 2), there are two preferred PLMN lists in the USIM: one configurable by the user and the other by the operator. In addition, there is a RAT (GSM or UTRA) associated to each PLMN in the preferred lists: for instance PLMN 1, RAT = 3G; PLMN 2, RAT = 2G; PLMN 3, RAT = 3G, etc.

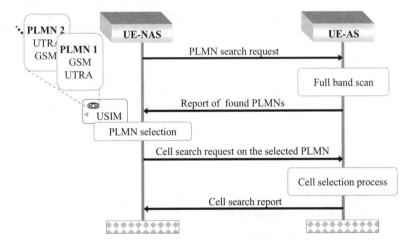

Figure 11.2. *UE NAS-AS interaction in PLMN search/selection and cell selection procedures*

11.3.2. *Phases in the cell selection process*

Frequency band scan

According to CDMA principles, multiple cells may use the same frequency carrier and therefore the cell search process requires the UE to find a cell within every single cell carrier in the UTRA/FDD spectrum: around 300 possible downlink carriers in Europe in the band 2,110-2,170 MHz [TS 25.101]. Two cases are possible:

– *Initial cell selection.* In the case of initial cell selection, the UE has no prior knowledge of the available RF carriers (UMTS, GSM, etc.). By using the available radio access technologies (GSM, UTRA FDD, TDD, etc.) in turn, the UE scans all the available carrier frequencies one by one, and for each carrier it searches for the strongest cell. Having found this cell, the UE reads its system information in order to find out which PLMNs are available. If the PLMN selected initially by the UE is amongst the available PLMNs, the search for a stronger cell on the other carrier frequencies is stopped.

– *Stored information cell selection.* The process of cell selection can be speeded up by using the information stored when the UE was last switched on. This information may include carrier frequencies, cell scrambling codes, etc. Apart from the way cell information is provided, the remaining steps of cell selection are the same as those used by the initial cell selection procedure. The process of cell selection in idle mode uses the stored information cell selection first. The initial cell selection approach is only used when the stored information method is unsuccessful.

Synchronization (cell search)

To be able to camp on a cell and listen to its common channels, the cell timing and scrambling codes are required. The cell search process described in Chapter 10 is then carried out so that the cell broadcast information in the BCH can be decoded.

BCH decoding

Decoding the BCH gives information about which PLMN the detected cell belongs to and makes it possible to obtain parameters to evaluate whether the cell is suitable or not;

Evaluation of the suitable cell criteria

The choice of a suitable cell for the purpose of receiving normal service is referred to as "normal camping". There are various requirements that a cell must satisfy before a UE can perform normal camping on it:

– It should be a cell of the selected PLMN or a PLMN considered as "equivalent" (see Chapter 8).

– It should not be barred. The PLMN operator may decide not to allow UEs to camp on certain cells, e.g. when the cell runs out of radio resources. Barred cell information is broadcast on the BCH. The barred cell status may in fact change dynamically; hence the UE needs to regularly check the BCH system information for this parameter.

– It should not be in an LA which is in the list of forbidden LAs.

– It should fulfill the cell selection criteria "S".

NOTE.– an *acceptable cell* providing limited service such as originating emergency calls requires only the cell not to be barred and the criteria "S" to be fulfilled.

Cell reselection

Initially the UE looks for a cell that satisfies the above 4 constraints by checking cells in descending order of the received signal strength. If a *suitable cell* is found, the UE camps on it and performs any registration as necessary. When camped on a cell the UE regularly looks to see if there is better cell in terms of a *cell reselection criterion*, and if there is, the better cell is selected. Also if one of the other criteria changes (e.g., the current cell becomes barred), or there is a downlink signaling failure a new cell is selected. This is called *cell reselection*.

PLMN reselection

The UE normally operates on its home PLMN (HPLMN). However, a visited PLMN (VPLMN) may be selected if, for instance, the UE loses coverage or if the user is roaming in a foreign country. When in a VPLMN, the UE searches for a PLMN according to the following two modes of operation: automatic (this mode utilizes a list of PLMNs in a priority order, the highest priority PLMN which is available and allowable is selected) and manual (here the UE indicates to the user which PLMNs are available).

Figure 11.3 illustrates the aforementioned phases involved in the cell selection process.

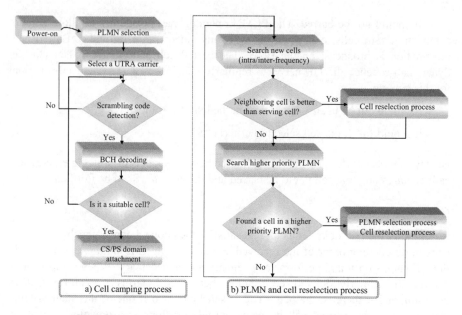

Figure 11.3. *Example of cell camping and cell reselection processes*

11.3.3. *"S" cell selection criterion*

For the purpose of selecting the best cell, the UE calculates the "S" selection criterion from the received level for each cell and from cell selection information broadcast on the BCH. This is done by calculating the parameters S_{qual} and S_{rxlev} given in dB as follows:

$$S_{qual} > 0 \text{ and } S_{rxlev} > 0, \text{ for UTRA/FDD mode} \tag{11.1}$$

$$S_{rxlev} > 0, \text{ for UTRA/TDD mode} \tag{11.2}$$

where:

$$S_{qual} = Q_{qualmeas} - Q_{qualmin} \tag{11.3}$$

$$S_{rxlev} = Q_{rxlevmeas} - Q_{rxlevmin} - P_{compensation} \tag{11.4}$$

The parameters in expressions [11.3] and [11.4] are described in Table 11.2. S_{qual} guarantees a good quality level in the downlink, whereas S_{rxlev}, which is similar to the *C1* criterion in GSM enables a certain radio quality balance in the two directions of uplink and downlink. These quantities also determine the coverage limit as well as the boundary between two adjacent cells for cell selection.

Parameter	Description
$Q_{qualmeas}$	CPICH E_c/N_0 measurement averaged over at least 2 samples
$Q_{qualmin}$	Minimum required quality level in the cell (broadcast on BCH)
$Q_{rxlevmeas}$	RX level value measurement based on CPICH RSCP (dBm). Averaged over at least 2 samples.
$Q_{rxlevmin}$	Minimum required RX level in the cell (dBm) broadcast on BCH
$P_{compensation}$	max(UE_TXPWR_MAX_RACH − P_max, 0)
UE_TXPWR_MAX_RACH	Maximum power level the UE is authorized to transmit when accessing the cell on RACH (broadcast on BCH) (dBm)
P_max	Maximum power level the UE can transmit according to its TX power class (dBm)

Table 11.2. *Parameters involved in the evaluation of "S" cell selection criteria*

11.4. Cell reselection process

By having camped upon the best cell using the cell selection criterion, the UE enters into the cell reselection mode. Cell reselection is the process of the UE ensuring that it is always on the best cell. Periodically, the UE repeats measurements of neighboring cells and then calculates the best cell for reselection using the cell reselection criteria "R". According to the specifications [TS 25.133], the UE should be able to monitor up to 3 frequency carriers containing as much as 64 cells. A maximum of 32 cells can be present in one of the 3 carriers (see Figure 11.4).

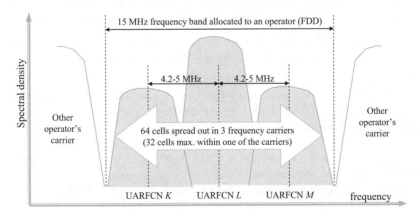

Figure 11.4. *Maximum number of cells to be monitored by the UE in UTRA/FDD*

As shown in Figure 11.5, unlike the CELL_DCH state where the handover procedure applies, the cell reselection process takes place in idle RRC, CELL_PCH, URA_PCH and CELL_FACH states (in which there is no dedicated channel allocated to the UE).

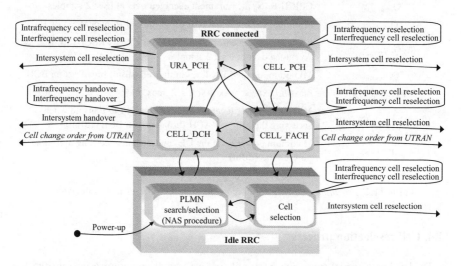

Figure 11.5. *Handover and cell selection/reselection procedures in UE RRC modes*

11.4.1. *Types of cell reselection*

The following cell reselection scenarios are possible in UTRA/FDD:

– *intrafrequency cell reselection*, in the case where the new selected cell is in the same frequency as the serving cell. Intrafrequency measurements are required;

– *interfrequency cell reselection*, in the case where the new selected cell is in a frequency carrier different to that of the serving cell. Interfrequency measurements are required;

– *intersystem cell reselection (inter-RAT)*, in the case where the new selected cell is part of a RAT (GSM, UTRA/TDD) different to that of the serving cell. Intersystem measurements are required.

NOTE.– intersystem cell reselection is also possible when explicitly ordered by the UTRAN based on the process: *Inter-RAT cell change order from UTRAN* [TS 25.133]. This applies to PS services in the sense UMTS → GPRS.

11.4.2. *Measurement rules for cell reselection*

Based on system information messages (SIB3/4), the network indicates to the UE the type of measurements to be performed on the serving and neighboring cells. This enables the network to somehow indirectly control the cell selection/reselection processes and therefore the load and coverage for a given cell.

1. Intrafrequency measurements:

– if $S_{qual} > S_{intrasearch}$, the UE shall not perform intrafrequency measurements since the radio quality in the serving cell is still good;

– if $S_{qual} \leq S_{intrasearch}$, the UE shall perform intrafrequency measurements;

– if $S_{intrasearch}$ is not broadcast the UE shall perform intrafrequency measurements.

2. Interfrequency measurements:

– if $S_{qual} > S_{intersearch}$, the UE shall not perform interfrequency measurements since the radio quality in the serving cell is still good;

– if $S_{qual} \leq S_{intersearch}$, the UE shall perform interfrequency measurements;

– if $S_{intersearch}$ is not broadcast the UE shall perform interfrequency measurements.

3. Intersystem measurements when camped on a UTRA/FDD cell:

– if $S_{qual} > S_{searchRAT,GSM}$, the UE shall not perform intersystem measurements on GSM neighboring cells since the radio quality in the serving cell is still good;

– if $S_{qual} \leq S_{searchRAT,GSM}$, the UE shall perform intersystem measurements on GSM neighboring cells;

– if $S_{searchRAT,GSM}$ is not broadcast the UE shall perform intersystem measurements on GSM neighboring cells.

Parameters $S_{intrasearch}$, $S_{intersearch}$ and $S_{searchRAT,GSM}$ are broadcast in system information and are read in the serving cell. The UE is also informed if a *Hierarchical Cell Structure* (HCS) is used in the network. In this case, specific measurements rules are applied (see [TS 25.304]).

11.4.3. *"R" ranking criterion*

The cell reselection procedure is based on a cell ranking criterion "R" where the UE calculates values R_s and R_n on the serving and neighboring cells, respectively:

$$R_s = Q_{meas,s} + Q_{hyst1,s} \qquad [11.5]$$

$$R_n = Q_{meas,n} - Q_{offset1,s,n} \qquad\qquad [11.6]$$

Parameter $Q_{hyst1,s}$ in expression [11.5] prevents a UE moving in the border of contiguous cells to alternatively select these cells, whereas parameter $Q_{offset1,s,n}$ in expression [11.6] may advantage or disadvantage one cell with respect to another. Table 11.3 gives more details about the parameters in expressions [11.5] and [11.6].

Note that the value of $Q_{meas,s}$ and $Q_{meas,n}$ in expressions [11.5] and [11.6] can be estimated from CPICH RSCP or CPICH Ec/No measurements as indicated by the network in SIB3/4 messages. When the CPICH Ec/No measurement is used, parameters $Q_{hyst2,s}$ and $Q_{offset2,s,n}$ replace respectively $Q_{hyst1,s}$ and $Q_{offset1,s,n}$.

NOTE.– in expression [11.6], the parameter TEMPORARY_OFFSET is added in case HCS is used in the network, thus giving: $R_n = Q_{meas,n} - Q_{offset1,s,n}$ – TEMPORARY_OFFSET. Such parameter penalizes a given cell during a certain time. For instance, this may prevent a fast-moving UE to camp on a cell of small size. Note also that when HCS is used, another criterion named "H" is applied together with "R" in order to decide whether cell reselection shall be performed or not [TS 25.304].

Parameter	Description
$Q_{meas,s}$	CPICH RSCP or CPICH E_c/N_0 measurement averaged over at least two samples in the serving cell
$Q_{hyst1,s}$	Hysterisis parameter giving a margin in respect to $Q_{meas,s}$. This is broadcasted on the BCH
$Q_{hyst2,s}$	Same as $Q_{hyst1,s}$ when the CPICH E_c/N_0 measurement is used
$Q_{meas,n}$	Same as $Q_{meas,s}$ but applied to one of the neighboring cells
$Q_{offset1,s,n}$	Cell broadcast parameter giving an offset between the serving and a neighboring cell when the quality measure is set to CPICH RSCP
$Q_{offset2,s,n}$	Same as $Q_{offset1,s,n}$ but used when the quality measure is set to CPICH E_c/N_0

Table 11.3. *Parameters used in the "R" ranking criterion*

11.4.4. *Phases in the cell reselection process*

The cell reselection process can be summarized in the following six steps:

– evaluation of criterion "S" on the candidate cells for reselection;

– ranking of the cells fulfilling the criterion "S" according to R_s and R_n values;

– if a GSM neighboring cell is the best ranked (i.e. with the highest R_n value), this cell is selected;

– if a UTRA/FDD neighboring cell is the best ranked and if CPICH RSCP is used, this cell is selected;

– if a UTRA/FDD neighboring cell is the best ranked and if CPICH E_c/N_o is used, a second ranking is applied. This time criterion "R" is evaluated using $Q_{meas,s}$ and $Q_{meas,n}$ estimated from CPICH E_c/N_o measurements and using parameters $Q_{hyst2,n}$ and $Q_{offset2,s,n}$ in expressions [11.5] and [11.6]. The cell with highest R_s or R_n value will be selected;

– in all the above cases the neighboring cell is reselected if it is better ranked than the serving cell during a time interval $T_{reselection}$ indicated in SIB3/4. Additionally, more than 1 second shall elapse since the UE camped on the current serving cell.

11.5. Handover procedures

The handovers in UTRA/FDD can be classified into two groups: soft and hard handover. In the case of soft handover the UE is simultaneously connected to several Node Bs, while in the case of a hard handover the UE is connected to one Node B at a time. For dedicated channels a soft handover is typically performed, while for shared and common channels a hard handover is the only possibility. A special case of soft handover is the softer handover whose principle is studied in Chapter 5.

From the UE perspective, the candidate cells the UE monitors are classified into three groups [TS 25.331]:

– cells belonging to the *active set*, i.e. the cells involved in soft handover (UTRA/FDD cells currently assigning a downlink DPCH to the UE). Intrafrequency measurements are performed on the active set which may contain as much as 6 cells;

– cells belonging to the *monitored set*, i.e. cells which are not included in the active set but which are monitored by the UE from the list sent by the UTRAN (CELL_INFO_LIST). Intrafrequency, interfrequency and intersystem measurements are performed on the cells of the monitored set;

– cells belonging to the *detected set*, i.e. cells which are neither in the active nor in the monitored set. Only intrafrequency measurements in the CELL_DCH state are made on the cells of the detected set.

Similarly to cell reselection, the handover procedure requires the UE to monitor total 64 UTRA/FDD cells spread out into three different frequency carriers: at most 32 cells on the same carrier frequency as the serving cell, and at least 32 cells on two

UTRA/FDD carriers in addition to the serving cell. If the UE supports GSM radio technology, it shall further monitor maximum 32 GSM cells indicated by the network in a separate list.

The above three categories lead to the following classification of handovers:

– *intrafrequency handover*, in the case where current communication is transferred or shared by a neighboring cell using the same frequency carrier as the cell(s) in the active set. Soft handover is a typical example of intrafrequency handover;

– *interfrequency handover*, in the case where current communication is transferred to a cell using a frequency carrier different to that of the cells in the active set. Interfrequency handovers are of hard handover type;

– *intersystem handover*, in the case where current communication is transferred to a neighboring cell belonging to a different radio access technology to the serving cell. This happens typically between a UTRA/FDD and a GSM cell. An intersystem handover is always of hard handover type.

NOTE.– intersystem handover is generally initiated by the network based on measurements reported by the UE. The standard specifications, however, makes it possible to execute intersystem handover without taking into account any measurement: this process is called *blind handover*.

11.5.1. *Phases in a handover procedure*

As illustrated in Figure 11.6, three phases are distinguished within the handover process: a measurement phase, an algorithm evaluation and decision phase, and an execution phase. During the measurement phase the UE monitors the signal quality and signal strength of the active and neighboring cells following the instructions supplied by the RNC and reports the measurement results back to the RNC. These results are then used by RNC to evaluate a handover algorithm (e.g. comparisons against predefined thresholds). The final decision regarding whether to execute or not a handover is done in the RNC. Note that the actual handover evaluation procedure is not defined in the technical specifications. Thus, it is up to each manufacturer to decide how the network reacts on the measurement reports the UEs are transmitting.

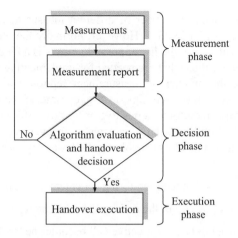

Figure 11.6. *Phases in a handover procedure*

11.5.2. *Intrafrequency handover*

The most representative example of intrafrequency handover is the soft handover which has special importance in UTRA/FDD due to its close relationship to power control. As discussed in Chapter 5, this has a direct impact on the amount of interference the system can tolerate and hence, on the overall system capacity.

NOTE.– soft handover is not the only possibility within the intrafrequency handover category. Intrafrequency hard handover is also possible and is needed when Node Bs involved in the handover are controlled by different RNCs and inter-RNC soft handover is not possible; for instance, when "Iur" interface is not implemented.

11.5.2.1. *Example of soft handover algorithm*

Soft handover is a *Mobile Evaluated Handover* (MEHO). Indeed, when in the CELL_DCH state, the UE continuously measures the strength of pilot signals (primary CPICH) of the serving and neighboring cells from the list provided by the RNC on current carrier frequency. The UE then compares the measurement results with the handover thresholds provided also by the RNC and returns the measurement reports to the RNC when triggers are fulfilled. The final decision will be made by the RNC who orders the UE to add or remove cells from its active set. The cell load is balanced between different cells by adjusting the CPICH reception level. If the CPICH power is reduced, part of UEs handover to other cells, on the contrary, if the CPICH power is increased, the cell will invite more UEs to handover to its own cell.

As previously described, a number of different measurement-based events are defined in specifications [TS 25.331]. However, normally the UE does not need to report all possible events, since the list of possible events is a kind of "toolbox" used by the network to choose those that are needed for implementing handover function. In order to decrease the number of measurement reports and avoid unnecessary handovers, a hysteresis (Hyst) and time-to-trigger timer ΔT can be implemented. Also, for each cell that is monitored, a positive or negative offset can be added to the measurement quantity before the UE evaluates if an event occurred. Figure 11.7 shows and example of soft handover involving the following measurement reporting events:

– event 1A (ADD). A cell in the monitoring set enters the reporting range;

– event 1B (REMOVE). A cell in the active set leaves the reporting range;

– event 1C (CHANGE BEST). A non-active cell becomes better than the active cell.

The following parameters are involved in this example:

– *Rep_Th*: reporting threshold for soft handover;

– *Hyst_event1A*, *Hyst_event1B*, *Hyst_event1C*: addition hysteresis for events 1A, 1B and 1C, respectively;

– *Meas_Sign*: CPICH Ec/No filtered measurement;

– *Best_Ss*: is the strongest measured cell in the active set.

In the example, a moving UE engaged in a communication in CELL_DCH state performs measurements on the primary CPICH of *cell 1* in the active set and of *cell 2* and *cell 3* which are supposed to belong to the monitored set. The size of the active set is assumed to be 2. Parameter *Meas_sign* is calculated from the measurements performed on the CPICH 2 (*cell 2*). If *Meas_Sign_cell_2* > (*Best_Ss* − *Rep_Th* + *Hyst_event1A*) for a period ΔT, and since the active set is not full, the cell is added to the active set (event 1A).

As time passes, the quality of *cell 1* is degraded and a measurement *Meas_sign* performed on CPICH 3 shows that *Meas_sign_cell_3* > *Meas_sign_cell_1* + *Hyst_event1C* for a period of ΔT. Thus, *cell 1* in the active set is replaced by the strongest candidate cell, i.e. cell 3 in the monitored set (event 1C).

Finally, after a certain time, *Meas_Sign* measured on CPICH 3 is such that *Meas_Sign_cell_3* < *Best_Ss* − *Rep_Th* − *Hyst_event1B* for a period of ΔT, so cell 3 is removed from the active set and only cell 2 is maintained therein (event 1B).

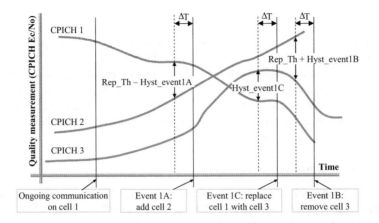

Figure 11.7. *Example of soft handover algorithm (adapted from [TR 25.922])*

11.5.2.2. *Timing adjustment of candidate cells*

In the downlink, soft handover gain is acquired with MRC (see Chapter 5) by the UEs. However, UTRA is an asynchronous system network, so it is necessary to adjust the transmission timing among the active sets – this enables the coherent combining in the Rake receiver to obtain multipath diversity. Otherwise, it will be difficult for the UE to combine the receiving signals from the different Node Bs. With this in mind, the UE shall perform additional measurements to determine the time difference between candidate cells to be added into the active set as illustrated in Figure 11.8. The measurement is the SFN-CFN observed time difference [TS 25.215]. The measurement result is sent to the RNC via the serving Node B. The candidate Node B shall adjust the downlink timing in steps of 256 chips based on the information it received from the RNC.

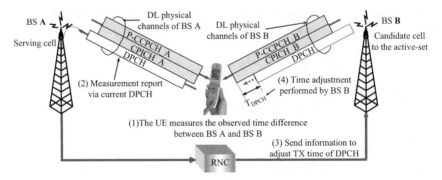

Figure 11.8. *Time adjustment process between two base stations*

11.5.2.3. *Power control during soft handover*

A given UE in overlapping cells that use the same pair of frequency carriers can dramatically increase the amount of uplink interference in the cells where its transmission power is not controlled. The same situation occurs in the downlink since only the transmission power of the serving Node B is controlled. This is the main motivation behind the soft handover process where closed power control loops are established with all the cells within the active set prior to the handover of the current communication from one cell to another.

Uplink power control in soft handover

In the uplink, each cell within the active set estimates the value of SIR_{est} from the different paths of the uplink DPCCH sent by the UE. Then, they generate individually the TPC commands indicating to the UE to increase or reduce its power. Thus, the UE receives different commands from different Node Bs.

As discussed in Chapter 10, two algorithms apply in the uplink. Figure 11.9 illustrates *Algorithm 1* where a TPC command results in a step size of + 1 or – 1 dB. Soft symbol decision is performed on each of the TPC commands transmitted by the n Node Bs based on a reliability figure W_i ($i = 1, ..., n$) assigned to each symbol. When a TPC command considered to be *reliable* is such that $TPC\_cmd_m = - 1$, the procedure is stopped and the UE decides to reduce its transmitted power. Otherwise, the UE shall consider that its power should be increased.

Algorithm 2 is also adapted to the soft handover case in order to emulate smaller step sizes for power control. The algorithm is carried out into two steps. First, the UE determines a temporary TPC command, TPC_temp_i, for each of the n sets of 5 TPC commands: $TPC\_temp_i = 1$ if all 5 hard decisions are "1" and $TPC\_temp_i = - 1$ if all decisions are set to "0". Otherwise, $TPC\_temp_i = 0$.

Then, the UE derives a combined TPC command for the fifth slot, TPC_cmd, as a function of all the n temporary power control commands of TPC_temp_i:

$$- TPC\_cmd = 1 \text{ if } \frac{1}{N} \sum_{i=1}^{N} TPC\_temp_i > 0.5 ;$$

$$-TPC\_cmd = -1 \text{ if } \frac{1}{N} \sum_{i=1}^{N} TPC\_temp_i < -0.5 .$$

Otherwise, TPC_cmd is set to 0.

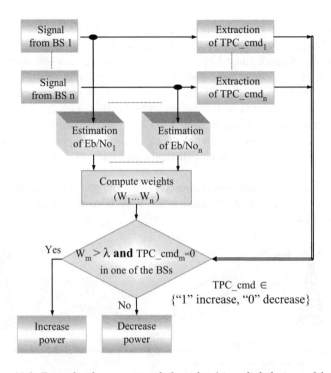

Figure 11.9. *Example of power control algorithm 1 in uplink during soft handover*

Downlink power control in soft handover

In the downlink, the UE uses a Rake receiver to combine all signals from the different Node Bs and issue a single power command that is transmitted in uplink. Note that Node Bs can receive the command differently and hence adjust their powers in opposite directions. The error rate of the PC commands can be reduced if the command is repeated over several PC groups (slots). This also results in Node Bs adjusting their power less often. This approach corresponds to DPC_MODE = 1, where the UE repeats the same TPC command over 3 slots (see Chapter 10). Another solution consists of synchronizing Node Bs transmitted powers (*power balancing*). Node Bs involved in the soft handover average the transmission code power levels and send the result to the RNC – averaging, for example, 750 TPC commands (500 ms). Based on this information, the RNC derives a reference power value and sends it to the cells under its control.

Power control based on Site Selection Transmit Diversity (SSDT)

Power control on soft and softer handover in the downlink can be conducted based on *Site Selection Transmit Diversity* (SSDT). The UE selects one of the cells from its active set to be "primary" and all the other cells are classed as "non-primary". The main objective is to transmit on the downlink from the best cell, thus reducing the interference caused by multiple transmissions in a soft handover mode. A second objective is to achieve fast site selection without network intervention, thus maintaining the advantage of the soft handover. In order to select a primary cell, each cell is assigned a temporary identification and the UE periodically indicates a primary cell identification ("ID") to the connecting cells. The non-primary cells selected by the UE switch off the transmission power. The primary cell identity code is delivered via uplink FBI field (see Figure 11.10).

NOTE.– the SSDT technique was proposed to be removed from specifications and is presented here for the sake of consistency with current *Release 99* documents [3GPP RP-050144].

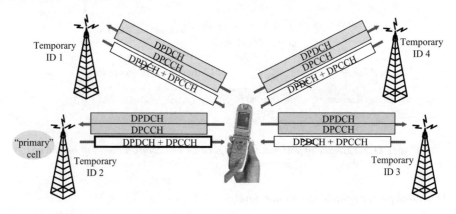

Figure 11.10. *Principle behind SSDT*

11.5.3. *Interfrequency handover*

Interfrequency handover is a hard handover between different UTRA/FDD carriers. This scenario is possible in Europe since most UMTS operators have two or three FDD carriers available. From the network architecture perspective, the inter-frequency handover may happen within the same Node B, or between two different Node Bs belonging to the same or to different RNCs.

By using a hierarchical cell structure, a better coverage can be obtained with micro cells having different frequency channels than the macro cells overlying the micro cells (see Figure 11.11). More effective load sharing techniques can also be implemented.

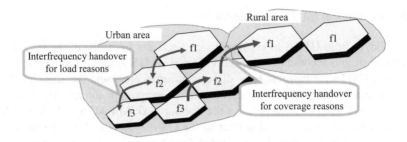

Figure 11.11. *Situations where interfrequency handover can apply*

In contrast to the soft handover, the interfrequency handover is a *Network Evaluated Handover* (NEHO): the RNC evaluates the possibility for the handover based on the frequency allocation, neighboring cell definition, cell layers, etc. The final decision is also taken by the RNC. The RNC may also ask the UE to start interfrequency measurements and perform periodic reporting.

Similarly to the soft handover, there are no "standard" algorithms defining when the interfrequency handover shall be conducted. Here we give some possible situations:

– the RNC rejects the radio access bearer set-up due to the serving cell's present load;

– the outer loop power control in the downlink cannot maintain the target BER/FER with maximum downlink power for real-time RABs;

– the UE reports quality deterioration from interfrequency measurement results. The reporting can be periodic or event triggered by events 2A, 2B, 2C, 2D, 2E and 2F [TS 25.331];

– the RNC rejects branch addition during soft handover due either to hard blocking or to the present downlink load of the target cell.

The nature of interfrequency handover causes temporary disconnection of the RAB (real-time) without any perceptible effect in the case of non-real-time RABs. Another disadvantage of the interfrequency handover lies on the need for the UE to have a second receiver or to support a *compressed mode* operation in order to measure other frequencies when communicating on a dedicated physical channel.

11.5.4. *Intersystem UMTS-GSM handover*

Handover from UTRA to GSM and vice versa has been one of the main design criteria taken into account in the specification of UMTS systems. In major European countries, UMTS deployment basically covers urban areas and therefore handovers to GSM are required to provide continuous coverage. On top of that, handovers GSM ↔ UMTS can be used to lower the loading in GSM/UMTS cells. In this latter case speech users can be handed over to the GSM system and data to the UMTS system with QoS renegotiation if needed (see Figure 11.12). This enables GSM networks to be used to give fallback coverage to UTRA/FDD. The subscribers can thus experience seamless services, by naturally taking into account the lower data rate capabilities in GSM when compared to UMTS maximum data rates (up to 2 Mbps).

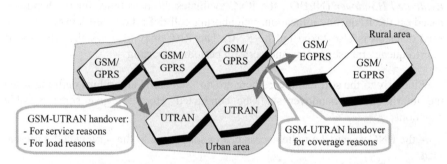

Figure 11.12. *Situations where intersystem handover can apply*

The intersystem handover is a hard handover and therefore causes temporary disconnection for real-time services. This is also a NEHO type and the actual algorithm is the property of the manufacturer. The same functional principles apply for the interfrequency handover, including periodic or event triggered reporting based on intersystem measurements with events 3A, 3B, 3C, 3D [TS 25.331]. Like in the case of interfrequency handovers, also intersystem (inter-RAT) measurements require the use of dual receivers or *compressed mode*, as described later in this chapter.

11.6. Measurements in idle and connected RRC modes

11.6.1. *Measurements in RRC idle, CELL_PCH and URA_PCH states*

From the layer 1 perspective, the RRC idle, CELL_PCH and URA_PCH states are equivalent when performing the measurements. Due to the *Discontinuous Reception*

(DRX) cycles, the UE possesses a certain level of flexibility to schedule intrafrequency, interfrequency and intersystem measurements as illustrated in Figure 11.13. As we previously mentioned, the standards do not specify when such measurements shall be performed but rather, they define the requirements these measurements shall fulfill [TS 25.133]. Measurement scheduling shall be optimized in such a way that the UE remains in *sleep* state most of the time during inter-paging gaps. It is up to the UE manufacturer to find the best trade-off between fulfilling standard requirements and prolonging the UE battery life.

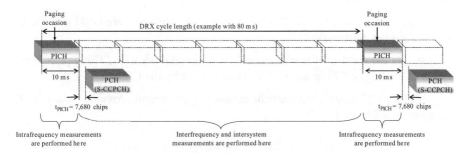

Figure 11.13. *Measurements in RRC idle, CELL_PCH and URA_PCH*

The length of the DRX cycle depends on the size of the cells and is broadcast in system information messages. The possible values for the DRX cycle are depicted in Table 11.4. For every DRX cycle the quality of the serving cell is monitored according to the "S" criterion. However, neighboring cells are only measured if the rules for intrafrequency, interfrequency and intersystem measurements described in previous sections are fulfilled. The list containing the scrambling codes and frequency carriers of the neighboring cells to monitor are indicated to the UE within SIBs 11/12.

DRX cycle length						
80 ms	160 ms	320 ms	640 ms	1.28 s	2.56 s	5.12 s

Table 11.4. *Possible values for a DRX cycle length [TS 25.133]*

11.6.2. *Measurements in CELL_FACH state*

Similarly to the DRX cycles, in the CELL_FACH state there are *measurement occasions* cycles where intrafrequency, interfrequency and intersystem measurements can be made. The cycles are of length T_{meas} and they are calculated from:

$$T_{meas} = (N_{FDD} + N_{GSM}).N_{TTI}.M\_REP.10 \qquad\qquad [11.7]$$

where:

– N_{FDD} is 1 if there are interfrequency FDD cells in the neighboring list. Otherwise N_{FDD} is 0;

– N_{GSM} is 1 if the UE is UMTS/GSM dual mode and there are GSM cells in the neighboring list. Otherwise N_{GSM} is 0;

– N_{TTI} is the number of frames in each measurement occasion. This equals to the length of the largest TTI in the S-CCPCH monitored by the UE;

– $M\_REP = 2^K$ is the measurement occasion cycle length. Possible values for K are given in Table 11.5.

N_{TTI}	1	2	4	8
K	3, 4, 5, 6	2, 3, 4, 5	2, 3, 4	1, 2, 3

Table 11.5. *Possible values for K as function of the TTI in current FACH*

The *FACH Measurement Occasion* of N_{TTI} frames will be repeated every $N_{TTI} \times M\_REP$ frames. Measurements are performed once every T_{meas} as illustrated in Figure 11.14. As in paging idle states, the actual scheduling of intrafrequency, interfrequency and intersystem measurements in CELL_FACH depends on the terminal manufacturer and is based on the requirements defined in [TS 25.133].

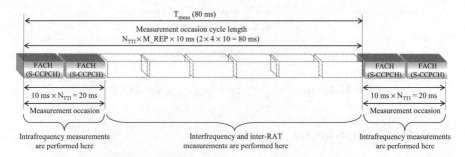

Figure 11.14. *Measurements in the CELL_FACH state.*
Example with $N_{TTI} = 2$, K = 2, $N_{REP} = 4$, $N_{TDD} = N_{GSM} = 0$, $N_{FDD} = 1$

11.6.3. *Measurements in the CELL_DCH state: the compressed mode*

In the CELL_DCH state, a dedicated physical channel (DPCH) is allocated to the UE in both directions UL/DL. Following the FDD principle, each channel uses a different frequency carrier, which makes the RF design more complex when the UE attempts to perform measurements in other frequency bands. This is needed when interfrequency and intersystem measurements must be made during an ongoing physical connection. The 3GPP specifications propose two techniques two deal with this problem:

– implementation of two receivers in the UE;

– compressed mode.

Implementing two receivers enables the UE to monitor cells in other frequency bands without interrupting the current communication in UL/DL. The price to pay is an increased cost of the RF section in the terminal as well as higher power consumption. The other technique concerns the so-called "compressed mode", where the reception and in some cases the transmission are stopped for a certain time period, thus enabling the UE to measure the other frequencies. This gap is achieved by compressing the data inside a given 10 ms frame, while degrading as little as possible the QoS (see Figure 11.15).

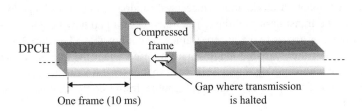

Figure 11.15. *Diagram of compressed mode principle*

The RNC controls which frames should be compressed and indicates this information to the UE within RRC control messages. For measurement purposes in the compressed mode, a *Transmission Gap Pattern Sequence* (TGPS) is defined, which consists of alternating transmission gap patterns 1 and 2, and each of these patterns comprises one or two transmission gaps. The parameters giving the structure, length, position and repetition of each gap pattern are defined by the UTRAN (see Figure 11.16). The UTRAN also indicates to the UE what each pattern sequence shall be used for, either to perform interfrequency or intersystem measurements. Additionally, the UTRAN may request the UE to perform internal measurements in order to optimize the time "compressed mode" that shall be initiated and finished (events 6A-6G).

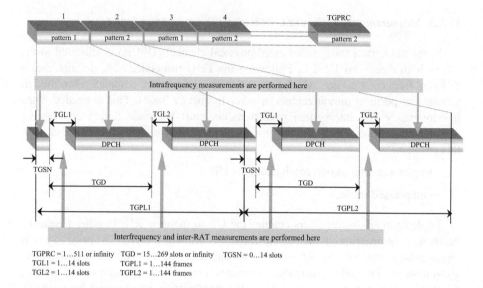

TGPRC = 1...511 or infinity   TGD = 15...269 slots or infinity   TGSN = 0...14 slots
TGL1 = 1...14 slots   TGPL1 = 1...144 frames
TGL2 = 1...14 slots   TGPL2 = 1...144 frames

Figure 11.16. *Measurements in the* CELL_DCH *state using the compressed mode*

The *Transmission Gap Length* (TGL) is the number of consecutive idle slots during the compressed mode and is measured in slots. Slots from N_{first} to N_{last} are not used for the transmission of data. The transmission gap may be generated within a single frame, or across two adjacent frames. The allowable TGLs are 3, 4, 5, 7, 10, and 14 slots. All the available lengths may be used when the gap is created across two frames by choosing N_{first} and TGL such that at least 8 slots in each radio frame are transmitted. Examples of single and double frame compressions are shown in Figure 11.17.

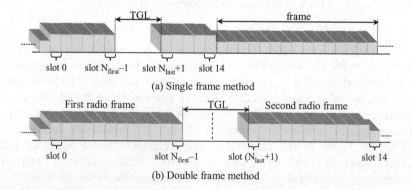

Figure 11.17. *Compressed mode based on single and double frame methods [TS 25.212]*

Compressed mode in the uplink

The UE might require uplink compressed mode when monitoring frequencies which are close to the uplink transmission frequency, i.e. frequencies in the TDD or GSM 1,800/1,900 bands. Avoiding transmission in these bands reduces the received interference caused by the UE's own transmissions. In the case of fixed duplex frequency space in the RF section of the UE, the UL compressed mode should take place in synchrony with the DL compressed mode. There is only one compressed mode frame structure type in the uplink (see Figure 11.18). In this frame structure, no information on either data or control is transmitted during the transmission gap. In the case of the DL compressed mode, on the other hand, some control information may be transmitted during the transmission gap as explained below.

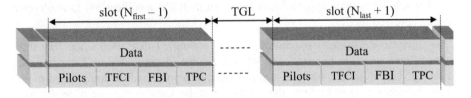

Figure 11.18. *Compressed mode frame structure in uplink [TR 25.212]*

Compressed mode in the downlink

Compressed mode in the downlink is used to perform:

– GSM/GPRS measurements while in UTRA/FDD mode (i.e. in intersystem handover situations);

– UTRA/TDD mode measurements while in UTRA/FDD mode (i.e. in intersystem handover situations);

– UTRA/FDD mode measurements on frequencies which are different to the serving cell carrier frequency (i.e. interfrequency handover situations).

Two different frame structure types, i.e. A and B, are used in the downlink compressed mode. These are shown in Figure 11.19. The pilot symbols of the last compressed slot (i.e. slot number N_{last}) are transmitted inside the transmission gap for both type A and B structures. The channel estimation obtained by using these pilots and the channel estimation obtained by using the pilot symbols of the following uncompressed slot (i.e. Slot N_{last} + 1) can be used to estimate the interim fading characteristics by methods such as linear interpolation. In frame structure B, the TPC of the slot N_{first} is transmitted to provide better power control.

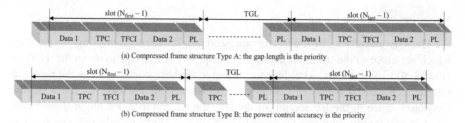

Figure 11.19. *Compressed frame structure in downlink [TS 25.212]*

11.6.3.1. *Compressed mode techniques*

The following three approaches are defined in 3GPP specifications to implement the compressed mode:

– *compressed mode by puncturing (rate matching)*. This is used when very short transmission gaps are required, and when a change in the spreading factor is to be avoided. The algorithm for puncturing is used *only* in downlink compressed mode. The algorithm used to perform this operation is the same as the one normally used for rate matching the user data to the air interface data rate [TS 25.212];

– *compressed mode by reducing the spreading factor by 2*. During compressed mode, the spreading factor (SF) is reduced by 2 during one or two radio frames. As the air interface chip rate is constant, this results in the same amount of payload data being sent in half the time. Doubling the data rate requires the transmit power to be doubled as well in order to maintain the same BER and/or FER at the receiver. The method applies to uplink and downlink and does not work if the normal SF in use is 4, since SFs lower than this are not allowed;

– *compressed mode by higher layer scheduling*. During compressed mode higher protocol layers set restrictions on the choice of the transport format combinations in a compressed frame. The maximum number of bits that will be delivered to the physical layer during the compressed radio frame is then known and a transmission gap can easily be generated. The major advantage of this method is that there should be little or no change to the operation of lower layers besides the discontinuity in transmission. However, some delay is added in the transmission chain, which means that its usage is restrained to non-real-time services.

NOTE.– compressed mode by puncturing is not used in real networks and it was proposed to be removed from specifications [3GPP RP-050144].

On the downlink, different UE scrambling and channelization codes can be used in compressed mode from those used in normal mode. The use of alternate codes is at the request of the UTRAN and can be useful if the primary code(s) is/are close to other codes in use. If the UE is ordered to use a different scrambling code in compressed mode, then there is a one-to-one mapping between the scrambling code used in normal mode and the one used in compressed mode [TS 25.213].

It should be noted that more power is needed during the compressed mode and therefore, the radio coverage can be affected. Moreover, the fast power control is not active during the silent period and the effect of the interleaving is decreased. All this means that a higher Eb/No target is required, which affects also the radio capacity. Therefore, the compressed mode should be activated by the RNC only when there is a real need to execute an intersystem or interfrequency handover. This can be done for instance by monitoring the downlink transmission powers for each user, or with the help of UE measurements.

11.6.3.2. Intersystem measurements in GSM

Performing intersystem measurements on UTRA carriers when in GSM active mode does not require any operation similar to the compressed mode. Indeed, the TDMA principle on which the GSM is based naturally creates idle gaps between the reception of two consecutive bursts. Thus, measurements on UTRA FDD and UTRA TDD can be scheduled therein.

The initial synchronization to a GSM cell is based on the FCCH (*Frequency Correction Channel*) and the SCH (*Synchronization Channel*). The FCCH helps to find the SCH and the successful reception and demodulation of the SCH gives all the precise information on the slot boundaries, as well as enough information to deduce the position of the demodulated timeslot within the GSM hyperframe. These channels are always transmitted on timeslot 0 of the GSM frame, always separated by exactly one TDMA frame (8 timeslots) and on the same frequency (see Figure 11.20). The base station transmits the FCCH and SCH on the BCCH carrier which is the carrier on which the *Broadcast Control Channel* (BCCH) is transmitted.

When in active GSM mode, the terminal performs RSSI measurements on serving and neighboring BCCH carriers and also identifies the BSIC on the SCH channel. The BSIC (*Base Station Identity Code*) gives the identity of a given base station similar to the scrambling code in the UTRA networks. The BSIC shall be reconfirmed periodically after its initial decoding.

Figure 11.20 shows the time relation between a UTRA/FDD superframe and a GSM BCCH-51 multiframe. The time difference, denoted by T_{offset}, between a given UTRA/FDD cell and each of the best GSM neighbors is another parameter that the UE shall measure and report to the network. Such parameter is defined as the time difference between a UTRA/FDD superframe (frame 4,096, with SFN = 0) and a GSM BCCH-51 multiframe. The accuracy of T_{offset} is ± 20 chips [TS 25.133]. For the UTRAN, the knowledge of T_{offset} is paramount since this is used to schedule the time when the frames shall be compressed so that the UE can perform the necessary interfrequency and intersystem measurements.

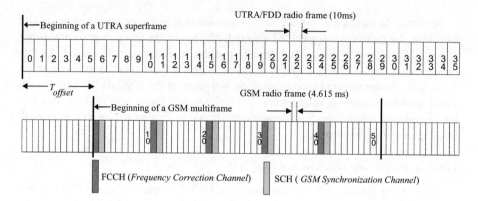

Figure 11.20. *Relationship between UTRA/FDD superframe and GSM BCCH-51 multiframe*

Chapter 12

UTRA/TDD Mode

12.1. Introduction

UTRA/TDD is the TDD (*Time Division Duplex*) mode of UTRA and is part of the radio access technologies in the IMT-2000 framework defined by the ITU. UTRA/TDD is based on a combination of wideband CDMA and TDMA, also known as TD-CDMA. The technical characteristics of UTRA/FDD and UTRA/TDD have been harmonized by the 3GPP community with respect to carrier spacing, chip rate and frame length. Interworking with GSM radio networks is also ensured. A similar harmonization process was carried out by 3GPP to harmonize UTRA/TDD with the radio standard TD-SCDMA (*Time-Duplex Synchronous CDMA*). This was originally developed by the *China Academy of Telecommunications Technology* (CATT) and adopted by ITU and 3GPP as part of the UMTS *Release 4* specifications. TD-SCDMA is also referred to as *low-chip rate UTRA/TDD* since it uses a chip rate of 1.28 Mcps in 1.6 MHz bandwidth rather than 3.84 Mcps in 5 MHz bandwidth [TR 25.834].

This chapter focuses on the air interface aspects of UTRA/TDD and more precisely on layer 1. Note that the key ideas behind upper radio layers in UTRA/TDD mode (RRC, RLC, BMC, PDCP, MAC) are equivalent to those described for the UTRA/FDD mode in previous chapters.

12.2. Technical aspects of UTRA/TDD

UTRA/TDD is a combination of TDMA and CDMA using TDD. TDD makes it possible to transmit traffic in uplink and downlink using the same frequency band –

it does not require pair bands as in UTRA/FDD. As depicted in Figure 12.1, in TDD, uplink and downlink radio signals are transmitted in the same frequency channel but at different time slots. Users are differentiated by the CDMA codes allocated within that time slot. Table 12.1 compares the main radio characteristics of UTRA/TDD with those of UTRA/FDD.

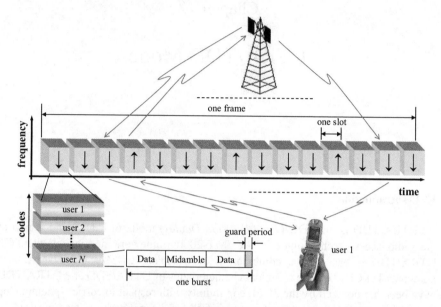

Figure 12.1. *UTRA/TDD principle*

12.2.1. *Advantages of UTRA/TDD*

The main benefits of the UTRA/TDD air interface can be summarized as follows:

– spectrum efficiency. UTRA/TDD employs only one frequency band for both uplink and downlink. This feature is particularly important in situations where a paired spectrum is not available or where the radio resources are limited;

– increased flexibility in spectrum usage, especially for asymmetric services;

– radio path reciprocity between uplink and downlink. In TDD operation mode, fast-fading is correlated between uplink and downlink because they use the same frequency. This property can be exploited in UTRA/TDD by using smart antennae in combination with joint detection in order to increase capacity and spectrum efficiency.

UTRA/TDD is especially helpful in situations where data traffic tends to be asymmetric, often requiring little uplink throughput, but significant bandwidth for downloading information (e.g. mobile Internet). The duplex switching point may be changed and the capacity from uplink to downlink moved or vice versa, thus utilizing spectrum optimally (see Figure 12.2).

With UTRA/TDD duplexers and isolation techniques between uplink/downlink carriers are not needed since transmit and receive channels are never active simultaneously and cannot interfere with each other. This advantage is, however, negligible for multi-mode terminals supporting UTRA/TDD as well as UTRA/FDD and/or GSM.

	UTRA/FDD	UTRA/TDD
Multiple access method	CDMA	CDMA and TDMA
Duplexing	FDD	TDD
Detection	coherent, based on pilot symbols	coherent, based on midambles
Chip rate	3.84 Mcps	3.84 Mcps (1.28 Mcps optional)
Spreading factor	between 4 and 512 (DL) between 4 and 256 (UL)	1 or 16 (DL) 1, 2, 4, 8 or 16 (UL)
Bandwidth	5,000 kHz	5,000 kHz (1,600 kHz optional)
Frequency spectrum (Europe)	1,920-1,980 MHz (UL) 2,110-2,170 MHZ (DL)	1,900-1,920 MHZ (UL/DL) 2,010-2,025 MHz (UL/DL)
RF modulation	QPSK	
Pulse shaping	Root Raise Cosine, $\alpha = 0.22$	
Power control rate for dedicated channels	500 Hz (UL and DL)	≤ 750 closed loop (DL) 100 or 200 Hz in open loop (UL)
Frame length	10 ms	
Slot length	$10/15 \approx 0.667$ ms	
Intrafrequency handover	hard handover	soft handover
Node Bs synchronization	asynchronous	synchronous

Table 12.1. *Radio characteristics of UTRA/FDD and UTRA/TDD*

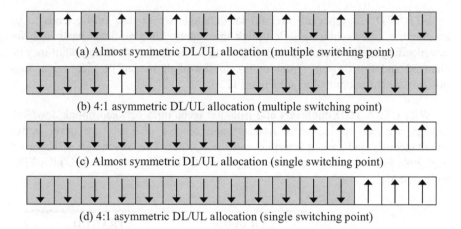

(a) Almost symmetric DL/UL allocation (multiple switching point)

(b) 4:1 asymmetric DL/UL allocation (multiple switching point)

(c) Almost symmetric DL/UL allocation (single switching point)

(d) 4:1 asymmetric DL/UL allocation (single switching point)

Figure 12.2. *Examples of switching point configurations within UTRA/TDD frame structure*

12.2.2. *Drawbacks of UTRA/TDD*

Using TDD duplexing mode within UTRA/TDD also carries some disadvantages:

– interference between uplink and downlink;

– need for accurate synchronization between base stations;

– lower coverage and mobility compared with UTRA/FDD.

Similarly to UTRA/FDD, where separation between uplink and downlink carriers is needed, UTRA/TDD requires the insertion of a *dummy* "guard period" within every single time slot in order to avoid mutual interference between uplink and downlink signals. The guard period length depends on the round trip time needed for a radio signal to propagate between the emitter and the receiver source. The longer the distance, the longer the guard period. This limitation can, however, be mitigated with additional techniques such as *timing advance*, which is studied later on in this chapter.

System synchronization is needed in order to have identical asymmetry for all the users in a cell. However, even if Node Bs are synchronized, inter-cell interference occurs if asymmetry is different in adjacent cells (see Figure 12.3). Interference between Node Bs can be reduced by appropriate network planning, whereas interference between UEs can be attenuated by power control mechanisms and the *Dynamic Channel Allocation* (DCA) technique.

UTRA/TDD uses a discontinuous data transmission scheme based on "bursts" as in the GSM radio interface. This contrasts with UTRA/FDD where data is transmitted continuously in a frame with average output power denoted by P. If we consider that one slot out of 15 is used during a transmission by the UE, the average output power in a UTRA/TDD frame is $P/15$. In order to have a similar performance to that of the UTRA/FDD mode (in terms of Eb/No), a UTRA/TDD terminal requires to increase its transmit peak power by 12 dB. This may have a negative impact on the link budget. Clearly, this problem is reduced if more slots within a frame are used.

The aforementioned limitations of UTRA/TDD restrain its usage to scenarios with small cell coverage (micro and picocells) and with low mobility. This is probably one of the reasons why UTRA/TDD radio networks are not expected to be commercially deployed in the European market in the short-term.

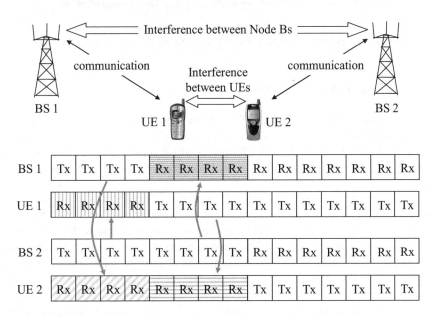

Figure 12.3. *Interference scenarios in UTRA/TDD in case where Tx/Rx is not coordinated*

12.3. Transport and physical channels in UTRA/TDD

From the functional perspective, most of the transport channels present in the UTRA/FDD physical layer are used in UTRA/TDD as shown in Figure 12.4. An exception is the USCH (*Uplink Shared Channel*) that is functionally similar to the DSCH in the downlink: this is shared in the uplink by several UEs carrying dedicated control or traffic data. This channel is mapped onto the PUSCH (*Physical USCH*). The other physical and transport channels in UTRA/TDD air interface keep the same functional role as in UTRA/FDD.

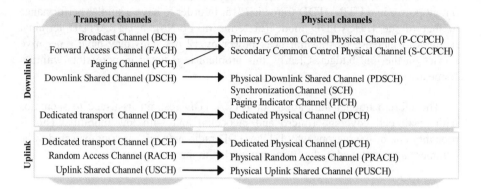

Figure 12.4. *Mapping of transport channels onto physical channels in UTRA/TDD mode*

12.3.1. *Physical channel structure*

A physical channel in UTRA/TDD is defined by a frequency carrier, a time slot, a time frame (where the slot is allocated), a channelization code, a scrambling code and a burst type. Each UTRA/TDD frame (TDMA frame) of length 10 ms is divided into 15 time slots, each of which may be allocated to either uplink or downlink.

Each time slot is 2,560 chips long (1 chip equals 1/3.84 MHz ≈ 0.26 μs). Several burst allocated to the same user can be transmitted within the same slot. Similarly, several bursts belonging to different users can be transmitted within the same slot. In both cases, each burst uses a different channelization code. The structure of a time slot contains two data fields: one midamble field and one guard period field (see Figure 12.5). The data fields result from the operations applied to the transport channels: multiplexing, interleaving, coding and spreading. Training sequences compose the midamble field. The guard period (GP) is used to mitigate channel impairments such as delay spread and propagation delay.

12.3.1.1. *Burst types*

Three different burst types are defined in UTRA/TDD and they are illustrated in Figure 12.5. They differ according to the length of each field. For instance, burst type 1 comprises less space for data transmission but a longer midamble field. A longer midamble enables a more accurate channel response estimation and therefore better detection performance, especially in long delay spread propagation channels.

The midambles are constructed by one single cyclic periodic code named "basic code". Different cells use different periodic basic codes.

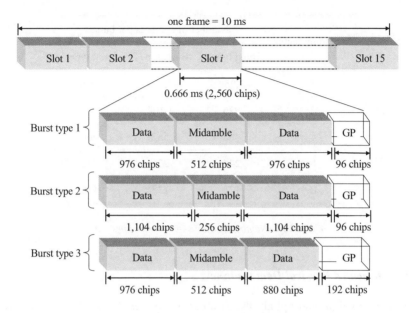

Figure 12.5. *Frame structure and burst types*

12.3.1.2. *Spreading and modulation*

The number of data bits N_{bits} carried out within the data fields of the bursts is calculated by the *Spreading Factor* (SF) and the length of the data field L_{data} given in chips [TS 25.221]:

$$N_{bits} = 2 \times L_{data} / SF \text{ where } SF \in \{1, 2, 4, 8, 16\}$$

NOTE.– the number of complex symbols per burst is half the number of bits (see Table 12.2). The symbols are generated from a pair of two subsequent interleaved and encoded data bits by mapping the bits to I and Q branches. The data symbols are spread by applying a complex channelization code from a real *Orthogonal Variable Spreading Factor* (OVSF) code. These codes follow the same structure as in the UTRA/FDD mode. In contrast to UTRA/FDD, the maximum value allowed for SF is 16, which enables the implementation of *joint detection* receivers with moderate complexity. In order to reduce inter-cell interference, the spread data symbols are further scrambled with a pseudo-random cell specific scrambling code whose length is 16 [TS 25.223].

Spreading factor	Number of symbols per burst		
	Burst type 1	Burst type 2	Burst type 3
1	1,952	2,208	1,856
2	976	1,104	928
4	488	552	464
8	244	276	232
16	122	138	116

Table 12.2. *Number of data symbols per burst type*

12.3.2. Dedicated Physical Data Channels

In the downlink and uplink, the *Dedicated Physical Data Channels* (DPDCH) are used to carry the data bits generated by the DCH as well as control information generated at layer 1 (see Figures 12.6a and 12.6b).

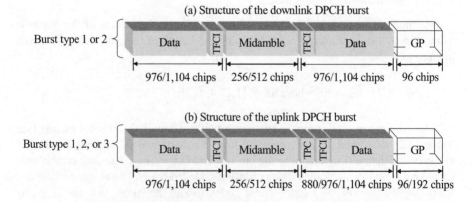

Figure 12.6. *Burst structure of the DPCH in a) downlink and b) uplink*

Besides the bits in the data fields, TFCI (*Transport Format Combination Indicator*) symbols are used in order to enable proper decoding, deinterleaving and demultiplexing on the physical layer. They can be omitted for simpler services such as fixed rate speech. In the uplink slot structure there is also a field with *Transmit Power Control* (TPC) symbols which enable closed loop power control for the downlink DPCH. For the downlink DPDCH, SF = 16 and SF = 1 are used. Higher data rates are possible when using different channelization codes in parallel following the so-called multicode operation. In the downlink, multiple parallel physical channels can be transmitted by using different channelization codes, with SF = 16. In the uplink, a UE shall use a maximum of two physical channels per timeslot simultaneously, each one using a different channelization code. For the uplink, transmission with SFs in the range of 16 to 1 is possible because this enables smaller PAR and therefore lower battery consumption than multicode transmission.

Tables 12.3a and 12.3b give some examples of DPCH slot formats among the 20 possible formats in the downlink and 90 formats in the uplink [TS 25.221].

a) Examples of downlink time slot formats

SF	Midamble length	Guard period	Bits/slot	TFCI bits /slot	Data bits /slot
16	512 chips		244	0	244
16	256 chips	96 chips	276	16	260
1	256 chips		4,416	32	4,384

b) Examples of uplink time slot formats

SF	Midamble length	Guard period	Bits/slot	TFCI bits /slot	TPC bits /slot	Data bits /slot
16	512	96 chips	244	0	0	244
8	512	96 chips	470	16	2	452
2	512	196 chips	1,618	32	2	1,584

Table 12.3. *Examples of DPCH a) downlink and b) uplink time slot formats*

12.3.3. *Common physical channels*

Table 12.4 summarizes some parameters for UTRA/TDD common physical channels.

12.3.3.1. *Downlink and uplink shared channels*

The PDSCH and the PUSCH have the same burst type structure and use the same spreading factors as the DPCH in the downlink and the uplink, respectively. The main difference with dedicated channels is that they are *shared*, i.e. they are allocated on a temporary basis (in a frame-by-frame basis in the case of the DSCH). They also apply the same mechanisms for power control.

The UEs in the cell know whether they have to decode the DSCH via in-band TFCI or midamble signaling [TS 25.224]: there is no an associated DPCH as in the PDSCH used in UTRA/FDD. In the uplink, the decoding of the USCH is communicated via higher layer signaling.

12.3.3.2. *Primary and Secondary Common Control Channels*

The *Primary Common Control Physical Channel* (P-CCPCH) is used in downlink to convey the data bits of the broadcast transport channel (BCH). The P-CCPCH uses a predefined channelization code and midamble facilitating decoding to the UEs once synchronization is achieved. The channel is in fact time multiplexed with the SCH – it is always transmitted in the first SCH time slot within a frame. Transmitted with a fixed power, the P-CCPCH is also used by the UEs as a reference for measurements (beacon channel).

The *Secondary Common Control Physical Channel* (S-CCPCH) carries common control or user data information from the PCH or/and the FACH. Multiplexing of the PCH and the FACH onto the S-CCPCH is possible by the use of the TFCI bits.

Transport channel	Physical channel	Burst type	Spreading Factor	TFCI field	TPC field
BCH	P-CCPCH	1	16	no	no
FCH	S-CCPCH	1 or 2	16	yes	no
PCH				no	no
DSCH	PDSCH	1 or 2	1 or 16	yes	no
USCH	PUSCH	1, 2 or 3	1, 2, 4, 8 or 16	yes	yes
RACH	PRACH	3	8 or 16	no	yes

Table 12.4. *Slot structure and spreading parameters of common physical channels*

12.3.3.3. *Paging Indicator Channel (PICH)*

The *Paging Indicator Channel* (PICH) enables the UE to prolong its battery autonomy with the help of the *Discontinuous Reception* (DRX) mechanism. DRX applies in idle RRC states where the UE is waiting for paging messages. The PICH contains the *Paging Indicators* (PI) generated at the physical layer. Figure 12.7 shows the structure of the PICH burst. The burst comprises N_{PIB} bits and a midamble. The number of bits depends on the burst type: $N_{PIB} = 240$ for type 1 and $N_{PIB} = 272$ for type 2. A set of these bits forms a PI. The number of PIs per frame is determined by the upper layer and may be 15, 17, 30, 34, 60, or 68. The adjacent bits to the midamble $b_{N_{PIB+1}},...,b_{N_{PIB+4}}$ are left unused.

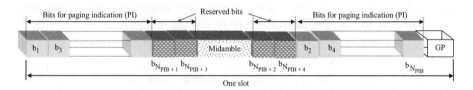

Figure 12.7. *Structure of the PICH burst*

When waiting for paging messages, the UEs have to detect their own PI and only if this indicates the presence of a paging message (the PI is set to "1") then the UE will start decoding the PCH. The time relationship between the PICH and the PCH is illustrated in Figure 12.8. The value $N_{GAP} > 0$ in frames and N_{PCH} are configured by higher layers. Note that the PCH is divided into PCH blocks comprising N_{PCH} paging sub-channels per block. The UE only monitors the sub-channel assigned by higher layers, thus making the DRX more efficient. The assignment of UEs to paging sub-channels is independent of the assignment of UEs to PIs.

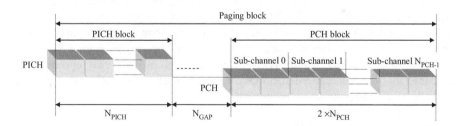

Figure 12.8. *PICH and PCH timing association*

12.3.3.4. *Physical Random Access Channel (PRACH)*

The *Physical Random Access Channel* (PRACH) carries RACH messages using a burst type 3. For the PRACH, only SF values of 8 and 16 are allowed in order to ensure the reception of RACH messages even in cells with large radius. A larger guard period as in burst type 3 is used for the PRACH because of the need to cope with the missing timing information during the initial access phase.

Before starting the random access process, the UE shall know the *access sub-channels* which contain available time slots that may be used to transmit the PRACH message. The messages are randomly transmitted in one or more time slots by noting that collisions may occur if two or more UEs transmit at the same time and with the same code. The UE shall also have information about the initial PRACH power for the first message. These parameters are broadcast in the BCH.

The random access procedure can be summarized as follows:

– the UE randomly selects an access sub-channel from the list of available sub-channel indicated by the network and calculates the permitted slots and frames. Then, a slot within the selected sub-channel is randomly selected;

– the UE randomly selects a spreading code from the list communicated by upper layers. From the selected code, the UE derives the midamble needed to compose the PRACH burst (the scrambling code was already determined during the synchronization process);

– the UE sets the PRACH power level according to the parameters communicated by the network and transmits the message within the selected slot/frame.

12.3.3.5. *Synchronization Channel (SCH)*

The *Synchronization Channel* (SCH) comprises two channels: the *Primary Synchronization Channel* (P-SCH) and the *Secondary Synchronization Channel* (S-SCH). The P-SCH uses a complex 256 chips long code which is common to all the cells within the radio network. The S-SCH can be chosen from a set of 12 complex codes whose length is 256 chips as well. These two channels are constructed from generalized hierarchical Golay codes [TS 25.223] that provide good aperiodic autocorrelation properties.

Three codes of the S-SCH are transmitted in parallel with the P-SCH within the same *slot* (see Figure 12.9). There are 32 sets of 3 codes called groups and each group is associated to 4 different scrambling codes. Every cell in the system is part of one of these 32 groups and uses a cell-specific scrambling code and two corresponding midamble codes from which long and short training sequences are constructed.

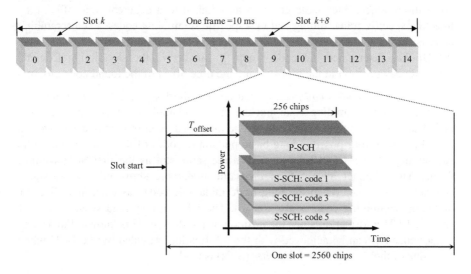

Figure 12.9. *Example of SCH structure: "case 2" with k = 1. S-SCH uses codes 1, 3 and 5*

The SCH is inserted into either one or two slots per frame depending on higher layer configuration. Two cases are possible:

– case 1: where the SCH and the P-CCPCH are transmitted in one slot with number k, $k = 0 ,...,14$;

– case 2: where the SCH is transmitted over two slots with numbers k and $k + 8$, $k = 0,...,6$. The P-CCPCH is transmitted only on slot number k.

The use of two slots within the 7 slot space was defined for facilitating cell monitoring and measurements in the case of intersystem handover with GSM and UTRA/FDD radio access technologies.

The UTRA/TDD synchronization process can be summarized in three steps:

– *slot synchronization.* During the first step of the cell search procedure, the UE uses the P-SCH (common to all cells) to find a cell and the beginning of the SCH. A cell can be found from the peaks detected at the output of a matched filter;

– *code group identification and slot synchronization.* In the second step, the UE uses the S-SCH to identify one of the 32 code groups. This is achieved by correlating the received signal with the S-SCH codes at the peak positions detected in the first step. Cells within different code groups transmit their synchronization sequences with a specific time t_{offset} with respect to the slot boundary (see Figure 12.9). The code group can be uniquely identified by detection of the maximum correlation values. Each code group is thus linked to a different time offset t_{offset}, thus to a specific frame timing and to 4 specific scrambling codes. Each scrambling code is further associated to a specific short and long basic midamble code. When the UE has determined the code group, it can derive the slot timing from the detected peak position in the first step and the time offset of the found code group;

– *scrambling code identification*. During the third step the UE determines the exact downlink scrambling code and the basic midamble code used by the found cell. The long basic midamble code can be identified by correlations made over the P-CCPCH with four possible long basic midamble codes of the group found in the second step. The P-CCPCH is located in the same slot as the first SCH within a frame. This channel always uses the same midamble shift and a pre-assigned channelization code. When the long basic midamble code has been identified, the downlink scrambling code is also known. The UE can now read system and cell-specific BCH information since the position of the P-CCPCH is known. Final frame synchronization can be achieved with the information provided by the BCH where the value of the *System Frame Number* (SFN) is read.

12.4. Service multiplexing and channel coding

From the functional point of view, there is no difference between UTRA/TDD and UTRA/FDD multiplexing and channel coding techniques. A transport channel serves as interface between the physical layer and MAC. The *Transmission Time Interval* (TTI) defines the delivery period of one or several transport blocks from MAC to the physical layer. The TTI value is chosen from the set {10 ms, 20 ms, 40 ms, 80 ms}.

CRC attachment is performed to each transport block separately. These blocks are concatenated and then segmented in order to achieve optimum coding block size according to the associated channel coding scheme. Four coding schemes are applicable and they are given in Table 12.5. After encoding, a first interleaving step is applied to each transport channel independently, which distributes the bits over the whole TTI of that transport channel. Radio frame segmentation makes it possible to distribute data equally to all the frames of the respective TTI. The rate matching algorithm is then applied by means of repeating or puncturing certain bits. The multiplexing of all bits follows this operation where the data bits of all transport channels are sequentially accommodated into one stream called CCTrCH. The resulting data stream is segmented onto single portions to be mapped separately onto physical channels. The segmented information goes through a second interleaving operation before the data information is finally mapped onto the physical channels.

Transport channel	Channel coding	Coding rate
BCH, PCH, RACH	convolutional coding	1/2
DCH, DSCH, FACH, USCH	convolutional coding	1/2, 1/3
	turbo coding	1/3
	no coding	

Table 12.5. *Channel coding schemes used in UTRA/TDD transport channels*

Figure 12.10 shows an example of multiplexing and channel coding in the downlink with two logical *Dedicated Traffic Channels* (DTCH) carrying speech data at 12.2 kbps each. They are transmitted in parallel with a logical *Dedicated Common Control Channel* (DCCH) containing signaling at 2 kbps. MAC inserts a header of 16 bit to the signaling block. Two DCHs are used to convey the information of these three logical channels. One of the DCH uses a TTI = 20 ms and rate matching is based on puncturing with rate 5%. The other DCH uses TTI = 40 ms and skips the rate matching operation. Convolutional channel coding is applied to the two DCHs with rate 1/3. The DCHs are finally mapped onto two DPCHs with a midamble whose length is 512 chips and SF = 16.

12.4.1. *Examples of UTRA/TDD user bit rates*

Similar user bitrates as those in UTRA/FDD can be achieved with UTRA/TDD mode. Table 12.6 gives some examples of maximum user bit rates according to the time slots allocated to a single user. The bit rates were calculated by neglecting RLC and MAC headers as well as tail, CRC, TFCI and TPC bits. A rate coding of 1/2 is considered and SF = 16.

Number of allocated codes to one user	Number of allocated slots			
	1	5	10	14
1 code with SF = 16	12.2 kbps	61 kbps	122 kbps	170.8 kbps
4 codes with SF = 16	48.8 kbps	244 kbps	488 kbps	683.2 kbps
8 codes with SF = 16	97.6 kbps	488 kbps	976 kbps	1,366.4 kbps
16 codes with SF = 16	195.2 kbps	976 kbps	1,952 kbps	2,732.8 kbps

Table 12.6. *Maximum user bit rates in UTRA/TDD air interface*

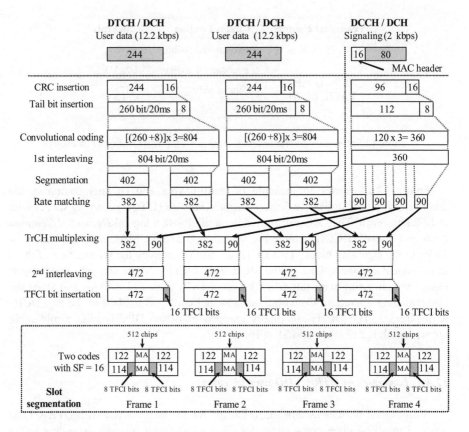

Figure 12.10. *Example of channel coding and multiplexing in downlink*

12.5. Physical layer procedures in UTRA/TDD

12.5.1. *Power control*

Power control in UTRA/TDD is used to limit the interference level within the system and thus reducing inter-cell interference as well as UE power consumption. It is performed on a frame basis and is implemented differently for the uplink and the downlink directions. Table 12.7 summarizes the characteristics of the power control methods defined in the 3GPP specifications for UTRA/TDD.

	Uplink	**Downlink**
Method	Open loop	Closed loop
Rate	Variable 1-7 slots (when 2 SCH slots are present in the frame) 1- 14 slots (when 1 SCH slot is present in the frame)	Variable, depends on the asymmetric (DL/UL) allocation in the frame (100 - 750 Hz)
TPC step size	---	1 dB, 2 dB or 3 dB

Table 12.7. *Power control methods used in UTRA/TDD*

12.5.1.1. *Uplink power control*

Uplink power control uses an open loop mechanism by exploiting the reciprocity of the uplink and downlink frequency channel with the assumption that the pathloss effects are similar in both directions. Open loop power control is applied to the PRACH, PUSCH and the DPCH. Note that the P-CCPCH can be used in the system as beacon channel, since its transmission power and midamble are known. The UE can thus measure the received power of this physical channel and calculate the pathloss.

The power level of the PRACH preamble used in the access procedure is calculated from [TS 25.331]:

$$P_{\text{PRACH}} = L_{\text{PCCPCH}} + I_{\text{BTS}} + C_{\text{RACH}} \qquad [12.1]$$

The power for the DPCH and the PUSCH is obtained from the expression:

$$P_{\text{UE}} = \alpha L_{\text{PCCPCH}} + (1-\alpha)L_0 + I_{\text{BTS}} + \text{SIR}_{\text{TARGET}} + C_{\text{DPCH}} \qquad [12.2]$$

In expressions [12.1] and [12.2], P_{PRACH} and P_{UE} denote the UE transmission power in dBm; L_{PCCPCH} is the pathloss measured from the P-CCPCH in dB; L_0 is the average value of the pathloss; I_{BTS} is the interference level in dBm measured by Node B and broadcast via the BCH; α is a weighting factor that takes into account the delay between the downlink pathloss estimation and the actual uplink transmission; $\text{SIR}_{\text{TARGET}}$ is the target value of the SIR signaled to the UE by the network together with C_{RACH} and C_{DPCH} which are constant values.

The open loop power control rate depends on the number of slots used by the SCH within one frame. For instance, when one slot is used (case 1) the power is adjusted once every 15 *slots*, i.e. at a rate of 100 Hz. Conversely, when two slots are used (case 2), the UE transmit power is adjusted once every 7.5 slots (average), i.e. at a rate of 200 Hz.

12.5.1.2. Downlink power control

Closed loop power control is used for downlink transmission and applies to the DPCH and the PDSCH. The UE compares the estimated SIR to an SIR target value. For all physical channels allocated to the same UE in a frame, a common power control command is generated and sent to Node B. This in turns can change its transmit power for all the physical channels addressed to the UE. As in UTRA/FDD, a quality-based outer loop power control mechanism adjusts the SIR target.

The closed loop power control rate is variable and depends on the asymmetric path used within a frame. However, TPC commands are sent to the UTRAN a maximum of once every 2 slots, i.e. at a 750 Hz rate.

12.5.2. Downlink transmit diversity

Transmit diversity reduces the impact of fading by offering multiple independent copies of the digitally modulated signal at the receiver. This results in increased receiver performance. Table 12.8 summarizes the transmit diversity techniques defined for UTRA/TDD mode. They concern two approaches: open loop and closed loop.

Physical channel	Open loop		Closed loop
	TSTD	SCTD	
P-CCPCH, S-CCPCH, PICH	no	yes	no
SCH	yes	no	no
DPCH	no	no	yes
PDSCH	no	yes	yes

Table 12.8. *Transmit diversity techniques defined for UTRA/TDD physical channels*

Open loop techniques involve no feedback from the receiver regarding channel conditions. The methods specified concern *Time Switched Transmit Diversity* (TSTD) and *Space Code Transmit Diversity* (SCTD) [TS 25.224]. TSTD is only applied to the SCH and the idea is to alternate transmission of the S-SCH and the P-SCH on a slot basis between two uncorrelated antennae. With SCTD, bits are grouped by blocks and then encoded so that two fields of bits are obtained at the output, which are spread and scrambled before they are transmitted via two uncorrelated antennae, respectively.

Closed loop transmit diversity applies to the baseband signal after spreading and scrambling operations. The signal is then split into two branches where complex weighting factors w_1 and w_2 are respectively applied in order to adjust phase and amplitude. The complex signal is then transmitted via two uncorrelated antennae 1 and 2. In contrast to UTRA/FDD, no feedback bits (FBI) are sent back to Node B in the closed loop scheme defined for UTRA/TDD mode. It is up to Node B to estimate w_1 and w_2 on a slot-by-slot basis, based on the channel impulse response estimated from the midambles of the uplink DPCH. The way of estimating w_1 and w_2 yields two different closed loop transmit diversity techniques [TS 25.224].

12.5.3. *Timing advance*

Timing advance is a mechanism used in UTRA/TDD to increase the range of the cell by aligning the UE transmission with Node B reception. The UTRAN shall continuously measure the time difference between the first channel path detected in the uplink (for PRACH, DPCH and PUSCH channels) and the time when, in theory, the second slot should arrive by assuming that the propagation time is zero. The measurement is referred to as *RX Timing Deviation*. Once the measurement is performed, the UE may be commanded to adjust its transmission timing by the UTRAN in ± 4 chip steps [TS 25.224]. In this way, the individual distance related transmission delay at Node B is compensated.

12.5.4. *Dynamic channel allocation*

Dynamic channel allocation (DCA) enables an effective strategy of interference avoidance. It enables, for instance, the deployment of UTRA/TDD for coordinated as well as for uncoordinated operations: interference in certain time slots, originated from neighboring cells can be efficiently mitigated. For each time slot, a long- and short-term recording and statistical evaluation of interference measurement is applied. Based on this measurement, the RNC creates a priority list of all time slots, which is updated dynamically. It allocates and reallocates the physical channels, especially for packet services which ensure a fast reaction to varying interference conditions. The highest priority is assigned to the time slot which exhibits the lowest interference level [TR 25.922].

12.5.5. *Handover*

In contrast to UTRA/FDD, there is no soft handover procedure in UTRA/TDD but only hard-handover. The handover is based on measurements performed on neighboring cells from a list transmitted by the UTRAN giving the characteristics of each cell including frequency, codes, and midamble. The actual handover procedure depends on the network manufacturer. Handover from one UTRA/TDD cell to another requires the UE to measure the receive power of the beacon channels of the neighboring cells. The cells within the UTRA/TDD network of the operator are typically synchronized which enables the UE to know where to search for midambles. For a handover to a neighboring UTRA/FDD cell, the Ec/No is measured on the CPICH. In the case of handover to a GSM cell, the received power level of the GSM broadcast (beacon) channel is measured. The final decision to execute the handover is taken by the UTRAN that exchanges information with the target cell. It should be noted that a method like compressed mode used in UTRA/FDD is not needed by the UEs in UTRA/TDD mode to perform intersystem or interfrequency measurements.

12.6. UTRA/TDD receiver

A Rake receiver is typically used to despread and it recovers the original signal by exploiting the orthogonality of the different codes. In practice, however, the received spreading codes are not completely orthogonal and the correlation process cannot be so efficient. As a result, *Multiple Access Interference* (MAI) is generated in the receiver: the desired signal does not significantly distinguish itself from interfering users whose effect can be mathematically modeled as increased background noise. One effective way to eliminate MAI is to use a *joint detection* approach which is an optimal multi-user detection receiver that extracts all CDMA signals in parallel.

Implementing multi-user detection receivers in UTRA/FDD is difficult due to the high number of codes: the implementation complexity is an exponential function of the number of codes. The efficiency of the joint detection receiver in UTRA/TDD is based on the limited number of codes employed (16). However, in contrast to UTRA/TDD, MAI can be compensated in UTRA/FDD mode by the fast power control mechanism used in both downlink and uplink directions.

In the example of Figure 12.11, N users in the uplink are detected in parallel by the receiver of Node B. The training sequence within the midamble enables the receiver to estimate the quality of the radio channel given the periodic properties of the basic code. By using a specific algorithm, all the CDMA channels can be extracted in parallel and the interference caused by undesired CDMA channels (MAI) can be removed.

Joint detection in combination with smart antennae may increase the capacity and the spectrum efficiency of UTRA/TDD and compensate somehow the inaccuracy of the fast power control mechanism inherent in UTRA/TDD.

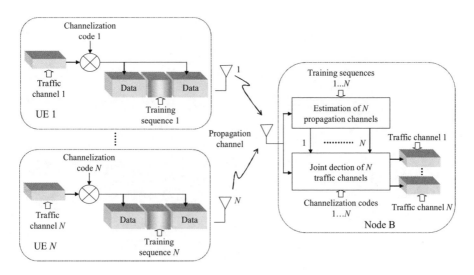

Figure 12.11. *Simplified receiver in Node B based on joint detection*

Chapter 13

UMTS Network Evolution

13.1. Introduction

The third generation (3G) of mobile communication systems, to which UMTS belong, is becoming a significant step forward in the convergence of mobile technologies and the Internet. Compared with the 3G, we can imagine the fourth generation providing more bandwidth, better QoS and personalization of services across heterogenous access networks. In the path towards this evolution, the 3GPP has defined a number of *Releases*, i.e. a group of technical specifications whose milestones are summarized in Table 13.1.

The detailed technical developments added to the GSM/GPRS network architecture were predominantly to support new services that required the definition of a new radio access network as well as updates on the core network. These were defined in the *Release 99* of 3GPP specifications published in March 2000 which have become the basis for most of the current commercially deployed UMTS networks. In March 2001 *Release 4* was standardized, which incorporated minor improvements to *Release 99*. More significant changes are provided by *Release 5* published in March 2002 where features such as *High Speed Downlink Packet Access* (HSDPA), *IP Multimedia Sub-system* (IMS) and IP UTRAN were defined. Within *Release 6*, the *Uplink Enhanced Dedicated Channel* (E-DCH), also known as *High Speed Uplink Packet Access* (HSUPA), is specified as well as the support of multicast and broadcast services through the *Multimedia Broadcast/Multicast Services* (MBMS) feature. In *Release 6* is also defined the integration of 3G mobile networks and WLAN which is a hot topic in wireless communication because the higher mobility of UMTS can be complemented with the higher data rate of WLAN in hotspot areas. Besides MBMS, UMTS *Release 6* also contains [TS 21.906 *R6*]: 3GPP and 3GPP2 harmonization in

IMS network architecture, emergency calls in the packet domain, radio network sharing between operators, push services and presence.

Release 7 is currently in development and brings enhanced features like *Multiple Input Multiple Output* (MIMO) antennae offering theoretical peak rates above 14 Mbps, while improving the average cell throughput. IMS is further enhanced in *Release 7* with the focus on improving QoS, service enabler and delivery for multimedia packet-based services.

	1ˢᵗ version	Main feature enhancements
Release 99	March 2000	– Basis for most current commercially deployed UMTS networks
Release 4	March 2001	– Enhancements applied to the CS domain: introduction to the *MSC server*, *Circuit-Switched Media Gateway* (CS-MGW) and IP transport of CN protocols – Enhancements of UTRAN QoS – Introduction to the *Transcoder Free Operation* (TrFO) concept – Definition of low chip rate UTRA/TDD mode (TD-SCDMA)
Release 5	March 2002	– Enhancements applied to the PS domain: introduction to the *IP Multimedia Sub-system* – Introduction to HSDPA (*High Speed Downlink Packet Access*) – Enhancements of GERAN so that this can be aligned with the UTRAN – Proposal of IP transport in the UTRAN – Introduction to the wideband AMR codec (AMR-WB)
Release 6	September 2005	– Introduction to *Multimedia Broadcast/Multicast Services* (MBMS) – Enhancements to support WLAN interworking – Definition of the enhanced uplink dedicated channels (E-DCH) also known as *High Speed Uplink Packet Access* (HSUPA) – Definition of IMS stage 2 – 3GPP and 3GPP2 IMS architecture harmonization
Release 7	Stage 1: December 2005 Stage 2: end of 2006 Stage 3: mid 2007	– Introduction to *Multiple Input/Multiple Output* (MIMO) antennae concept for HSDPA – Proposals for UTRAN enhancements: introduction to the *continuous connectivity for packets data users* concept; definition of techniques to reduce delay between PS/CS connections – Proposals for CN and IMS enhancements: multimedia telephony, support of voice call continuity and *Policy and Charging Convergence* (PCC) – Introduction to the *Advance Global Navigation Satellite System*

Table 13.1. *Main features of 3GPP Releases within the UMTS system evolution path*

13.2. UMTS core network based on *Release 4*

In *Release 4*, distributed concepts of H.248/Megaco have been adopted with the final goal to provide speech and multimedia services over the CS domain which are neither limited nor restricted to the fixed bandwidth and functionality limitations of 64 kbps TDM technology. For this purpose, a "Bearer Independent" core network architecture has been introduced [TS 23.205 *R4*]. As shown in Figure 13.1, the focus has been the separation of the MSC onto independent components of *MSC server* and *Circuit-Switched Media Gateway* (CS-MGW). The same applies to GMSC, which is functionally split into GMSC server and CS-MGW. The MSC server comprises all the mobile-specific call, service and mobility management functions and uses H.248 to interact with the CS-MGW for CS services, including enhancements to provide the allocation of switching and transcoding functionality for voice and data communications. On the other hand, the CS-MGW performs the requested media conversion (it contains, for example, the *Transcoding and Rate Adaptation Unit*, TRAU) and provides the bearer control and the payload processing such as codec and echo canceller.

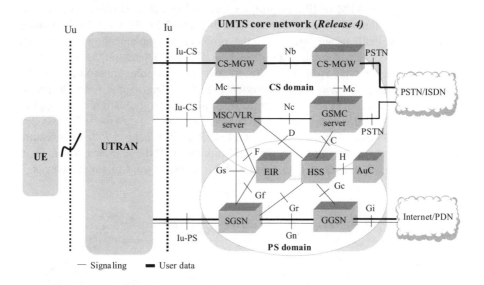

Figure 13.1. *UMTS core network architecture based on Release 4*

Communication between the (G)MSC server and the CS-MGW is made possible via interface "Mc" based on the H.248/IETF Megaco protocol, whereas the interface "Nc" performs network-to-network-based call control between the MSC server and the GMSC server. The "Nc" interface uses an evolvement of the ISUP protocol named *Bearer Independent Call Control* (BICC). Finally, in the "Nb" interface bearer control and user data transport are performed between CS-MGWs. It should be noted that the NAS/AS protocols used by the UE to communicate with the CS domain supporting this network architecture are the same as for *Release 99*.

The main advantages of the *Release 4* network architecture are:

– the possibility to offer better transport efficiency since different transport resources besides the typical TDM/64 kbps can be used as ATM and IP. With respect to the "Nc" interface, an ATM or IP transport enables the physical separation of the call control entities from the bearer control entities – hence, the name of *Bearer Independent Call Control* (BICC) architecture. On the other hand, using ATM or IP for user control and transport, and bearer control in the "Nb" interface, enables the passage of compressed speech at variable rates through the CS domain. Note also that *Release 4* standards enable the use of ATM and IP technologies in the transport layers of "Iu-CS", "F", "D" and "C" interfaces [TS 29.202 *R4*];

– the local transcoding function (TRAU) can be placed in the MGW. Transcoding into A/μ-law format is only necessary when passing the network border; due to an ATM or IP transport, compressed speech at variable bit rates can be carried out through the CS domain (see Figure 13.2). This results in a more efficient bandwidth usage. However, the problem of implementing two speech codecs (i.e. coder/decoder pairs) at both ends in "Tandem Operation" remains – with the inconvenience that speech quality is degraded. Double transcoding can be avoided by using *Tandem Free Operation* (TFO) or *Transcoder Free Operation* (TrFO) principles. TFO [TS 22.053] introduced in previous standard releases applies to UMTS and GSM networks and is well suited for "tunneling" the radio-encoded speech on PCM links. On the contrary, TrFO [TR 25.953 *R4*] is a *Release 4* feature where there is no constraint to use the PCM link on the "Nb" interface, and thus, in addition to the advantages of TFO, there is also a saving of transmission resources.

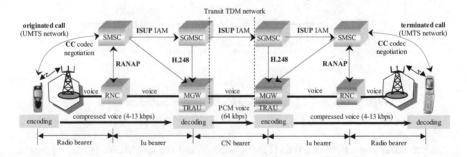

Figure 13.2. *Illustration of saving transmission resources within Release 4 architecture*

13.3. UMTS core network based on *Release 5*

In *Release 5* a set of new entities, dedicated to the handling of the signaling and user traffic flows related to voice and multimedia-over IP applications, was introduced in the PS domain. This set, called *IP Multimedia CN Subsystem* (IMS) [TS 23.228, *R5*], enables operators to converge their networks from the traditional parallel CS and PS domains towards an "all-IP" operation. IMS enables person-to-person services which facilitate shared applications like browsing, whiteboards, games and two-way radio sessions. The first applications are expected to concern Push-to-talk over Cellular (PoC), Presence, Instant Messaging and many other interactive applications evolving towards Voice and Video over IP.

The key architectural principles of IMS are:

– separation of transport, call/session control and services. IMS further enables users and applications to control sessions and calls between multiple parties;

– possibility for a user of a single terminal to establish and maintain several connections simultaneously;

– introduction of new signaling for the control of multimedia sessions: the *Session Initiation Protocol* (SIP) defined by IETF, which allows operators to offer multiple applications simultaneously over multiple access technologies including GPRS or other wireless or fixed network technologies;

– secure authentication and confidentiality based on the *IP Multimedia Services Identity Module* (ISIM) application;

– QoS control enabling, for instance, mobile voice over IP (VoIP) users to access legacy telephony with voice quality as that provided in the CS domain;

– roaming between IMS, so that the users may access their services from any UMTS/GSM network based on the IMS architecture;

– provision of enhanced mechanisms (compared with the PS domain of *Release 99*) for charging purposes containing information on time, duration, volume and type of data sent/received, and participants.

A conceptual view of the key elements in the IMS is shown in Figure 13.3. The motivation for this architecture is to provide maximum flexibility and independence from the access technologies: access, transport and control functions are separate. The new functional entities defined for IMS are:

– the *Serving-Call State Control Function (S-CSCF)* which holds user-specific data services (downloaded from the HSS) and performs session control. It handles the SIP requests and coordinates the establishment of associated bearers in both the home and the visited networks;

– the *Interrogating-CSCF (I-CSCF)* which is the main entrance of the home network: it interrogates the HSS during mobile terminated communications set-up in order to determine the appropriate S-CSCF;

– the *Proxy-CSCF (P-CSCF)* which performs the bridging of the signaling (by acting as a firewall) between the UE and the S-CSCF. Its main task is to select the I-CSCF of the home network of the user;

– the *Media Gateway Control Function (MGCF)* which converts the SIP message into appropriate ISDN messages and forwards them to the external ISDN network. The *IMS Media Gateway Function* (IM-MGW) at the edge facing the external ISDN network is controlled by the MGCF.

In addition, many other interworking functions and entities shown in Figure 13.3 are defined for interconnection with legacy networks such as PSTN, GSM/GPRS, UMTS, etc. They concern the IM-MGW, the *Breakout Gateway Control Function* (BGCF) and the *Signaling Gateway Function* (SGW). Note that the name of an HLR in IMS is changed into HSS (*Home Subscriber Server*).

In Figure 13.3, the "Mc" interface is based on the H.248 protocol, whereas the "Mg", "Mi", "Mj", "Mm", "Mr" and "Mw" interfaces use SIP. The transport within these interfaces can be based either on IPv6 or IPv4.

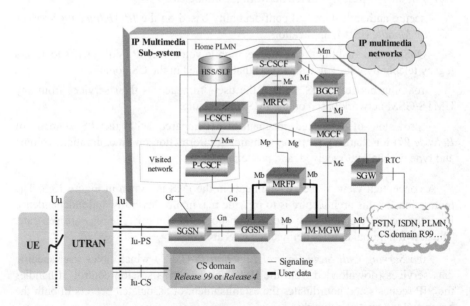

Figure 13.3. *IP multimedia subsystem (IMS) architecture*

13.4. Multimedia Broadcast/Multicast Service (MBMS)

Resources for dedicated transmission are limited in a UMTS network, especially for video services. Rather than setting up a PDP context and associated RABs for every single user among thousands willing to watch the same TV program, the MBMS (*Multimedia Broadcast/Multicast Service*) architecture enables a more efficient way of using radio network and core network resources. This example shows the need to transmit simultaneously time critical events to several users, which might cause congestion. MBMS is an alternative way to solve this problem by providing unidirectional point-to-multipoint services in which data is transmitted from a single source to multiple recipients in one cell by using the same radio resources [TS 22.146, *R6*]. Examples of MBMS service can be real-time broadcasts of events for which a subscription may or may note be required, including video clips from national sports, weather and traffic information, news; etc. Hence, MBMS is hence well suited for streaming and store-and-playback (in the UE) type services.

13.4.1. *Network aspects*

As shown in Figure 13.4, an MBMS feature requires the introduction of a central media server within the operators' UMTS network architecture: the *Broadcast/Multicast Service Center* (BM-SC). This serves as the entry point for content delivery services that want to use MBMS. The UE interacts with this node which is responsible for setting up and controlling MBMS transport bearers. The BM-SC is also used to schedule and deliver MBMS transmissions. The UTRA radio interface is also affected since techniques for optimizing transmission such as point-to-multipoint transmission, selective combining and transmission mode selection between point-to-multipoint and point-to point bearer are needed. Despite this fact, the impact of MBMS feature on the UE's architecture remains small. Several categories of UEs can, however, be specified according to their capabilities including, for instance, low tier terminals providing limited simultaneous operation of HSDPA and MBMS. On the contrary, high tier terminals can provide HSDPA and voice calls which are possible during MBMS contents reception.

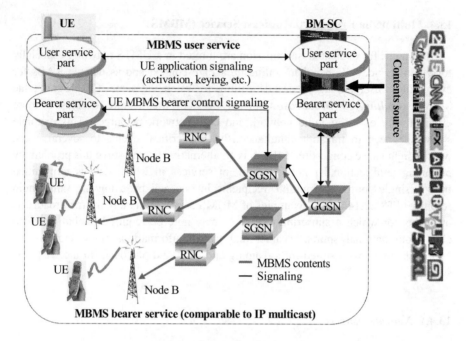

Figure 13.4. *MBMS architecture*

13.4.2. *MBMS operation modes*

MBMS operates in two modes: the broadcast mode and the multicast mode. In broadcast mode, the MBMS contents (text, audio, picture, video, data) are delivered to users within a requested service area and with a requested QoS. This mode does not require a subscription and data is transmitted in the cells of the service area irrespectively of the number of users that are interested in receiving the session.

In multicast mode, the MBMS contents (text, audio, picture, video, data) are delivered to users within a requested service area with a requested QoS. In contrast to the broadcast mode, in multicast mode interested users have to register with the MBMS service in order to receive data and will be authenticated according to a subscription. Radio resources are used only in cells with registered users that can be charged for the MBMS services using different tariff models.

Figure 13.5 gives a bird's eye view of the typical phases during an MBMS session for both broadcast and multicast operation modes. The session starts when information of a given MBMS service is sent to a service provisioning server. This information known as "service announcement" contains information related to the service and how UEs may access it. Such service announcements can be sent to UEs by means of SMS or MMS mechanisms or they can be delivered over a special MBMS announcement channel. An end-user willing to access the service from the reception of the service announcement may follow the workflow depicted in Figure 13.5. Note that broadcast services require no further action, whereas multicast services need to send a *session join* request to the network based on the parameters extracted from the service announcement. Before transmission can start, the BM-SC shall send a *session start* request to the GGSN, which in turn coordinates with the SGSN the set-up of CN and radio access bearers with the required QoS. At this stage, the UEs of the associated MBMS service group are notified that the service is about to deliver the content. End-users who want to leave the MBMS multicast service send a *service leave* request to the network in order for this to remove the user from the related MBMS service group.

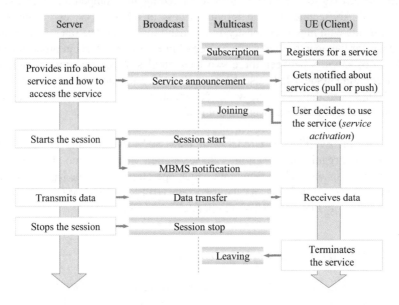

Figure 13.5. *Typical flow of MBMS service provision*

13.4.3. *MBMS future evolution*

In *Release 7* MBMS features will benefit from a better air interface enabling higher bit rates. Providing MBMS services via WLAN is also part of the work in this *Release*. Note, however, that today very few operators have shown interest in commercially offering MBMS services before the 2007 time frame. One reason for this is the increasing interest in developing TV over mobile techniques with other non-3GPP technologies such as DVB-H and DMB. The performance of MBMS is not yet comparable to DVB-H and DMB. Thus, MBMS may coexist with them as a complement or as an alternative. However, we should consider that the DVB-H and DMB approaches are subjected to licensing and regulatory aspects.

13.5. UMTS-WLAN interworking

There is an increasing demand for high speed wireless access to the Internet by using existing WLAN (*Wireless Local Access Network*) technologies such as IEEE 802.11b. Public and private WLAN hotspots in dense areas are in direct competition with UMTS operators, but still lack the services and billing integration as well as greater mobility. This fact is motivating mobile operators to also provide this alternative access to their users, instead of letting the competition take over the market. WLAN technology can provide access to UMTS services in deployment environments with high user density and demand for higher data rates. This requires, however, for the operator that some degree of interworking exists in order to provide mobility of services between these two technologies. Authentication, authorization and accounting issues shall also be centralized for the operator. By coupling the WLAN access to the UMTS network, the operator can provide a unified service to its customers, even when they are roaming abroad.

Besides interworking of UMTS with IEEE 802.11b operating in the 2.4 GHz ISM, other organizational entities have started looking into UMTS-WLAN interworking aspects such as Bluetooth which operates in the same band. Also, ETSI Project BRAN is defining a standardized HIPERLAN/2-UMTS interworking architecture. Note that HIPERLAN/2 operates on 5 GHz bands.

13.5.1. *UMTS-WLAN interworking scenarios*

UMTS-WLAN interworking is a 3GPP work item named "WLAN" and is part of *Release 6*. The operator may offer different levels of services to its users depending on the level of integration between the WLAN and the UMTS network. For this purpose, six scenarios have been identified by 3GPP SA WG2 [TR 22.934, *R6*], each of them proposing an increased level of services to the users (see Figure 13.6). Note that *Release 6* specifications only consider a *basic* integration with 3GPP networks by focusing mostly on scenarios 2 and 3. Other aspects will be covered in *Release 7* (scenarios 4 and 5) such as QoS management in WLAN, charging correlation (i.e. pricing based on bearer characteristics), service continuity (i.e. moving between access networks) and seamless services (i.e. without the user noticing interruptions).

Scenario 1: common billing and customer care

In this scenario, the end-user receives a common bill for both his/her 3GPP and WLAN access. This is the simplest scenario since there is no impact on the 3GPP architecture or on the terminals. Authentication is still done via a username/password combination and security is not shared. Only Internet access service is offered in the WLAN access network.

Scenario 2: 3GPP access control and charging

In scenario 2 of the UMTS-WLAN interworking, the WLAN network uses the 3GPP network for access control and charging. Authentication, authorization and accounting (AAA) are provided by a *3GPP AAA server*. Authentication is USIM-based (or GSM-SIM-based) instead of based on username/password as in scenario 1. Scenario 2 enables pre-paid, volume-based, time-based charging. As for scenario 1, only Internet access service is provided through the WLAN Access Network.

Scenario 3: access to 3GPP PS services

In Scenario 3, the end-user has access to his/her 3GPP services as if he/she were connected directly through the UTRAN. The services are PS-based and may include: IMS, MBMS, location services, presence-based services, messaging, etc. All 3GPP-related traffic goes through a *Packet Data Gateway*, a new node that shall be incorporated into the CN. Internet access service can be provided either by the WLAN or through the 3GPP network. Note that service continuity is not provided during roaming/handovers.

Scenario 4: service continuity

Contrary to scenario 3, in scenario 4 services are preserved when handing over occurs between WLAN and the UTRAN: the end-user does not have to disconnect and re-connect every application when he/she is roaming between the UMTS/WLAN access networks. Nevertheless, the service continuity is not seamless: there may be dropped packets, temporary loss of data – but at least the connection is preserved. The QoS of the target access network may be different from the source access network.

Scenario 5: seamless services

Within scenario 5, the user is able to move from WLAN to the UTRAN without experiencing any *remarkable* drop in services (e.g. a VoIP or multimedia session is maintained without any loss of frames). The QoS in the target access network may be different from the source access network.

Scenario 6: Access to 3GPP CS services

The user is able to access traditional CS services (voice, fax/modem, CS data, CS multimedia, etc.) through the WLAN. The services should operate seamlessly when switching from WLAN to the UTRAN. No use case has been found for this scenario, so it has not been studied any further by SA2.

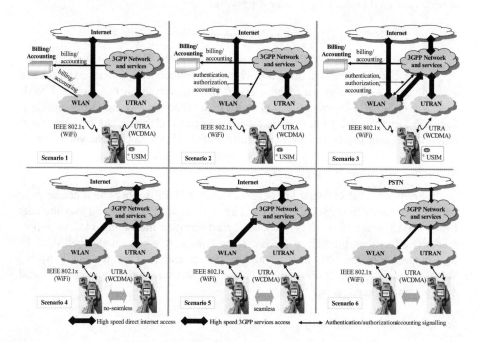

Figure 13.6. *UMTS-WLAN interworking scenarios*

13.5.2. *Network and UE aspects*

UMTS-WLAN interworking requires the definition of new functional entities in the UMTS architecture. Some of them are:

– the *WAG (WLAN Access Gateway)*, which is the entry point to the UMTS PLMN from the WLAN access network;

– the *PDG (Packet Data Gateway)*, which gives access to UMTS PS services;

– the *3GPP AAA Server*, which performs authentication, authorization and accounting;

– the *3GPP AAA Proxy*, which proxies the 3GPP AAA server in the VPLMN.

As far as the mobile terminal architecture is concerned, the impact on its hardware and software architecture depends on the scenario adopted by the operator for UMTS-WLAN service provision. Similarly to PLMN selection rules for dual-mode UMTS-GSM, terminals may be applied to WLAN-enabled terminals [TS 22.234, *R6*]. Both automatic and manual network selection shall be possible. Generally speaking, the terminal must be able to listen/emit on both WLAN and UMTS frequencies. More complex hardware/software architecture is required in case the terminal is capable to access both the WLAN and UTRAN PS and CS services at the same time.

Note that a competing architecture to WLAN interworking has been defined in 3GPP GERAN, referred to as *GERAN Generic Access*. An "independent" group was even created for this purpose: UMA (*Unlicensed Mobile Access*). In this architecture, the WLAN is connected to the 3GPP CN through an interface similar to the A/Gb in the GSM A/Gb in such a way that the WLAN appears like a BSS to the CN. Regarding the impact on the terminal, the UMA client sits below the user interface of the phone, thus ensuring any application on the handset (push-to-talk, presence, ringtones) is supported seamlessly between the GSM/GPRS and UMA/Wi-Fi access networks. In this case, the applications are delivered to the subscriber on UMA exactly as they are delivered over the GSM or UMTS network.

13.6. UMTS evolution beyond *Release 7*

There are several research and development works underway around the world to identify technologies and capabilities intended to materialize the vision of the 4G. The main items of such a work include reduced latency in 3GPP networks, higher user data rates, improved capacity and coverage, and reduced overall cost for the operator. As part of the UMTS network architecture evolution, the packet-switched technology will be enhanced to cope with the rapid growth in IP data traffic. Until 2010, the current UMTS network architectures will progressively migrate towards a packet-based architecture optimized to carry all QoS classes of traffic and the legacy CS traffic will increasingly be carried over the PS domain. During the second decade of this century, such migration will be completed and the large majority of the traffic will be carried over the PS domain while the CS domain will be only used for small proportions of legacy devices. It is also expected that IP-based 3GPP services will be provided through various access technologies including mechanisms to support seamless mobility across heterogenous access networks.

The UMTS system evolution does not stop at *Release 7*. The migration and enhancements of both the network architecture and the radio interfaces are analyzed in the 3GPP work items called *High Speed Packet Access Evolution* (HSPA+), *System Architecture Evolution* (SAE) and *Long Term Evolution* (LTE). Their commercial deployment is expected to follow the time frame as in the order given here.

13.6.1. *HSDPA/HSUPA enhancements*

The radio performance of HSDPA and HSUPA (E-DCH) described in Chapter 14 are being improved in the so called *HSPA evolution*, or HSPA+. Such evolution defines a framework for enhancing HSDPA/HSUPA towards LTE and SAE performance targets. Areas for improvement are the latency, throughput and spectrum efficiency based on current 5 MHz bandwidth. HSPA+ shall also define constraints in terms of acceptable hardware and software changes in the current UMTS elements: UE, Node B, RNC, SGSN and GGSN. Backward compatibility with *Release 99* and HSDPA/HSUPA terminals shall be preserved without any performance degradation. The HSPA+ work is in the early stages of the 3GPP [3GPP RP-060296] using issues which remained open in *Release 7* as a starting point and keeping as the main focus the PS-only operation mode. The items for improvement concern CS/PS set up procedures, continuous connectivity, MIMO, gaming, advanced receivers, etc.

13.6.2. *System Architecture Evolution*

System Architecture Evolution (SAE) is a SA2 work item started in December 2004 in the 3GPP. SAE aims at rationalizing the system architecture with the ultimate goal of providing full service delivery within the PS domain and therefore enable the evolution towards an all-IP network. The evolved architecture defined by SAE shall improve basic system performance: communication delay, quality and connection set-up time. This shall also efficiently support a wide range of service in the PS domain by ensuring mobility, service continuity, access control, privacy and charging between heterogenous radio access technologies.

A global view of the SAE architecture is depicted in Figure 13.7. The new functional elements introduced by this architecture model are [TS 23.882, *R7*]:

– the *MME (Mobility Management Entity)*, which is responsible for managing UEs mobility and security parameters including temporary identities, authentication and user authorization;

– the *UPE (User Plane Entity)*, which is in charge of triggering/initiating paging when downlink data arrives for the UE. It also stores and controls the UE parameters of the IP bearer service or network internal routing information;

– the *3GPP anchor*, which anchors the user plane for mobility between 2G/3G access networks and the LTE access system;

– the *SAE anchor*, which anchors the user plane for mobility between 2G/3G access networks and non-3GPP access systems.

Implementation of the above new functional entities requires the definition of associated new interfaces S1, S2, S3, S4, S5a, S5b, S6, S7 and SGi.

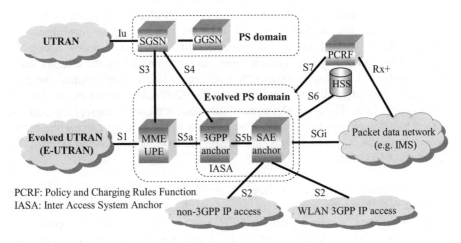

Figure 13.7. *Possible architecture for SAE*

13.6.3. *Long Term Evolution (LTE)*

Long Term Evolution (LTE) refers to the technology developments targeting capacity and data rate enhancements of UTRA as well as the optimization of the UTRAN towards an *Evolved UTRAN* (E-UTRAN) architecture by 2010. Some of the main objectives of LTE are [TR 25.913]:

– scaleable bandwidth in uplink and downlink: 1.25 MHz, 2.5 MHz, 5 MHz, 10 MHz and 20 MHz;

– increased peak data rate: 100 Mbps in downlink and 50 Mbps in uplink with 20 MHz spectrum allocation in both downlink and uplink;

– reduced latency in the user plane: less than 10 ms round trip delay between the UE and the RNC;

– reduced latency in the control plane:

  - transition time of less than 100 ms from a idle state, such as *Release 6* Idle Mode, to an active state such as *Release 6* CELL_DCH,

- transition time of less than 50 ms between a dormant state such as *Release 6* CELL_PCH and an active state such as *Release 6* CELL_DCH;

– improved spectrum efficiency: 3 to 4 times higher than HSDPA (*Release 6*) and 2 to 3 times higher than HSUPA (*Release 6*);

– support for inter-working with existing 3G systems and non-3GPP specified systems.

In order to achieve 100 Mbps peak rates the 3GPP decided to adopt *Orthogonal Frequency Division Multiplexing* (OFDM) technology in the downlink (see [TR 25.814, *R7*]). OFDM is implemented with *frequency domain adaptation* approach providing large performance gains in situations where the channel varies significantly over the system bandwidth. Similarly to HSDPA, information about the downlink channel quality is obtained from the feedback reported by the UEs to Node B scheduler. The scheduler dynamically selects the most appropriate data rate for each downlink spectrum to be allocated to a given user by varying the output power level, the channel coding rate and/or the modulation scheme (QPSK, 16-QAM and 64-QAM).

In the uplink direction, the *Single Carrier FDMA* (SC-FDMA) solution was selected, which works out how to provide better signal *Peak-to-Average Ratio* (PAR) properties in comparison to OFDM when operating in the uplink (see [TR 25.814, *R7*]). A scheduler in Node B dynamically assigns a unique time frequency interval to the UE for transmission of user data. This feature ensures intracell orthogonality and maximizes coverage. Coding and modulation schemes used in uplink are similar to the downlink transmission.

It should be noted that the standardization of LTE by the 3GPP community is still ongoing and it is not clear yet which technical solutions will be finally adopted and become part of the standard. Note also that when LTE becomes available on the market, this will certainly have to coexist with UMTS operators that deployed HSPA. With this in mind, the 3GPP community should consider the possibility to adapt the LTE technical solutions to HSPA evolutions (HSPA+).

Chapter 14

Principles of HSDPA

High-Speed Downlink Packet Access (HSDPA) is a radio technology included in UMTS 3GPP *Release 5* specifications intended to increase the user peak data rates and quality of service in the downlink. In fact, the maximum data rate to a user in present UTRA networks achieved in ideal conditions is limited to 2 Mbps. The introduction of HSDPA to the UMTS specifications will potentially offer up to 10 Mbps downlink data rates in the same 5 MHz bandwidth, while remaining backward compatible with the UEs of *Release 99*. Quality of service is improved with dynamic adaptive modulation and coding, multi-code operation, fast scheduling and physical layer retransmissions.

HSDPA is well suited for providing interactive (e.g. Internet browsing), streaming (e.g. video on demand) and background end-user services. The doors of the "3.5G" are opened by this radio technology. However, HSDPA is just a first step in the evolution of UTRA: the second step will be to enhance uplink data rates, improve uplink capacity and reduce uplink delay [TS 25.309, R6]. Although HSDPA applies to both UTRA/FDD and UTRA/TDD variants, this chapter will explore exclusively the key technical aspects behind the HSDPA concept in the context of UTRA/FDD. Finally, note that the principles of HSDPA are also part of the evolution of cdma2000 networks.

14.1. HSDPA physical layer

In *Release 5*, the UTRA physical layer was enhanced with the introduction of a new transport channel called *High-Speed Downlink Shared Channel* (HS-DSCH). Consequently, four possible transport channels can then be used for user data transfer in the downlink: the DCH, the FACH, the DSCH and the HS-DSCH. The key characteristics of these channels and those of their corresponding physical channels are given in Table 14.1. On the other hand, Figure 14.1 summarizes the transport-channel to physical-channel mapping within *Release 5*.

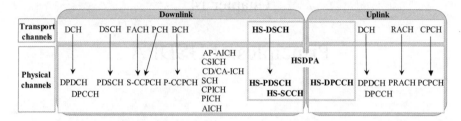

Figure 14.1. *Transport-channel to physical-channel mapping in UTRA/FDD Release 5*

	Downlink transport channels			
	DCH	**FACH**	**DSCH**	**HS-DSCH**
TTI (ms)	10, 20, 40, 80	10, 20, 40, 80	10, 20, 40, 80	2
Type of channel coding	turbo coding convol. coding	turbo coding convol. coding	turbo coding convol. coding	turbo coding
Code rates	1/2, 1/3	1/2, 1/3	1/2, 1/3	1/3
CRC size	0, 8, 12, 16, 24	0, 8, 12, 16, 24	0, 8, 12, 16, 24	24
HARQ	yes, in RLC	yes, in RLC	yes, in RLC	yes, in phy layer
	Downlink physical channels			
	DPCH	**S-CCPCH**	**PDSCH**	**HS-PDSCH**
Spreading factor	variable, 4-512	variable, 4-256	variable, 4-256	fixed, 16
Codes per user	8 codes (max.)	1 code	8 codes (max.)	15 codes (max.)
Modulation	QPSK	QPSK	QPSK	QPSK, 16-QAM
Power control	fast	slow	fast	slow
TX diversity	OL STTD and CL modes 1 and 2	OL STTD	OL STTD and CL modes 1 and 2	OL STTD and CL mode 1
Soft handover/ macrodiversity	yes	no	associated DPCH only	associated DPCH only

Table 14.1. *Key parameters of downlink transport/physical data channels in Release 5*

14.1.1. *HS-DSCH transport channel*

The coding chain of the HS-DSCH is depicted in Figure 14.2 [TS 25.212, *R5*]. After receiving a transport block from MAC, the physical layer adds CRC bits, code block segmentation, turbo encoding, rate matching, interleaving and constellation rearrangement before the data is sent to the physical channel(s). The physical layer spreads and scrambles the data and then maps it to one or more QPSK or 16-QAM constellation. Bits are taken in groups of 4 and used to obtain the appropriate QAM symbol [TS 25.213, *R5*]. Before this symbol lookup, a bit rearrangement is performed. There are four possible rearrangements – this gives the system the chance to transmit bit streams in such a way that all bits experience the same average level of error after combining successive retransmissions in the receiver side.

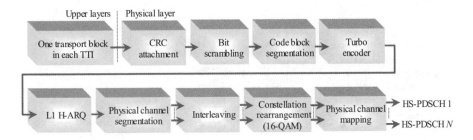

Figure 14.2. *Operations associated to HS-DSCH in the transmitter side*

As the DSCH, the HS-DSCH can be shared in time by several users attached to a Node B. However, the HS-DSCH distinguishes itself from DSCH by the following:

– the *Transmission Time Interval* (TTI) is always 2 ms (mapped to a radio sub-frame of 3 slots). This enables short transmit delays between packets while the channel is shared by multiple users. It also enables better tracking of the time varying radio conditions and fast multiple retransmissions in the case of receive errors;

– the number of transport blocks and the number of HS-DSCHs per TTI is always one;

– only one interleaving operation is applied over 2 ms time periods;

– the adoption of the *Hybrid Automatic Repeat reQuest* (H-ARQ) mechanism in the physical layer;

– only turbo coding is used for channel coding based on the *Release 99* 1/3 turbo encoder scheme, though other coding rates can be obtained with H-ARQ;

– support for 16-level *Quadrature Amplitude Modulation* (16-QAM) in addition to QPSK modulation.

14.1.2. *Mapping of HS-DSCH onto HS-PDSCH physical channels*

The HS-DSCH is mapped onto one or several *High Speed Physical Downlink Shared Channels* (HS-PDSCHs). The structure of an HS-PDSCH is composed of sub-frames of 3 slots (7,690 chips) each. Channelization coding and scrambling are applied as shown in Figure 14.3.

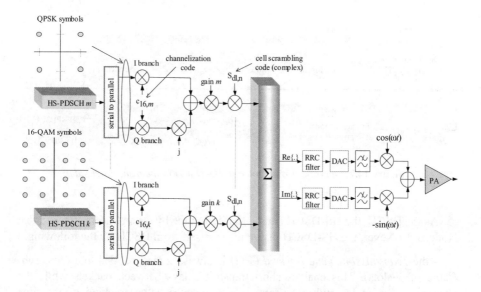

Figure 14.3. *Simplified transmission chain in Node B for HS-PDSCH*

The channelization codes have a fixed spreading factor, SF = 16. This enables a maximum of 15 parallel codes for user traffic and signaling while leaving one for other required control and data bearers. Although the available HS-PDSCHs are primarily shared in the time domain, it is also possible to share the code resources using code multiplexing, in which case several users share the code resources within the same TTI, as shown in Figure 14.4. Multi-code transmissions are allowed: multiple channelization codes (HS-PDSCHs) can be assigned to the UE in the same TTI, depending on the UE capability. The same scrambling code sequence is applied to all the channelization codes. Besides QPSK, the 16-QAM modulation can be used by the HS-PDSCH, thus increasing by 2 the peak rate. Since noise also appears at amplitude variations, 16-QAM is prone to interference and more sophisticated detection techniques may be required as well as high linear power amplifiers.

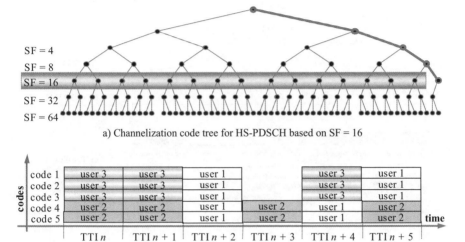

a) Channelization code tree for HS-PDSCH based on SF = 16

b) 5 channelization codes are allocated to 3 users based on a time-code multiplexing scheme

Figure 14.4. *Example of time and code multiplexing of HS-PDSCH shared by 3 users*

14.1.3. *Physical channels associated with the HS-DSCH*

The HS-DSCH is associated in the downlink with one or several *High-Speed Shared Control Channels* (HS-SCCHs) and in the uplink to the *High-Speed Dedicated Physical Control Channel (uplink) for HS-DSCH* (HS-DPCCH). The sub-frame structure of all these channels is shown in Figure 14.5.

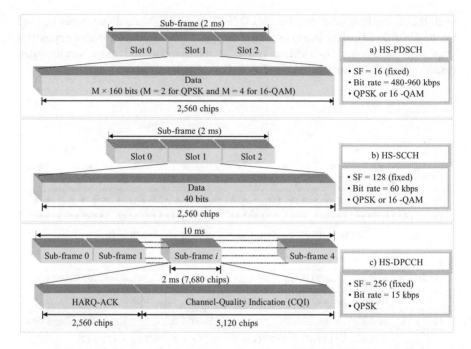

Figure 14.5. *Sub-frame structure of HS-PDSCH, HS-SCCH and HS-DPCCH*

HS-SCCH structure

The HS-SCCH is a fixed rate (60 kbps, SF = 128) downlink physical channel used to carry the downlink signaling related to the HS-DSCH transmission (see Figure 14.5b). The power of the HS-SCCH is controlled by Node B. This channel conveys:

– the set of channelization codes for the current HS-PDSCH(s);

– the modulation scheme that is being used (QPSK or 16-QAM);

– the size of the transport block;

– the H-ARQ process identifier;

– the *Redundancy Version* (RV) and constellation re-arrangement parameters;

– the UE identity, i.e. the *HS-DSCH Radio Network Identifier* (H-RNTI).

The above information provides timing and coding information, thus enabling the UE to listen to the HS-DSCH at the correct time and using the correct codes to enable successful decoding of UE data. The UE using an active HS-DSCH must be capable of receiving up to four parallel HS-SCCHs in order to determine if data is being transmitted to the UE in the next time period [TS 25.214, *R5*].

HS-DPCCH structure

The structure of the HS-DPCCH is presented in Figure 14.5c. It is a fixed rate (15 kbps, SF = 256) uplink channel that carries the acknowledgements (positive (ACK) or negative (NACK)) of the packet received on HS-PDSCH and also the *Channel Quality Indication* (CQI). The CQI are estimated and then transmitted by the UE in steps of 2.0 ms according to the network configuration and a repetition scheme can be applied. The HS-DPCCH is always accompanied by a DPCCH in the uplink. The power of the HS-DPCCH is defined as an offset compared to this channel.

Associated DPCH

On top of the HS-PDSCH, the HS-SCCH and the HS-DPCCH, every UE has an associated *Release 99* dedicated physical channel (DPCH) in both the uplink and downlink directions (see Figure 14.6). The downlink associated channel carries the signal radio bearer for layer 3 signaling as well as power control commands for the uplink channel, whereas the uplink channel is used as feedback channel, carrying for instance the TCP and FBI bits. Other services such as speech can also be carried on the DPCH.

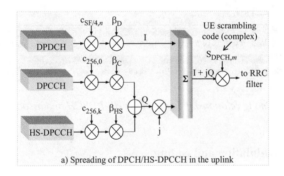

a) Spreading of DPCH/HS-DPCCH in the uplink

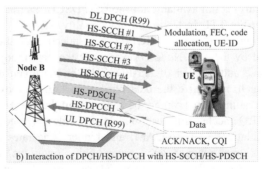

b) Interaction of DPCH/HS-DPCCH with HS-SCCH/HS-PDSCH

Figure 14.6. *Spreading of DPCH/HS-DPCCH and interaction with HS-SCCH/HS-PDSCH*

14.1.4. *Timing relationship between the HS-PDSCH and associated channels*

Figure 14.7 shows the timing offset between the uplink and downlink DPCHs, the HS-PDSCH, the HS-SCCH and the HS-DPCCH at the UE. An HS-DPCCH subframe starts a multiple of 256 chips after the start of an uplink DPCH frame associated to the downlink DPCH. Similarly, the HS-SCCH starts 2 slots (5,120 chips) before the start of the HS-PDSCH, thus providing the UE with the required information for decoding it. For every packet sent on the HS-DSCH, Node B expects a feedback from the UE via the HS-DPCCH after 7.5 slots (19,200 chips).

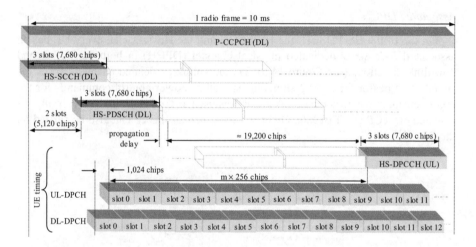

Figure 14.7. *Timing relationship between the HS-PDSCH and associated channels*

14.2. Adaptive modulation and coding

Fast power control is not available for the HS-PDSCH: its transmission power is hold constant over the TTI. Instead, this functionality has been replaced by a fast *Adaptive Modulation and Coding* (AMC) scheme. Indeed, according to the value of the CQI estimated by the UE, Node B determines the best modulation technique, transport block size and number of HS-PDSCHs for the given downlink channel conditions in order to maximize the data rate to the UE. Moreover, as previously mentioned, the spreading factor can not be changed but the effective coding rate can – due to the H-ARQ mechanism described in the following section. In this way, higher order modulations (e.g. 16-QAM) and higher code rates (e.g. 3/4) will be assigned to users experiencing good radio link conditions. Similarly, lower modulation schemes (e.g. QPSK) and lower code rates (e.g. 1/4) will be used in poor radio conditions to maintain the error rate. Table 14.2 gives examples of user data rates from different combinations of modulation and coding rates.

Modulation	Effective code rate	Data rate (1 HS-PDSCH)	Data rate (5 HS-PDSCHs)	Data rate (15 HS-PDSCHs)
QPSK	1/4	120 kbps	0.6 Mbps	1.8 Mbps
QPSK	1/2	240 kbps	1.2 Mbps	3.6 Mbps
QPSK	3/4	360 kbps	1.8 Mbps	5.4 Mbps
16-QAM	1/2	480 kbps	2.4 Mbps	7.2 Mbps
16-QAM	5/8	600 kbps	3.0 Mbps	9.0 Mbps
16-QAM	3/4	720 kbps	3.6 Mbps	10.8 Mbps

Table 14.2. *User data rates on top of Layer 2 obtained from different coding rates and modulation schemes (including overhead)*

14.3. Hybrid Automatic Repeat Request (H-ARQ)

In a typical ARQ mechanism, the receiver sends an acknowledge (ACK) message to the sending station when a data block has been successfully received, e.g. by CRC checksum comparison. When the checksum calculated by the receiver does not match the checksum included within the transmitted data block, the receiver will send a negative acknowledge (NACK) to the sender and discard the erroneous block. This so-called *Stop and Wait* (SAW) method is not very efficient for two reasons. First, the sender may retransmit the erroneous block and second, the transmitter is inactive until it gets a response. Hybrid ARQ is a combination of ARQ and forward error correction (FEC) aiming at minimizing retransmissions: the erroneous blocks are kept and are used for a combined detection with retransmission. Moreover, in order to avoid waiting times, N parallel SAW-ARQ processes are alternatively used within the same channel in separate TTIs. The delay between the original and the first retransmission is in the order of 12 ms.

The H-ARQ mechanism employed in the HSDPA concept is located in the physical layer, thus enabling fast retransmissions. Two retransmission strategies are used: *chase combining* and *incremental redundancy* (IR). With chase combining, the UE sends a NACK to the sending Node B when detecting a block with errors. Rather than discarding the erroneous block, it will be stored. In the case where the retransmitted block is also received in error, the previous block and the current block are combined weighted by the SNR estimation. Each time a block is resent, the same coding scheme is used.

The IR scheme is similar to chase combining with the difference that retransmitted data is coded by using additional redundant information in order to improve the chances that the block will be received either without errors or with enough errors removed which will make it possible to combine it with previous blocks, thus enabling error correction. This can result in fewer retransmissions than for chase combining and is particular useful when the initial transmission uses high coding rates (e.g. under poor radio conditions or high velocity scenarios). However, it results in higher memory requirements for the UE [FUR 02].

Implementation of H-ARQ in the physical layer

H-ARQ can be seen as a complement to AMC: an initial estimate for the redundancy required for reliable transmission is obtained with AMC. With H-ARQ, fine tuning of the effective code rate is achieved. In contrast to AMC, the H-ARQ mechanism is not based on channel quality measurements and therefore inherent CQI estimation delays and errors can be avoided.

The 3GPP H-ARQ scheme is part of the physical layer coding chain (see Figure 14.2). This is based on the *Release 99* rate matching algorithm. However, an architecture based on 2 separate stages of rate matching has been adopted in order to facilitate the incremental redundancy operation [TS 25.212, *R5*]. The overall rate matching architecture is illustrated in Figure 14.8.

The first rate matching is used to adjust the number of available coded bits at Node B to the *Virtual IR* (VIR) buffer size. The maximum number of soft bits available in the VIR buffer is signaled by higher layers for each H-ARQ process (8 max.). The received bits are stored in the VIR buffer and retained for possible subsequent repeat transmissions. Each repeat transmission may produce some soft-bits in positions of the VIR buffer already containing soft bits from a previous transmission attempt. Thus, the new soft bits are simply added to the existing soft values. The second rate matching stage matches the capacity of the VIR buffer to that of the physical channels assigned to the HS-DSCH given the redundancy version (RV) selected.

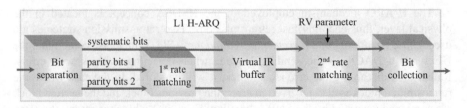

Figure 14.8. *Overall H-ARQ architecture in the physical layer*

14.4. H-ARQ process example

Figure 14.9 illustrates an example of H-ARQ process:

– Node B schedules data on the HS-DSCH to two users UE1 and UE2 according to packet prioritization and resource availability;

– prior to sending data on the HS-DSCH, Node B transmits the HS-SCCH two slots in advance of the HS-DSCH;

– the UEs monitors the set of HS-SCCH signaled by the network on every TTI (2 ms). In decoding the UE identity field within the first slot of the HS-SCCH, the UEs know to whom the data block on the HS-DSCH is destined. If it turns out to be for UE1, it decodes the remaining information in the HS-SCCH by giving the parameters required to decode the HS-DSCH: modulation scheme, multi-code set, H-ARQ process control, etc.;

– the H-ARQ starts after decoding the HS-SCCH and receiving the first transport block 1. in order for the serving Node B to know if the block was detected with errors or not, the UE1 sends a CRC-based ACK/NACK response on the HS-DPCCH. Whilst waiting for the UE1 feedback, Node B takes the opportunity to send blocks 2 and 3 so that several H-ARQ processes are active in parallel;

– from the procedure described in [TS 25.214, R5], the UE1 estimates and then sends a report on the channel quality (CQI) to Node B by choosing a transport format (modulation, code rate and Tx power offset) such that a target BLER is met. For the next transport block sent, Node B determines the transport format according to the recommended transport format and possibly on power control commands of the associated DPCH;

– the UE keeps on monitoring exclusively the HS-SCCH used in the immediately preceding sub-frame, looking for an indication that there is about to be some data destined to it.

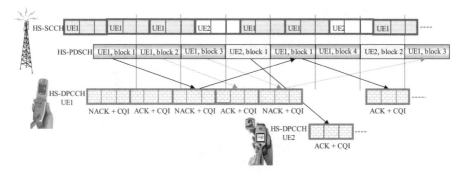

Figure 14.9. *Example of H-ARQ process with two users*

14.5. Fast scheduling

For each TTI, the fast scheduling mechanism defines which UE(s) the HS-DSCH should be transmitted to and, together with the AMC, at which data rate. in order to enable the scheduler to quickly respond to the changes in the channel conditions and ensure that the UE is served while on a constructive fade, this mechanism resides in Node B. This contrasts with *Release 99* where the packet data scheduler is located at the RNC. The "best" packet scheduler is that which optimizes the cell capacity while fulfilling the QoS requirements (e.g. transfer delay and guaranteed bit rate). Several algorithms can be used in practical implementations. Some examples are: the *Round-Robin*, the *Maximum Carrier-to-Interference* (C/I) and the *Proportional Faire* schedulers [TR 25.848, *R4*].

In the *Round-Robin* scheduler users active at the same time are served in sequential order so they all get to experience the same delay and throughput. This scheme provides a high degree of fairness. However, since channel conditions are not taken into account, the packet data can be corrupted for certain users who experience a destructive fade.

The *maximum C/I* mechanism serves users with the highest C/I during the current TTI until the packet queue is empty (see Figure 14.10). This naturally leads to the highest system throughput, since the served users are the ones with the best channel conditions. Nevertheless, the users experience very different service quality. For instance, users at the cell edge will be largely penalized by experiencing excessive service delays and significant outage.

An alternative to the above approaches, is the *Proportional Fair* scheduler where all users have an equal probability of being served, although they may experience a very different average channel quality. The Proportional Fair scheduler provides a good trade-off between system capacity and fairness.

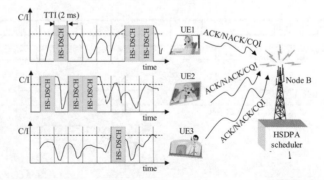

Figure 14.10. *Principle of maximum C/I scheduling mechanism*

14.6. New architecture requirements for supporting HSDPA

Although backguard compatible with *Release 99*, the deployment of HSDPA requires network upgrades. Figure 14.11 illustrates the HSDPA protocol architecture. Node B is the element affected the most. Important changes are expected in both its physical layer, as described in previous sections, and in layer 2 because a new high speed MAC (MAC-hs) entity will be needed. The software in the RNC must also be updated to manage HSDPA and non-HSDPA radio resources (e.g. hard handovers from a DCH to the HS-DSCH). No hardware modifications are foreseen on the RNC. With respect to the core network, the impact from the introduction of HSDPA is minor: the existing radio access bearers must be modified and re-dimensioned in order to cope with the higher bit rates provided by HSDPA.

14.6.1. *Impact on Node B: high-speed MAC entity*

Implementing a MAC-hs entity in Node B side results from the need to achieve low delays in the link control. In fact, the control of the H-ARQ mechanism is located in the MAC-hs, thus enabling the storage of NACK data blocks and the associated retransmission scheme without the inherent delays of an RNC-based ARQ implementation. In Node B, MAC-hs further manages the fast scheduling and users priority operations. Finally, MAC-hs selects the appropriate transport format and resources for the data transmitted in the HS-DSCH according to the aforementioned AMC mechanism.

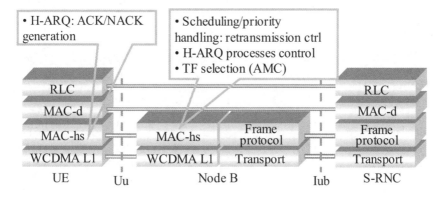

Figure 14.11. *HSDPA protocol architecture*

14.6.2. *Impact on the UE: HSDPA terminal capabilities*

As in *Release 99*, the HSDPA UEs tell the network, upon RRC connection set-up, their radio access capabilities. These capabilities determine, for example, the maximum number of HS-PDSCH multi-codes that the UE can simultaneously receive and the minimum inter-TTI time, which defines the minimum time between the beginning of two consecutive transmissions to this UE. 12 new categories have been specified by *Release 5*, as shown in Table 14.3. Similarly to *Release 99*, some guidance on radio access capability combinations is also defined resulting in different UE classes – this gives UE manufactures the freedom to differentiate from each other. UEs of categories 1 and 5 are the first to be commercially available. It should be noted that UEs belonging to the categories 1 to 10 support both QPSK and 16-QAM modulations, whereas categories 11 and 12 support exclusively QPSK modulations. Finally, note that UEs from category 10 support bit rates of up to 14 Mbps. This of course is a theoretical peak rate value unlikely to be achieved in practical implementations.

HS-DSCH category	Max. number of HS-PDSCH codes received	minimum TTI interval	Max. number of bits in a transport block per TTI	User data rate (Mbps)
category 1 (1.2 Mbps class)	5	3	7,298	1.22
category 2	5	3	7,298	1.22
category 3	5	2	7,298	1.83
category 4	5	2	7,298	1.83
category 5 (3.6 Mbps class)	5	1	7,298	3.65
category 6	5	1	7,298	3.65
category 7 (7 Mbps class)	10	1	14,400	7.20
category 8	10	1	14,400	7.20
category 9 (10 Mbps class)	15	1	20,251	10.13
category 10	15	1	27,952	13.98
category 11	5	2	3,630	0.91
category 12	5	1	3,630	1.82

Table 14.3. *Categories and classes of HSDPA UEs according to their physical layer capability*

14.7. Future enhancements for HSDPA

Additional performance improvements are foreseen for HSDPA with the introduction of *enhanced uplink techniques*, and *Multiple Input Multiple Output Antenna (MIMO)*.

14.7.1. *Enhanced UTRA/FDD uplink*

HSDPA focuses exclusively on providing higher capacity and reduced delays in the downlink. However, the uplink may be a future congestion point in terms of capacity and delay (contribution to overall round-trip delay) as well – higher data rates in the uplink is not a main target so far. With this in mind, a similar approach has been followed in *Release 6*, where a new transport channel called *Enhanced-DCH* (E-DCH) was introduced [TS 25.309, *R6*]. In terms of services, the priority is given to streaming, interactive and background traffic classes. E-DCH is also known as *High Speed Uplink Packet Access* (HSUPA). The premise is still to remain backward compatible with *Release 99* and *Release 5*.

The enhancement applied to the uplink includes similar features to those proposed for HSDPA with a focus on decreasing delay, improving coverage and increasing throughput for packet services with bit rates of up to 5.76 Mbps. The main features behind HSUPA are [TR 25.896, *R6*]:

– introduction of a new dedicated channel: fast DCH set-up;

– adaptive modulation and coding scheme;

– introduction of fast hybrid ARQ with soft combining in the uplink;

– shorter frame size (2 ms TTI) and improved QoS;

– fast Node B controlled scheduling – this is Node B that controls the set of transport format combinations to be used by the UE in the uplink.

Unlike HSDPA, HSUPA remains based on a dedicated channel and auxiliary signaling and traffic channels were introduced so that overall uplink capabilities are improved. Similar to HSDPA, fast retransmission based on H-ARQ principle were introduced in HSUPA at the physical layer for error recovery. With the scheduling function located at Node B improved coverage and capacity in the uplink can be achieved. HSUPA can operate with or without HSDPA.

Like HSDPA, new terminals are required in HSUPA whose capability has been standardized per category as depicted in Table 14.4. All categories support 10 ms TTI whereas categories 2, 4 and 6 may support 2 ms TTI as an option. Maximum peak rate is 2 Mbps in 10 ms TTI and 5.76 Mbps with 2 ms TTI.

E-DCH category	Max. number of E-DCH codes transmitted	Minimum SF	Support for 10 and 2 ms TTI E-DCH	Max. number of bits in a transport block within a 10 ms TTI	Max. number of bits in a transport block within a 2 ms TTI	Max. user data rate (Mbps)
category 1	1	4	10 ms only	7,296	--	0.73
category 2	2	4	10 ms and 2 ms	14,592	2,919	1.46
category 3	2	4	10 ms only	14,592	--	1.46
category 4	2	2	10 ms and 2 ms	20,000	5,837	2.92
category 5	2	2	10 ms only	20,000	--	2.00
category 6	4	2	10 ms and 2 ms	20,000	11,520	5.76

Table 14.4. *Categories of HSUPA UEs according to their physical layer capability*

14.7.2. Multiple Input Multiple Output antenna processing

With the MIMO concept, both Node Bs and the UEs are equipped with multiple antenna elements for data transmission and reception. This creates a form of antenna diversity which increases the data link robustness against fading channels [GES 03]. Moreover, since the same spreading codes (channelization and scrambling) can be re-used for each antenna branch, the peak rate can be increased proportionally to the number of elements in the antenna, provided that this number is the same in both the UE and Node B. Due to the costly implementation of MIMO within the UE, it was scheduled for *Release 7* specifications.

Appendix 1

AMR Codec in UMTS

Originally developed to be used in GSM by the ETSI, the *Adaptive Multi-Rate* (AMR) speech codec [TS 26.071] was approved within the 3GPP forum in 1999 to be mandatory for circuit- and packet-switched speech in UMTS networks. An AMR speech codec adapts the error protection level to the local radio channel and traffic conditions so that it always selects the optimum channel and codec mode to deliver the best combination of speech quality and system capacity. AMR uses *Multi-Rate Algebraic Code Excited Linear Prediction* (MR-ACELP) scheme based on two different synthesis filters. It converts a narrowband speech signal (from 300 to 3,400 Hz) to 13-bit uniform *Pulse Coded Modulated* (PCM) samples with 8 kHz sample rate. This leads to 20 ms AMR frames consisting of 160 encoded speech samples. This means that the codec can switch mode, i.e. source bite rate, every 20 ms. AMR has 8 coded modes in UMTS systems, whereas in GSM AMR uses either 6 or 8 modes. The eight source rates vary from 4.75 to 12.2 kbps. It also contains a low rate encoding mode, called *SIlence Descriptor* (SID), which operates at 1.8 kbps to produce background noise and a non-transmission mode.

The AMR codec dynamically adapts its error protection level to the channel error conditions. For instance, lower speech coding bit rate and more error protection schemes are used in bad channel conditions. This principle is illustrated in Figure A1.1 where AMR strives to change to the best curve associated to a given AMR mode. It has been shown that the degradation on the audio quality caused by a lower speech coding rate is compensated by increased robustness with the channel coding. Note, however, that this channel robustness is more beneficial in GSM than in UMTS due to the embedded fast power control used in WCDMA systems. Using a variable-rate transmission scheme also makes it possible to control the transmission power of the UE, a fact that is particularly useful when the UE suddenly attains its maximum transmit power: in CDMA: lower bit rates generally need lower transmit power and vice versa.

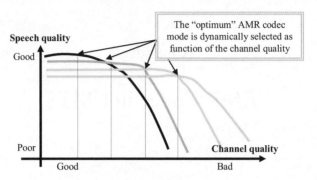

Figure A1.1. *AMR principle*

A1.1. AMR frame structure and operating modes

Figure A1.2 depicts the generic structure of the AMR frame. As observed in the figure, the frame is divided into a *header*, *auxiliary information* and *core frame*. The header contains the *Frame Types* and *Frame Quality Indicator* fields. The Frame Type can indicate the use of one of the eight AMR codec modes for that frame, a noise frame, or an empty frame. The Frame Quality Indicator indicates if the frame is good or bad. The *auxiliary information* part includes the *Mode Indication*, *Mode Request* and *Codec CRC* fields. The CRC field is used for the purpose of error-detection calculated over all the Class A bits in the AMR *Core frame*. The *Core frame* part is used to carry the encoded bits divided into A, B and C classes. In case of a comfort noise frame, comfort noise parameters, i.e. a SID frame, replace "class A" bits of the core frame while "class B" and "class C" bits are omitted.

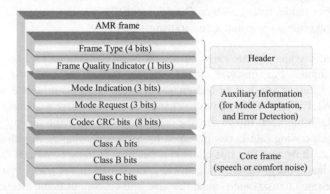

Figure A1.2. *Generic structure of the AMR frame*

Classification of the encoded bits according to their sensitivity to errors

AMR encoded bits are divided into three indicative classes according to their importance: A, B and C. The reason for dividing the speech bits into classes is that they can be subjected to different error protection in the network. Class A contains the bits that are most sensitive to errors and any kind of errors in these bits typically result in a corrupted speech frame which should not be decoded without applying appropriate error concealment. This class is protected by the CRC in *auxiliary information* field. Classes B and C contain bits where increasing error rates gradually reduce the speech quality, but the decoding of an erroneous speech frame is usually possible without a strongly perceptible quality degradation.

AMR operating modes

Table A1.1 depicts the 8 different modes (source bit rates) AMR can operate. It should be noted that some of these modes are equivalent to the speech codecs currently used in other mobile communication systems. For instance, the "AMR 12.20 kbps" mode is equal to the ETSI GSM called codec EFR (*Enhanced Full Rate Speech* [TS 06.60]). Similarly, the "AMR 7.40 kbps" mode is equivalent to the IS-641 codec used in the USA standard IS-136 (US TDMA). Finally, "AMR 6.70 kbps" mode is equivalent to the codec used in the PDC Japanese standard.

Frame type index	Frame content (AMR mode, comfort noise, or other)	ClassA bits	Class B bits	Class C bits
0	AMR 4.75 kbps	42	53	0
1	AMR 5.15 kbps	49	54	0
2	AMR 5.90 kbps	55	63	0
3	AMR 6.70 kbps (PDC EFR)	58	76	0
4	AMR 7.40 kbps (IS-136 EFR).	61	87	0
5	AMR 7.95 kbps	75	84	0
6	AMR 10.2 kbps	65	99	40
7	AMR 12.2 kbps (GSM EFR)	81	103	60
8	AMR SID	–	–	–
9	GSM EFR SID	–	–	–
10	TDMA EFR SID	–	–	–
11	PDC EFR SID	–	–	–
12-14	Future usage	–	–	–
15	No data to transmit/receive	–	–	–

Table A1.1. *AMR modes and relationship with AMR frame structure*

Based on the fact that voice activity in a normal conversation is about 40%, all AMR modes implement a *Voice Activity Detection* (VAD) algorithm that detects if each 20 ms-frame contains speech or not on the transmitting side. VAD works together with the *Discontinuous Transmission* (DTX) or *Source Controlled Rate* (SCR) [TS 26.093] techniques where RF transmission is cut during speech pauses. When the transmission is cut, "comfort noise" parameters are sent at a regular rate in AMR frames during discontinuous activity. These frames are known as SID (*SIlence Descriptor*) frames. The receiver decodes these parameters and generates locally a "comfort noise". Without this background noise the participants in a conversation, might think that their connection is broken during silence periods. The SCR technique for AMR in UMTS is mandatory and aims at prolonging the battery life (UE side) and reducing the interference.

A1.2. Dynamic AMR mode adaptation

The AMR mode adaptation in UMTS networks means using different AMR coding for the data stream. Mode adaptation can independently be applied in the uplink and the downlink. At any point in time, a different AMR mode can be used in each direction and this can be dynamically changed during a voice conversation.

Location of the AMR speech codec in UMTS networks

The AMR speech codec is located in the *Transcoder* (TC) function defined to be in the UMTS core network and as such, logically controlled by *Non-Access Stratum* protocols. From the transfer point of view, this means that all AMR coded data is going to be transmitted not only via Iub and air interface but also via Iu-interfaces. Note, however, that the AMR mode control that generates the AMR mode command cannot be located in the TC, since this control entity needs information from the air interface to make a decision about valid AMR modes – the AMR mode command is used to change the current AMR mode to the new one. The only element in the network which can provide this type of information is the UTRAN. Note that in GSM networks the control of the codec mode is provided by the BTS. This solution is not applicable in UTRA due to the soft-handover procedure defined for dedicated traffic channels. Therefore, the AMR mode control function is part of the RNC, and more precisely a part of layer 3 functionality. Within the radio interface, the rate on the speech connection is either decreased or increased depending on the new valid AMR mode by changing the valid *Transport Format* (TF) in the corresponding MAC-d entity (see Chapter 7).

AMR mode adaptation in the downlink

As shown in Figure A1.3, the RNC generates the AMR mode adaptation command based on existing radio conditions in the downlink as reported by the UE from radio quality measurements and from traffic volume measurements. The command is sent to the encoder inside the TC via the Iu interface.

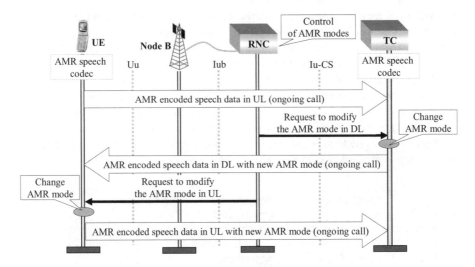

Figure A1.3. *Overview of AMR codec mode control during an ongoing voice call*

Uplink AMR mode adaptation

Two different alternatives for the AMR mode control in the uplink have been proposed:

– Based on the air-interface load, the RNC decides when to request the encoder in the UE to change the valid AMR mode and a new valid AMR mode is sent to the UE inside the AMR mode command message. When received by the UE, mode adaptation is made accordingly (see Figure A1.3). Within this approach, the UE does not have any rights to request the mode adaptation from the network nor to change the used AMR mode autonomously.

– In the second proposed alternative, the AMR codec control is not only included into the RNC but also into the UE. This enables the UE to change the valid AMR mode of the speech connection on uplink more quickly without requesting the mode change from the RNC first. For instance, when the level of the maximum transmission power is reached, the UE may change the valid AMR mode independently. The new mode can, however, be selected only from the valid *Transport Format Set* (TFS), which has been communicated to the UE by RRC from the RNC side. The changed AMR mode is discovered by the RNC from the TFCI bits in the uplink dedicated physical data channel.

A1.3. Resource allocation for an AMR speech connection

An AMR speech connection can be initiated either by the UE or the network. When the UE requests resources from the network, a first negotiation is made based on NAS procedures in order to configure the call connection. The CN will determine the QoS, needed which will be then indicated to the UTRAN inside the RANAP *RAB ASSIGNMENT REQUEST* message. Based on this request, RNC can define the requested RAB and associated *Radio Bearer*(s) (RB). Depending on whether the requested AMR base speech connection supports the concept of *Unequal Error Protection* (UEP) or *Equal Error Protection* (EEP), the RNC assigns either one or three RBs (including one or three DCHs), respectively, for the user plane (see Figure A1.4). In the control plane, RRC may allocate one or none signaling radio bearer according to the alternative method used to change the AMR mode.

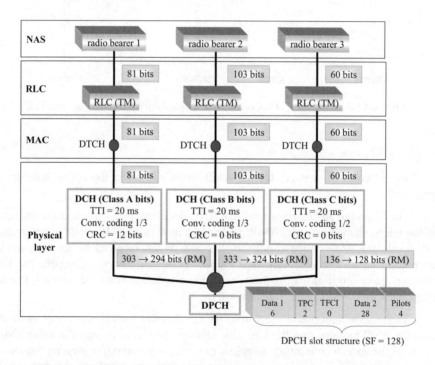

Figure A1.4. *Implementation of AMR 12.2 kbps mode in the radio interface*

A1.4. AMR wideband

The AMR wideband (AMR-WB) codec has been standardized in 3GPP and is part of the specification in *Release 5*. The codec is based on the same adaptive principles as the AMR narrowband. The AMR-WB comprises nine codec modes: 6.6 kbps, 8.85 kbps, 12.65 kbps, 14.25 kbps, 15.85 kbps, 18.25 kbps, 19.85 kbps, 23.05 kbps and 23.85 kbps. The encoder of the AMR-WB is able to code an audio signal with bandwidth between 50 and 7,000 Hz. A higher sampling rate is thus needed compared with the narrowband approach (16 kHz instead of 8 kHz) leading to a 14 bit samples with 16,000 samples/s. Wideband coding provides improved voice quality especially in terms of increased voice naturalness since it covers twice the audio bandwidth compared to the classical telephone voice bandwidth of 4 kHz [TS 26.190, *R5*].

Appendix 2

Questions and Answers

1. Radio access technologies defined in the IMT-2000 framework are	A	UTRA/FDD and UTRA/TDD only
	B	UTRA FDD/TDD, TD-SCDMA, cdma2000, UWC-136 and DECT
	C	UTRA FDD/TDD, TD-SCDMA, cdma2000, UWC-136, and WiFi
	D	UTRA FDD/TDD, TD-SCDMA, cdma2000, UWC-136, and Bluetooth
2. Radio access technologies defined in the IMT-2000 framework are based on CDMA except	A	UTRA/TDD and EDGE
	B	UTRA/TDD, TD-SCDMA, DECT and EDGE
	C	UTRA/TDD and UWC-136
	D	UWC-136 and DECT
3. Radio access technologies defined in the IMT-2000 framework use TDD duplexing except	A	UTRA/TDD and EDGE
	B	cdma2000 and DECT
	C	UTRA/FDD, UWC-136, cdma2000 and DECT
	D	UTRA/FDD, UWC-136 and cdma2000
4. A UMTS network comprises	A	A UTRAN and a core network based on GSM/MAP
	B	Several Node Bs, one RNC and a core network based on GSM/MAP
	C	One RNC, one MSC and one SGSN
	D	One RNC, a CS domain and a PS domain

5. GSM standardization is developed in ____ whereas that of UMTS is carried out in ____	A	ETSI, 3GPP
	B	ETSI, ETSI
	C	3GPP, 3GPP
	D	3GPP, 3GPP2
6. Maximum and theoretical data rates that can be achieved with GSM, HSCSD, GPRS, GPRS/EDGE and UMTS are respectively	A	14.4 kbps, 171.2 kbps, 171.2 kbps, 384 kbps and 2 Mbps
	B	9.6 kbps, 38.4 kbps, 115 kbps, 307 kbps and 2 Mbps
	C	14.4 kbps, 57 kbps, 171.2 kbps, 384 kbps and 2 Mbps
	D	None of these
7. In Europe, a UMTS license typically comprises	A	2×20 MHz for UTRA/FDD and 1×5 MHz for UTRA/TDD
	B	1×5 MHz for UTRA/TDD only
	C	2×20 MHz for UTRA/FDD only
	D	2×15 MHz for UTRA/FDD and 1×5 MHz for UTRA/TDD
8. Passing a video telephony call while at the same time surfing on Internet with a UMTS terminal	A	Is not possible
	B	Is only possible with Type 1 terminals
	C	Is only possible with Type 2 terminals
	D	None of these
9. A USIM is	A	A smart card
	B	A smart card with an application
	C	An application
	D	The same thing as a SIM card
10. A UMTS subscriber needs to dial ____ for a video telephony call and ____ for a voice call	A	MSISDN, MSISDN
	B	MSISDN, a special number that is not standardized
	C	MSISDN, IMSI
	D	IMSI, IMSI
11. ____ is not part of the information that can be stored in a USIM	A	Scrambling code ID
	B	Ciphering and integrity keys
	C	IMSI
	D	IMEI

12. With Type 2 UMTS/GSM terminals, it is possible to	A	Receive simultaneously user data from 3G and 2G networks
	B	Transmit simultaneously user data to 3G and 2G networks
	C	Switch automatically from 3G to 2G networks
	D	None of these
13. Real-time services are associated to traffic classes of type	A	Conversational and interactive
	B	Conversational and streaming
	C	Interactive and streaming
	D	Streaming and background
14. With identity___ the network can deny access to stolen UMTS terminals when registered in the database ___	A	IMSI, HLR
	B	MSISDN, HLR
	C	IMSI, AuC
	D	IMEI, EIR
15. The PS domain in UMTS networks typically comprises	A	One or several MSCs and GMSCs
	B	Only one SGSN and several GGSN
	C	One or several SGSNs/GGSNs and HLR/AuC/EIR databases
	D	Two or more RNSs
16. UTRAN typically comprises	A	One or several RNCs controlling one or several Node Bs
	B	Only one RNC controlling one or several Node Bs
	C	One or several RNCs controlling one Node B each
	D	Two or more RNSs
17. The BSS in GSM is functionally equivalent to the ___ in UMTS	A	UTRAN
	B	RNS
	C	RNC
	D	There is no equivalence
18. In contrast to GPRS, UMTS networks	A	Enable streaming-type services
	B	Introduce new SGSN and GGSN equipments in the PS domain
	C	Introduce the concept of "PDP context"
	D	None of these

19. A radio access bearer (RAB) can be seen as a "pipe" that conveys user data between	A	The UE and the RNC
	B	The UE and the GGSN
	C	The UE and a Node B
	D	The UE and the MSC or the SGSN
20. In order to access Internet services, it is mandatory for a UMTS terminal to	A	Support WAP
	B	Be of Type 2
	C	Use the PS domain
	D	None of these
21. A service which is not envisaged in 3GPP specifications for UMTS networks is	A	Circuit and packet data services
	B	SMS, EMS and MMS messaging services
	C	Location services
	D	None of these
22. Some procedures handled by non-access stratum in the control plane are	A	Mobility management and radio resource allocation
	B	Handover, SRNS relocation and authentication
	C	Mobility management, call control and authentication
	D	Cell reselection, macrodiversity and ciphering execution
23. Mobility management, call control and authentication functions are located in	A	The UE and the core network
	B	The UE, UTRAN and the core network
	C	The UE, Node B, the RNC and the core network
	D	The UE, the RNC and the core network
24. In UTRA radio interface, the TCP/IP header compression function is performed by	A	RRC
	B	RLC
	C	MAC
	D	PDCP
25. One of the roles of MAC is	A	Multiplexing one or several transport channels into one physical channel
	B	Multiplexing one or several logical channels into one transport channel
	C	TCP/IP header compression
	D	None of these

26. In the UE user plane, an example of the path followed by data streams generated by an IP application is	A	Application→SM→GMM →RRC→RLC→MAC→Phy layer
	B	Application→PDCP→RLC→MAC→Phy layer
	C	Application→RRC→PDCP→RLC→MAC→Phy layer
	D	Application→RRC→RLC→MAC→Phy layer
27. In the UE control plane, an example of the path followed by NAS messages addressed to the CS domain is	A	CM→MM →RRC→RLC→MAC→ Phy layer
	B	CM→MM →RLC→MAC→ Phy layer
	C	SM→GMM →RLC→MAC→ Phy layer
	D	SM→GMM →RRC→RLC→MAC→ Phy layer
28. Layer ___ controls paging, open loop power control and radio bearer allocation	A	MM
	B	NBAP
	C	RRC
	D	RLC
29. Due to the Iur interface in UTRAN	A	The UE can perform combined RA/LA updating
	B	Intersystem handover can be accelerated
	C	Macrodiversity can be supported when the implied Node Bs belong to different RNCs
	D	Two Node Bs can communicate to accelerate cell reselection
30. In UTRA, ciphering is performed according to the traffic class on either	A	RRC or PDCP
	B	RRC or RLC
	C	RLC or MAC
	D	RRC or MAC
31. Communication between RNC and Node B is enabled by	A	NBAP
	B	RANAP
	C	RNSAP
	D	RRC
32. A RAB request is enabled by	A	NBAP
	B	RANAP
	C	RNSAP
	D	RRC

33. Similarly to GSM radio networks, UTRAN incorporates	A	ATM in the transport network layer
	B	BSSAP protocol to communicate with the core network
	C	WCDMA on the radio interface
	D	A hierarchical architecture with base stations controlled by a base station controller
34. When signaling and user data flows are transmitted in parallel to the network, the UE needs	A	Separate physical channels for each signaling and data flow
	B	A logical channel where signaling and data flows are multiplexed
	C	Separate RLC entities for each signaling and data flow
	D	None of these
35. Macrodiversity is a CDMA feature applied to	A	UTRA/FDD and UTRA/TDD
	B	UTRA/TDD only
	C	UTRA/FDD only
	D	UTRA/FDD, UTRA/TDD and EDGE
36. In the physical layer of UMTS terminals, spread spectrum is performed	A	On each transport channel
	B	After RF modulation
	C	On each logical channel
	D	Before RF modulation
37. Based on CDMA principles, several users can communicate	A	Simultaneously but in different frequency carriers
	B	Within the same frequency carrier but at different time slots
	C	Simultaneously and in the same frequency carrier
	D	None of these
38. In UTRA/FDD, the reduction of the CDMA processing gain for high data rate services can be compensated by	A	Increasing the coding rate in the channel coding scheme
	B	Decreasing the transmit power
	C	Using m-sequences
	D	None of these
39. The main difference between GSM physical channels and UTRA/FDD physical channel is in	A	The usage of "frame" and "slot" time metrics
	B	The fact that TDD is used in GSM
	C	The fact that power control is not used in GSM
	D	None of these

40. Physical layer operations like spread spectrum, chip synchronization, multiplexing and channel coding are located in	A	RNC and the UE
	B	RNC, the UE and Node B
	C	Node B only
	D	Node B and the UE
41. In the UE transmission chain, spread spectrum operation results from	A	A long or short scrambling code only
	B	A channelization code only
	C	The combination of a channelization and a scrambling code
	D	None of these
42. The OVSF codes (channelization codes) are used for	A	Helping UTRAN to distinguish one UE from another in the UL
	B	Helping the UE to distinguish one Node B from another in the DL
	C	Helping UTRAN to decode the DPDCH and DPCCH of one UE in the UL
	D	Introducing the orthogonality in the UL only
43. Contrary to GSM radio interface, in UTRA/FDD	A	The process of "hard handover" does not exist
	B	Antenna sectoring is not applicable
	C	The concept of "soft handover" does not exist
	D	The technique of "frequency hoping" is not used
44. Similarly to GSM radio interface, in UTRA/FDD	A	QPSK RF modulation is used
	B	The chip rate is fixed
	C	The base stations are not synchronized
	D	The Tx power is controlled once every time slot
45. In UTRA/FDD uplink, user data can be conveyed by	A	FACH, PCH or DCH
	B	RACH or DCH
	C	FACH, BCH or DCH
	D	DCH only
46. In UTRA/FDD uplink, signaling messages can be conveyed by	A	RACH only
	B	DCH only
	C	RACH or DCH
	D	None of these

47. During a communication in UTRA/FDD uplink, when there is no upper layer data to transmit	A	The UE goes automatically into RRC idle mode
	B	DPDCH and DPCCH transmissions are stopped
	C	DPDCH is stopped and DPCCH is transmitted
	D	None of these
48. The Radio Access Bearer set-up procedure is always initiated by	A	The UE
	B	UTRAN
	C	The core network or UTRAN
	D	The core network
49. The Radio Bearer set-up procedure is initiated by	A	The UE
	B	UTRAN
	C	The core network or UTRAN
	D	The core network
50. A PDP context is	A	Activated and deactivated by the UE and modified by the external packet network
	B	Activated by the UE or by the SGSN and modified by the UE or by the external packet network
	C	Activated by the UE or by the SGSN and modified by the external packet network
	D	Activated, deactivated and modified by the UE
51. In the core network it is possible to combine CS/PS procedures related to mobility management when interface ___ is implemented	A	Gi
	B	Gs
	C	Gn
	D	Iur
52. In order to communicate with CS and PS domains, the UE needs	A	Zero RRC connections and one Iu connection per domain
	B	One RRC connection per domain and one Iu connection per domain
	C	One RRC connection and one Iu connection for both domains
	D	None of these

53. When the UE possesses multiple primary PDP contexts activated simultaneously	A	They are all associated to the same APN
	B	They all must use the same PDP address
	C	They all must use the same QoS parameters
	D	None of these
54. From the UE perspective, a PDP context is	A	Automatically created when attaching to the PS domain
	B	A dedicated "pipe" between the UE and the SGSN
	C	Initiated by Session Management NAS layer
	D	None of these
55. A Routing Area (RA) is a location area composed of a number of cells and	A	Controlled by the MSC
	B	Controlled by the MSC and the SGSN
	C	Contained within one LA
	D	None of these
56. A UTRAN Routing Area (URA) is a location area composed of a number of cells and	A	Contained within one RA
	B	Controlled by the MSC and the SGSN
	C	Contained within one LA
	D	None of these
57. If the UE transmits PS data while waiting for an incoming CS call, the service states with regard to the PS and CS domains are	A	PS-CONNECTED and CS-DETACHED
	B	PS-CONNECTED and CS-CONNECTED
	C	PS-CONNECTED and CS-IDLE
	D	PS-CONNECTED only
58. After detaching from the PS domain, user PS data transmission can start	A	Just after the UE attaches again to PS domain
	B	Just after a PDP context is set up
	C	At any time, without any other procedure
	D	None of these
59. Similarly to GSM, in UMTS the following security procedures are used	A	Ciphering, integrity and authentication
	B	Mutual authentication of both the UE and the network
	C	P-TMSI and TMSI allocation
	D	None of these

60. During soft handover, the UE	A	May receive user data from multiple Node Bs transmitting on different frequency carriers
	B	Is in macrodiversity state and transmits on different physical channels associated to different Node Bs
	C	Is in CELL_DCH state
	D	Is in CELL_PCH or CELL_FACH states
61. The RRC states where the UE can transmit and receive user data are	A	CELL_PCH and CELL_DCH
	B	CELL_DCH only
	C	CELL_PCH, CELL_FACH and CELL_DCH
	D	CELL_FACH and CELL_DCH
62. In CELL_PCH state, the UE	A	Transmits and receives user messages of small size
	B	Performs soft or hard handover
	C	Performs periodic LA and RA updating procedures
	D	Performs cell selection/reselection processes
63. In CELL_DCH state is not possible to	A	Perform intrafrequency measurements
	B	Report interfrequency measurements
	C	Receive paging messages
	D	None of these
64. A PLMN reselection process is performed	A	Systematically, just after the UE camped on a suitable cell
	B	When the UE is registered on a VPLMN
	C	In CELL_DCH state
	D	None of these
65. Intrafrequency hard handover in UMTS networks	A	Is necessary when "Iur" is not implemented between two RNCs event if their Node Bs transmit on the same frequency carrier
	B	Is necessary when "Gs" is not implemented in the network
	C	Happens when the UE is in RRC idle
	D	None of these

66. Following 3GPP specifications, intersystem handover	A	Is only possible for voice services
	B	Is only possible in the direction GSM→UMTS
	C	Is not possible when multiple PDP contexts are active
	D	None of these
67. If the UE transmits PS data while waiting for an incoming CS call, the service states with regard to UTRAN are	A	RRC connected (for PS domain) and RRC idle (for CS domain)
	B	RRC connected
	C	RRC idle
	D	None of these
68. Compressed mode transmission is used to	A	Perform inter-system and interfrequency measurements in CELL_DCH state
	B	Perform inter-system and interfrequency measurements in CELL_DCH and CELL_FACH states
	C	Perform inter-system and interfrequency measurements in all RRC states
	D	None of these
69. In UMTS, HSDPA brings	A	A different concept on the core network architecture
	B	A different radio access technology applied to UTRA/FDD only
	C	Enhancements on the DSCH channel of *Release 99*
	D	None of these
70. In 3GPP *Release 4* specifications	A	PS domain architecture is strongly impacted
	B	CS domain architecture is strongly impacted
	C	UE architecture is strongly impacted
	D	None of these
71. IMS (IP Multimedia Subsystem)	A	Was introduced in 3GPP *Release 4* specifications
	B	Strongly impacts UTRAN architecture
	C	Makes useless SGSN and GGSN entities
	D	Impacts UE architecture at NAS level

72. GERAN can be seen as	A	An evolution of UMTS networks towards 3GPP *Release 5* specs
	B	An enhanced version of UTRAN
	C	Improved GSM/GPRS/EDGE networks offering similar QoS as in UMTS networks
	D	None of these
73. Inter-operability between WLAN and UMTS networks	A	Is not possible
	B	Is only possible with IEEE.802.1x WLAN networks
	C	Is possible but is not standardized
	D	Is defined in 3GPP *Release 6* specifications

Answers

1.B, 2.D, 3.D, 4.A, 5.C, 6.C, 7.D, 8.D, 9.C, 10.A, 11.D, 12.C, 13.B, 14.D, 15.C, 16.A, 17.B, 18.D, 19.D, 20.D, 21.D, 22.C, 23.A, 24.D, 25.B, 26.B, 27.A, 28.C, 29.C, 30.C, 31.A, 32.B, 33.D, 34.C, 35.C, 36.D, 37.C, 38.A, 39.D, 40.D, 41.C, 42.C, 43.D, 44.C, 45.B, 46.C, 47.C, 48.D, 49.B, 50.D, 51.B, 52.D, 53.D, 54.C, 55.C, 56.D, 57.C, 58.B, 59.C, 60.C, 61.D, 62.D, 63.D, 64.B, 65.A, 66.D, 67.B, 68.A, 69.C, 70.B, 71.D, 72.C, 73.D.

Bibliography

Books

[LAG 00] LAGRANGE X., GODLEWSKI P., TABBANE S., *Réseaux GSM-DCS – des principes à la norme*, Hermès, 2000.

[LEE 98] LEE J. S., MILLER L. E., *CDMA Systems Engineering Handbook*, Artech House, 1998.

[PRO 95] PROAKIS J. G., *Digital Communications*, McGraw-Hill, 1995.

[PUJ 00] PUJOLLE G., *Les réseaux*, Eyrolles, 2000.

[RAH 98] RAHMAN M., *Guide to ATM systems and technology*, Artech House, 1998.

[RAP 96] RAPPAPORT T. S., *Wireless Communications, Principles and Practice*, Prentice Hall, 1996.

[SIM 94] SIMON M. K., OMURA J. K., SCHOLTZ R. A., LEVITT B. K., *Spread spectrum communications handbook*, McGraw-Hill, 1994.

[TAB 97] TABBANE S., *Réseaux mobiles*, Hermès, 1997.

[VIT 95] VITERBI A. J., *CDMA, Principles of Spread Spectrum Communications*, Addison-Wesley, 1995.

Articles

[AKH 99] AKHTAR S., ZEGHLACHE D., "Capacity evaluation of UTRA WCDMA interface", *Proceedings of IEEE Vehicular Technology Conference*, Amsterdam, Netherlands, p. 19-22, September 1999.

[BAH 74] BHAL L. R., COCKE J., JELINEK F., RAVIV J., "Optimal decoding of linear codes for minimizing symbol error rate", *IEEE Trans. on Inf. Theory*, vol. 20, p. 284-287, March 1974.

[BER 93] BERROU C., GLAVIEUX A., THITIMAJSHIMA P., "Near Shannon limit error-correcting coding and decoding: turbo-codes", *IEEE ICC*, p. 1064-1070, Geneva, May 1993.

[FUR 02] FURUSKAR A., PARKVALL S., PERSSON M., SAMUELSSON M., "Performance of WCDMA high-speed packet data", *Proceedings of IEEE Vehicular Technology Conference*, vol. 3, p. 1116-1120, May 2002.

[GES 03] GESBERT D., SHAFI M., SHIU D., SMITH P., NAGUIB A., "Performance of WCDMA high-speed packet data", *IEEE Journal on Selected Areas in Communications*, vol. 21, p. 281-302, April 2003.

[GIL 91] GILHOUSEN K. S., JACOBS I. M., PADOVANI R., VITERBI A. J., WEAVER JR., WHEATLEY III C. E., "On the capacity of a cellular CDMA system", *IEEE Trans. on Vehicular Technology*, vol. 40, p. 303-312, May 1991.

[GOL 68] GOLD R., "Maximal recursive sequences with 3-valued recursive cross-correlation functions", *IEEE Trans. Info. Theory*, vol. 4, p. 154-156, January 1968.

[GOL 92] GOLOMB S. W., "Shift register sequences", *Aegean Park Press*, 1992.

[HAG 89] HAGENAUER J., HOEHER P., "A viterbi algorithm with soft-decision output and its applications", *Proc. GLOBECOM'89*, p. 1680-1686, 1989.

[KLE 96] KLEIN A., KAWAS KALEH G., BAIER W., "Zero forcing and minimum mean square error equalization for joint detection in CDMA channel", *IEEE Trans. on Vehicular Technology*, vol. 45, p. 276-287, May 1996.

[KLE 97] KLEIN A., "Data detection algorithms specially designed for the downlink of CDMA mobile radio systems", *Proceedings of IEEE Vehicular Technology Conference*, Phoenix, AZ, p. 203-207, 1997.

[KOH 95] KOHNO R., MEIDAN R., MILSTEIN L. B., "Spread spectrum access methods for communications", *IEEE Communication Magazine*, vol. 33, p. 58-67, January 1995.

[KUM 96] KUMAR P. V., HELLESETH T., CALDERBANK A. R., HAMMONS A. R., "Large families of quaternary sequences with low correlation", *IEEE Trans. on Info. Theory*, vol. 42, March 1996.

[MOS 96] MOSHAVI S., "Multi-user detection for DS-CDMA communications", *IEEE Communication Magazine*, vol. 34, p. 124-136, October 1996.

[OJA 98] OJANPERA T., PRASAD R., "An overview of third-generation wireless personal communications: a European perspective", *IEEE Personal Communications*, p. 59-65, December 1998.

[PAR 00] PARKVALL S., "Variability of user performance in cellular DS-CDMA – long versus short spreading sequences", *IEEE Trans. on Communications*, vol. 48, p. 1178-1187, July 2000.

[RHE 96] RHEINSCHMITT R., TANGEMANN M., "Performance of sectorised spatial multiplex systems", *Proc. of 46th IEEE Vehicular Technology Conference*, p. 426-430, 1996.

[ROH 99] ROHANI K., HARRISON M., KUCHI K., "A comparison of base station transmit diversity methods for third generation cellular standards", *IEEE 49th Vehicular Technology Conference*, vol. 1, p. 351-355, 1999.

[SAR 80] SARWATE D. V., PURSLEY M. B., "Correlation properties of pseudorandom and related sequences", *Proceedings of the IEEE*, vol. 68, p. 593-619, May 1980.

[TAN 00] TANAKA S., HARADA A., SAWAHASHI M., ADACHI F., "Experiments on coherent adaptive antenna array diversity for wideband DS-CDMA mobile radio", *IEEE Journal on Selected Areas in Communications*, vol. 18, no. 8, p. 1495-1504, August 2000.

[VER 86] VERDU S., "Minimum probability of error for asynchronous Gaussian multiple-access channels", *IEEE Trans. Info. Theory*, vol. 32, p. 85-96, January 1986.

[WAL 23] WALSH J. L., "A closed set of normal orthogonal functions", *American J. Mathematics*, vol. 45, p. 5-24, 1923.

[YEE 93] YEE N., LINNARTZ J.-P., FETTWEIS G., "Multi-carrier CDMA in indoor wireless radio networks", *Proc. of IEEE PIMRC'93*, p. 109-113, Yokohama, Japan, September 1993.

Glossary

3G   *Third Generation*. Name given to cellular telecommunication systems provided with enhanced Quality of Service in comparison to Second Generation systems. In the ITU, these systems are defined within the IMT-2000 framework.

3GPP   *Third-Generation Partnership Project*. Collaboration partnership in charge of the maintenance and development of the technical specifications for GSM, GPRS, EDGE and UMTS mobile systems. Established in December 1998, the 3GPP involves regional standardization organisms: ARIB, CCSA, ETSI, ATIS, TTA and TTC.

3GPP2   *Third-Generation Partnership Project 2*. Collaboration partnership in charge of the maintenance and development of the technical specifications for CDMA2000 3G mobile system. Established in December 1998, the 3GPP2 involves regional standardization organisms like ARIB, CCSA, TIA, TTA and TTC.

16-QAM   *16-Quadrature Amplitude Modulation*. Modulation scheme based on amplitude modulation used to transmit 4 data bits per symbol. Used in HSDPA, it makes it possible to double the data rate of transmissions compared to QPSK used in *Release 99* physical channels.

AAL   *ATM Adaptation Layer*. Protocol layer in charge of accommodating higher layer data into "ATM cells" following a specific format and size.

Active Set   Set of Node Bs that have assigned a downlink DPCH to the UE in the case of the soft-handover procedure.

AICH   *Acquisition Indicator Channel*. Downlink physical common channel involved in RACH access procedure. It conveys the Acquisition Indicators (AI) used by the UTRAN to acknowledge the reception of an access preamble.

ALCAP   *Access Link Control Application Protocol*. Generic name for control plane protocols in the transport layer used to set up and release the data bearers in some UTRAN interfaces.

AMR	*Adaptive Multi Rate.* Audio codec standardized by ETSI which makes it possible to adapt the source and channel coding rates according to the radio channel conditions. AMR is adopted as the standard speech codec in UMTS networks, while it is optional in GSM.
ANSI-41	Set of specifications defining the core network used in AMPS, IS-136 (US TDMA) and the IS-95 (cdmaOne) mobile networks. In an evolved form, this is the core network used in the 3G systems cdma2000 and UWC-136.
AP-AICH	*Access Preamble-AICH.* Downlink physical common channel conveying Access Preamble Indicators in the CPCH access procedure.
APN	*Access Point Name.* Logical name enabling the identification of an external packet-data network to which the UE attempts connection. The APN is determined when the UE sets up a PDP context in the PS domain.
ARQ	*Automatic Repeat reQuest.* Error control technique that consists of automatically requesting the transmitter retransmission of a message in which errors where detected in the receiver side.
AS	*Access Stratum.* Functional layer where data and signaling bearers are realized in order to transport upper layers information between the UE and the core network.
ATM	*Asynchronous Transfer Mode.* Data transfer technology where data traffic is encoded into small fixed sized bytes called "cells" that are transmitted based on connection oriented technology.
AuC	*Authentication Centre.* Database in the GSM, GPRS and UMTS core networks where security parameters associated with each IMSI are stored.
BCCH	*Broadcast Control Channel.* Downlink control logical channel containing cell system information messages.
BCH	*Broadcast Channel.* Downlink common transport channel conveying BCCH messages.
BER	*Bit Error Rate.*
BLER	*BLock Error Rate.* Error rate measured for every transport block from the CRC check result.
BMC	*Broadcast/Multicast Control.* Layer 2 sub-layer in UTRA radio protocols that handles cell broadcast of short message services.
BSC	*Base Station Controller.* Equipment in charge of the administration of radio resources in GSM/GPRS radio access networks. The BSC controls one or several BTSs.
BSS	*Base Station System.* Term that groups one BSC and one or several BTSs in a GSM/GPRS radio access network.
BTS	*Base Transceiver Station.* Equipment containing devices for transmitting and receiving radio signals in the GSM/GPRS radio access network.

CAMEL   *Customized Application for Mobile network Enhanced Logic*. Extension
               of the *Intelligent Network* concept adapted to GSM/UMTS networks for
               personalized service creation. CAMEL enables access to these services
               even in situations of roaming.

CBC   *Cell Broadcast Center*. Entity in a GSM/UMTS network which handles
               CBS teleservice, i.e. the service enabling point-to-multipoint transmission
               of short messages to users located within the same cell or in different
               cells.

CC   *Call Control*. Operating between the UE and the MSC, this protocol
               handles mechanisms for establishing and releasing CS domain services.
               CC is part of the CM sub-layer in the NAS control plane.

CCCH   *Common Control Channel*. Bidirectional common logical channel
               containing signaling messages exchanged between the UE and the
               UTRAN. This channel is always mapped onto RACH and FACH
               transport channels.

CCTrCH   *Coded Composite Transport Channel*. In UTRA physical layer, this is a
               data stream resulting from encoding and multiplexing operations of one or
               several transport channels.

CD/CA-ICH   *CPCH Collision Detection/Channel Assignment Indicator Channel*.
               Downlink physical channel that conveys collision detection and channel
               assignment indicators in the CPCH access procedure.

cdma2000   Third generation mobile telecommunication standard within the IMT-
               2000 framework. The technical specifications of cdma2000 are developed
               in 3GPP2 and it remains compatible with 2G CDMA networks of the
               cdmaOne family.

CDMA   *Code Division Multiple Access*. Multiple access method that encodes data
               with special codes named spreading codes associated with each channel. It
               results in a multiplexing technique where multiple users can transmit
               within the same frequency carrier and at the same time.

Channelization
codes   Orthogonal codes of length SF generated from the OVSF approach which
               are used to spread the symbols in UTRA physical channels. In the
               downlink, they enable a UE to separate the spread signals of other users
               from its own signal information.

Chip   In CDMA terminology, a chip is every data unit that composes a
               spreading code. In UTRA, a set of SF chips is used to encode a symbol in
               the physical layer enabling spread spectrum since the chip rate (3.84
               MHz) is much larger than the symbol rate.

CM   *Connection Management*. Operating between the UE and the MSC, this
               layer 3 sub-layer is part of NAS. It is functionally divided into CC and SS.
               This handles the call set-up, maintenance and clearing as well as SMS and
               supplementary services.

CN   *Core Network.* Set of functional entities and equipments of UMTS networks responsible for enabling or denying access to users' subscribed services. The CN also keeps track of the UE's individual location, performs security functions, provides interworking with external networks and handles charging information. The CN is based on two separate domains: CS and PS which enable separate CS and PS service handling, respectively.

Common channel   Any channel not dedicated to a specific UE.

Compressed mode   Method used in UTRA/FDD in order to create gaps inside the DPCH frame structure where transmission is halted. In the downlink, this enables the UE to perform intersystem and interfrequency measurements.

CPCH   *Common Packet Channel.* Uplink common transport channel used for transmitting user data and signaling information. This channel is not used in real networks.

CPICH   *Common Pilot Channel.* Downlink common physical channel composed of a pre-established sequence of pilot symbols. Since it is continuously transmitted within a cell, the CPICH is used by the UEs to perform downlink quality measurements on serving and neighboring cells. It is also used to estimate the propagation channel impulse response.

CRC   *Cyclic Redundancy Check.* Algorithm that enables the detection of erroneous transport blocks in the UTRA physical layer.

CRNC   *Controlling Radio Network Controller.* One of the functional roles of the RNC enabling load and admission control of the cells under its responsibility.

C-RNTI   *Cell Radio Network Temporary Identifier.* Identity allocated by the CRNC to the UE upon accessing a new cell. This is used as UE ID on all common channel messages in air interface.

CS   *Circuit Switched.* Term used here to refer to all services provided via the CS domain based on circuit switched transfer mode.

CSCF   *Call Session Control Function.* Within an IP multimedia sub-system (IMS), this is a SIP server providing session establishment, modification and termination.

CS domain   Set of physical and logical entities in the UMTS core network that enable CS services offered by the network operator or by an external network.

CSICH   *CPCH Status Indicator Channel.* Downlink physical channel that indicates whether PCPCH channels are available in a cell.

CTCH   *Common Traffic Channel.* Downlink logical common channel used to transfer user plane information in a point-to-multipoint scheme.

DCA   *Dynamic Channel Allocation.* Method for automatically allocating resources that, based on measurements, determines which among the non-used channels possess the lower level of interference.

DCCH   *Dedicated Control Channel*. Dedicated logical channel transmitting dedicated control information in both directions between the UE and the UTRAN. This is only used when an RRC connection has been set up.

DCH   *Dedicated transport Channel*. Bidirectional transport channel intended to carry dedicated user or signaling data.

Dedicated channel Any channel dedicated to a specific UE.

DPCCH   *Dedicated Physical Control Channel*. Dedicated physical channel that in the DPCH coveys layer 1 control information.

DPCH   *Dedicated Physical Channel*. Bidirectional dedicated physical channel used for the transfer of dedicated signaling or user data. The channel is divided into two parts: DPDCH and DPCCH.

DPDCH   *Dedicated Physical Data Channel*. Bidirectional dedicated physical channel that in the DPCH conveys upper layer information.

DRNC   *Drift RNC*. Functional role of the RNC when the RNS acts as DRNS.

DRNS   *Drift RNS*. One of the roles of the RNS when it supports the SRNS in terms of resources and radio link handling via Iur interface. This situation typically happens when the UE is in macrodiversity involving Node Bs controlled by different RNSs (RNCs).

DRX   *Discontinuous Reception*. Technique used by the UE in order to prolong the autonomy of its battery. The key idea is to read PICH/PCH channels in well-defined time intervals denoted DRX cycles and to remain in "sleep mode" the rest of the time.

DS-CDMA   *Direct-Sequence Code Division Multiple Access*. Multiple access method based on CDMA principles where spread spectrum is achieved by directly multiplying each data symbol by the spreading code on a chip-by-chip basis.

DSCH   *Downlink Shared Channel*. Downlink transport channel that can be shared in time by multiples users for user data and signaling transfer. This is not currently used in real-life networks.

DTCH   *Dedicated Traffic Channel*. Bidirectional logical channel dedicated to one UE for the transfer of user data.

DTX   *Discontinuous Transmission*. Technique consisting of cutting RF transmission during speech pauses. It makes it possible to prolong UE's battery life and reduce interference within the cell.

Ec/No   Measurement of the energy per chip received compared with the total spectral density within the channel that satisfies RSCP/RSSI.

EDGE   *Enhanced Data Rates for GSM Evolution*. Technology used in GSM/GPRS networks with the purpose of increasing the data rate in the radio interface based on 8 states QPSK modulation (8-QPSK).

EIR   Equipment Identity Register. Database in the core network containing the IMEIs of terminals for which the access to the network can be refused.

FACH   Forward Access Channel. Downlink common transport channel enabling the transmission of small amounts of user data or signaling.

FBI   FeedBack Information. Field of bits in an uplink DPCCH used by the UE to send information to the UTRAN when SSDT or closed loop Tx diversity are implemented.

FDD   Frequency Division Duplex. Duplexing method where the radio communication between the terminal and the base station uses a paired spectrum, i.e. transmission and reception take place within different frequency carriers.

FDMA   Frequency Division Multiple Access. Multiple access method where each frequency carrier of the frequency spectrum within a cell can be assigned to only one user at a time. A frequency carrier represents a single channel.

FOMA   Freedom of Mobile Multimedia Access. NTT DoCoMo's third-generation mobile communication service based on early specifications of UMTS. It was launched in Japan as the world's first 3G commercial service in October 2001.

GERAN   GSM/EDGE Radio Access Network. GSM radio access network based on the EDGE technology and specified within the 3GPP. It supports an interface with the UMTS core network and offers access to 3G classes of service.

GGSN   Gateway GPRS Support Node. Gateway between the packet domain of the cellular network and an external PDP network (e.g. IP network). It performs the routing of incoming/outgoing packets towards the SGSN/PDP network.

GMM   GPRS Mobility Management. Protocol of the NAS part within layer 3. It is responsible for the handling of mobility, authentication and ciphering algorithm selection. It is located in the control plane within the UE and the SGSN.

GMSC   Gateway MSC. Gateway node used to interconnect the CS domain of the cellular network to an external network (PSTN, ISDN). It is responsible for routing calls from the PSTN/ISDN towards the MSC of mobile stations and vice versa.

GPRS   General Packet Radio Service. Technology that has been specified by ETSI for GSM phase 2+ to enable the support of packet-based services and interworking with external packet networks (e.g. IP-based networks). It has evolved and has been integrated as the PS core network domain to UMTS system.

GSM   Global System for Mobile communications. Digital cellular system specified by ETSI and based on the TDMA access method. It was first introduced in 1991 and became the *de facto* 2G wireless telephone standard in Europe. It is also used in other parts of the world.

GTP

GPRS Tunneling Protocol. Protocol responsible for tunneling (encapsulation in IP datagrams) of user data between SGSN and GGSN. It is also used in Iu-PS interface.

Handover

Procedure consisting of transferring an ongoing communication from one radio channel to another.

Hard handover

Type of handover where all current links are released before establishing new ones. It is typically performed when the UE moves to a new cell using frequency carriers that are different from those used in the cell it is leaving. This type of handover is also used when the UE moves from UTRA to GSM.

H-ARQ

Hybrid-Automatic Repeat reQuest. Technique used in HSDPA and combining the traditional ARQ and the turbo code error correction mechanism. It gives better performance than the simple ARQ but is more complex to implement.

HLR

Home Location Register. Functional unit located in the core network that contains subscriber information and current location of the UE. It is the central database of the mobile network.

HPLMN

Home PLMN. PLMN where the MCC and MNC of the PLMN identity are the same as the MCC and MNC of the IMSI. This is the nominal network of the UE where subscriber information is permanently located.

HSCSD

High Speed Circuit Switched Data. Dedicated circuit switched data communications technology which makes it possible for a TDMA-based network to allocate to a user several (up to four) concatenated slots each delivering 9.6 kbps or 14.4 kbps.

HSDPA

High Speed Downlink Packet Access. Radio access technology applied to UTRA/FDD and UTRA/TDD that enables downlink data rates of up to 10 Mbps. It is defined in 3GPP *Release 5* specifications.

HS-DPCCH

High Speed Dedicated Physical Control Channel (uplink) for HS-DSCH. HSDPA uplink physical channel associated with the HS-PDSCH that conveys signaling information consisting of the acknowledgement used in H-ARQ and downlink channel quality information.

HS-DSCH

High Speed Downlink Shared Channel. High speed downlink transport channel used by HSDPA that carries the user data with a peak rate reaching 10 Mbps with 16-QAM. It is associated with a DPCH and one or several HS-SCCH.

HS-PDSCH

High Speed Physical Downlink Shared Channel. Physical channel associated with the HS-DSCH. It uses a fixed spreading factor SF = 16; it supports multicode transmission and could use QPSK or 16-QAM modulations.

HS-SCCH *High Speed Shared Control Channel.* Downlink physical channel used to convey the necessary physical layer information to enable decoding of the data on HS-DSCH.

HSUPA *High Speed Uplink Packet Access.* Also known as *Enhanced DCH* (E-DCH), this is a 3GPP *Release 6* technology which makes it possible to improve the uplink air interface capacity utilization and end user experience just as HSDPA does in the downlink. HSUPA offers a theoretical uplink capacity of 5.7 Mbps.

IMEI *International Mobile Equipment Identity.* A number assigned to each mobile phone equipment and uniquely identifying it to the network (serial number). It can be found behind the battery in the phone or by dialing the sequence code *#06#.

IMS *IP multimedia sub-system.* Core network sub-system that makes it possible to connect the PS domain to an IP network providing multimedia services. It first appeared in *Release 5* specifications.

IMSI *International Mobile Subscriber Identity.* An international number consisting of a unique 15-digit code used to identify an individual subscriber on a mobile phone network. It is stored in the USIM and the HLR.

IMT-2000 *International Mobile Telecommunications 2000.* Term used by the International Telecommunications Union (ITU) to designate the global standard for 3G wireless communications.

IMT-DS *International Mobile Telecommunications Direct Spread.* Name given to UTRA/FDD radio access technology within the IMT-2000 family.

IMT-FT *International Mobile Telecommunications Frequency Time.* Name given in the IMT-2000 framework to the TDMA-based radio access technology used by the 3G DECT system.

IMT-MC *International Mobile Telecommunications Multi Carrier.* Name given in the IMT-2000 framework to the cdma2000 radio interface.

IMT-SC *International Mobile Telecommunications Single Carrier.* Name given in the IMT-2000 framework to the UWC-136 radio interface.

IMT-TC *International Mobile Telecommunications Time Code.* Name given in the IMT-2000 framework to the UTRA/TDD and TD-SCDMA radio interfaces.

IP *Internet Protocol.* Set of communication standards that control communications activity on the Internet. By extension, IP also designates data networks that are based on it.

ISCP *Interference Signal Code Power.* Interference measured at the receiver on pilot bits of the DPCCH after combining and despreading operations.

ISDN	*Integrated Services Digital Network*. Circuit switched communication network supporting simultaneous end-to-end digital transmission of different media (voice, data, video, text, fax) and signaling.
Iu	Interface enabling user data and signaling exchanges between the UTRAN and the core network. It is functionally split into Iu-CS and Iu-PS when the exchanges take place between the RNC and the MSC and between the RNC and the SGSN, respectively.
Iub	Interface enabling user data and signaling exchanges between the RNC and Node B.
Iur	Interface enabling user data and/or signaling exchanges between two RNCs.
IWF	*InterWorking Function*. Function in the UMTS/GSM network that enables communications between the PLMN and an external CS data network (e.g. the public telephone system). This is typically located in the MSC.
JD	*Joint Detection*. Multi-user detection technique where all spread user symbols that are active during the same time slot are simultaneously extracted while efficiently reducing Multiple Access Interference (MAI). It is well adapted for UTRA/TDD and TD-SCDMA systems.
LA	*Location Area*. Area controlled by a single MSC/VLR with the purpose of tracking the mobility of a UE for CS service provision.
LCS	*LoCation Services*. Generic term that groups all the 3GPP techniques that can be implemented in a UMTS/GSM enabling provision of services based on the geographical position of the UE.
Logical channel	Channel defined by the type of information to be transferred. It consists of a data stream to be transmitted over the radio interface and is located between RLC and MAC. Every logical channel is associated with a transport channel.
LTE	*Long-Term Evolution*. Technology developments in 3GPP defining the future evolution beyond *Release 7* of UTRA and the UTRAN. The LTE target is a high data rate, low latency, packet-optimized radio-access technology based on new air interfaces and architectures. LTE peak data rates could reach 100 Mbps in the downlink and 50 Mbps in the uplink.
MAC	*Medium Access Control*. Layer 2 sub-layer offering services to the RLC by means of logical channels which are mapped onto transport channels. It also selects appropriate transport formats for each transport channel.
Macrodiversity	Situation where the UE receives the same information via radio links (DPCHs) from different cells. The UE exploits this by combining the different radio paths to improve detection performance.
MAP	*Mobile Application Part*. Set of ETSI's standard functions and protocols enabling signaling exchanges between the internal entities within the

UMTS/GSM core network. It also enables roaming between different PLMNs.

MBMS	*Multimedia Broadcast Multicast Service*. Service architecture and procedures defined in *Release 6* which enable transmission of multimedia content to multiple users in a point-to-multipoint scheme. MBMS further enables the UTRAN to control and monitor the users receiving the data.
ME	*Mobile Equipment*. Functional part in the UE composed of the Terminal Equipment (TE) and the Mobile Termination (MT).
MExE	*Mobile Station Application Execution Environment*. Standardized framework enabling the execution of applications by defining "classmarks" which take into account the SW and HW capabilities of the mobile terminal.
MIMO	*Multiple Input Multiple Output Antenna processing*. Antenna technology in which multiple antennas are used at both the source (transmitter) and the destination (receiver). This is specified in *Release 7* and applicable to HSDPA with the purpose of increasing the peak data rate.
MM	*Mobility Management*. NAS sub-layer located in the UE and the MSC control planes in charge of controlling the mobility of the UE for CS service provision.
MNC	*Mobile Network Code*. Code used to differentiate networks within a same country. Its attribution is under the responsibility of the regulation authority of the country's network.
Monitored set	Neighboring cells that are not included in the active set but that are monitored by the UE from a list transmitted by the UTRAN.
MRC	*Maximal Ratio Combining*. Technique used by the Rake receiver that exploits the estimation of the propagation channel conditions in order to rotate back the received symbols so as to undo the phase rotation caused by the channel. The resulting symbols can then be simply added together to recover the energy across all delay positions.
MSC	*Mobile Switching Centre*. Switch in the CS domain responsible for connecting calls together by switching the digital voice data packets from one network path to another. The MSC also provides additional information to support UE service subscribers, including user registration, authentication and location updating.
MSIN	*Mobile Station Identification Number*. Part in the IMSI that identifies the mobile subscriber.
MSISDN	*Mobile Subscriber ISDN Number*. Standard international number that can be dialed from a fixed or mobile network to call a mobile subscriber. It is based on the ITU E.164 standard.

MT	*Mobile Termination.* Functional part in the UE which performs specific functions related to the signal transmission and reception in the radio interface.
NAS	*Non-Access Stratum.* Protocols that enable signaling and data exchanges between the UE and the MSC/SGSN and which are transparent to the UTRAN.
NBAP	*Node B Application Part.* Application protocol in the Iub interface in charge of the allocation and control of radio resources in Node B.
NMSI	*National Mobile Station Identifier.* Number part of the IMSI that identifies the UE within its home country. It consists of the MNC and the MSIN.
Node B	Equipment in the UTRAN responsible for radio transmission/reception in one or more cells to/from the UEs. It is under the control of one RNC and communicates with it via the Iub interface.
NSAPI	*Network layer Service Access Point Identifier.* Identity used by the UE and the core network that associates the user data streams to a given PDP context.
NSS	*Network Sub System.* Group of entities in a GSM network primarily responsible for interconnection to the PSTN, collecting charging information, maintaining subscriber profile and service information, wide area mobility management, etc.
OSA	*Open Service Access.* Set of 3GPP specifications that enable third party application development and deployment by means of open, secure and standardized access to the core network.
OVSF	*Orthogonal Variable Spreading Factor.* Spreading codes used in UTRA/FDD and UTRA/TDD formed from a code tree where all codes at a given SF are orthogonal to each other. They compose the channelization codes.
Paging	Act of seeking a UE within a known geographic area.
PCCH	*Paging Control Channel.* Downlink logical control channel that conveys paging information.
P-CCPCH	*Primary Common Control Physical Channel.* Downlink common control physical channel on which the BCH is mapped.
PCH	*Paging Channel.* Downlink common control transport channel carrying paging information.
PCPCH	*Physical Common Packet Channel.* Uplink common physical channel on which the CPCH is mapped.
PDCP	*Packet Data Convergence Protocol.* Sub-layer of layer 2 in the UE and RNC user plane that contains compression methods to obtain better spectral efficiency for IP packets to be transmitted over the radio interface. It is only used for PS services.

PDP context	A logical "pipe" handled by the UE, the SGSN and the GGSN which contains the session information when established for accessing a PS service including the UE's IP address and QoS parameters.
PDSCH	*Physical Downlink Shared Channel*. Downlink shared physical channel on which the DSCH is mapped.
Physical channel	In UTRA/FDD a physical channel consists of a channelization code, a scrambling code, a frequency and, in uplink, a relative phase between I and Q branches.
PICH	*Page Indicator Channel*. Downlink common physical channel associated with the PCH. The PICH carries paging indicators that are used by the UEs to get efficient sleep mode operation.
PLMN	*Public Land Mobile Network*. Any mobile communication network (e.g. UMTS or GSM) belonging to a mobile network operator offering telecommunications services over an air interface.
PRACH	*Physical Random Access Channel*. Uplink common physical channel on which the RACH is mapped.
Processing gain	Defined as the ratio in dB of the chip rate to the bit rate of the information over layer 2 (or also the ratio of the bandwidth of a spread spectrum to the data rate of the information).
PSCH	*Primary Synchronization Channel*. Downlink physical channel which are part of the SCH and are used by the UEs to achieve chip and slot synchronization. The PSCH is common to all Node Bs.
PS domain	Set of physical and logical entities in the UMTS core network that enable PS services offered by the network operator or by an external packet-data network.
PSTN	*Public Switched Telephone Network*. General term referring to any fixed network offering telephony services.
P-TMSI	*Packet Temporary Mobile Subscriber Identity*. Temporary identity allocated to a UE used to secure the transmission of data in the radio interface after registration to the PS domain.
PUSCH	*Physical Uplink Shared Channel*. Uplink common physical channel on which the USCH is mapped. This is specific to UTRA/TDD mode.
QoS	*Quality of Service*. Set of service performances which determines the degree of satisfaction of a user of a service. It is characterized by the combined aspects of performance factors applicable to all services, such as BER, maximum data rate, transfer delay, etc.
QPSK	*Quadrature (Quaternary) Phase Shift Keying*. Digital modulation technique used in UTRA where four different phase angles are used.
RA	*Routing Area*. Area controlled by a single SGSN with the purpose of tracking the mobility of a UE for PS service provision.

RAB   *Radio Access Bearer.* Service offered by the access stratum for the transfer of user data between the UE and the core network following the required QoS.

RACH   *Random Access Channel.* Uplink common transport channel intended to carry control information (e.g. RRC connection request) from the UE. It can also be used to send small amounts of packet data from the terminal to the network.

Radio bearer   Service provided by in the UTRA radio interface for the reliable transfer of user data between the UE and the UTRAN.

RANAP   *Radio Access Network Application Part.* Application protocol that handles all signaling within Iu-PS and Iu-CS interfaces.

Rake   A radio receiver in CDMA systems having multiple receptors called "fingers" which use different offsets of a common spreading code to receive and combine several multipath time-delayed signals.

RLC   *Radio Link Control.* Layer 2 sub-layer in the UTRA radio interface that provides segmentation and retransmission services for both user and control data.

RNC   *Radio Network Controller.* The network element in the UTRAN responsible for the control of the radio resources.

RNS   *Radio Network Subsystem.* Sub-network within the UTRAN consisting of one RNC and one or more Node Bs.

RNSAP   *Radio Network Subsystem Application Part.* Radio network application protocol in charge of the signaling in the Iur interface.

RNTI   *Radio Network Temporary Identifier.* Generic term grouping a set of temporary identities that may be allocated by the UTRAN to the UE after the establishment of an RRC connection.

RPLMN   *Registered PLMN.* A PLMN where the UE has successfully performed the attachment procedure. This can be the HPLMN or the VPLMN.

RRC   *Radio Resource Control.* Layer 3 sub-layer that controls signaling between the UE and the UTRAN including messages required to set up, modify and release protocol entities in layers 2 and 1.

RRC connection   Logical connection established between one UE and the RNC used by this later to track its position into the UTRAN and to handle common and dedicated resources for that UE.

RSCP   *Received Signal Code Power.* Measure of the received signal power after despreading performed on the CPICH or the PCCPCH.

RSSI   *Received Signal Strength Indicator.* Measure of the received signal power within the frequency carrier bandwidth.

SAE   *System Architecture Evolution.* Case study in 3GPP intended to develop a framework for an evolution and migration of current systems to a higher data rate, lower latency, packet-optimized system that supports multiple radio access technologies.

S-CCPCH   *Secondary Common Control Physical Channel.* Downlink common control physical channel on which the FACH and/or the PCH can be mapped.

SCH   *Synchronization Channel.* Downlink physical channel used by the UEs to get time synchronization with the serving cell. It comprises the P-SCH and the S-SCH.

Scrambling codes   Pseudo-random codes used to enhance correlation properties of spread signals in the UTRA physical layer. In the uplink, they enable Node B to separate spread signals from different users within the cell. Similarly, in the downlink, they enable the UE to separate spread signals from different Node Bs.

SF   *Spreading Factor.* In UTRA it indicates the number of chips that are spreading one data symbol.

SFN   *Cell System Frame Number.* UTRAN frame counter that makes it possible to schedule in time the system information blocks broadcast via the BCH.

SGSN   *Serving GPRS Support Node.* Node in the PS domain responsible for the UE mobility management and IP packet session management. It also routes user packet traffic to the appropriate GGSN, providing access to external packet data networks.

SIM   *Subscriber Identity Module.* Also known as a smart card, it is a chip that resides in GSM handsets and identifies the caller to the mobile network as a legitimate subscriber.

SIP   *Session Initiation Protocol.* Signaling protocol used in the IMS architecture which is responsible for setting up, modifying and releasing the session of a multimedia call.

SIR   *Signal-to-Interference Ratio.* Ratio between the strength of the received signal and the channel interference. In UTRA the SIR satisfies $(RSCP/ISCP) \times SF$.

SM   *Session Management.* NAS protocol layer located in the UE and SGSN control plane in charge of the activation and release of a PDP context.

SMS   *Short Message Service.* Teleservice enabling the transmission and reception of text messages.

Soft handover   A procedure used in UTRA/FDD where the radio links are added and abandoned in such a way that the UE always keeps at least one radio link with the UTRAN.

Softer handover	Special case of soft handover where the UE is in the overlapping cell coverage area of two adjacent sectors controlled by the same Node B.
Spread spectrum	Technique used to transmit a signal in a bandwidth much larger than the frequency content of the original signal information. The signal is spread from a code which is independently generated from the information message.
SRNC	*Serving RNC.* Name given to the RNC when the RNS to which it belongs plays the role of serving RNS (SRNS).
SRNS	*Serving RNS.* Functional role of the RNS when it takes charge of the RRC connection for a given UE.
S-RNTI	*SRNC Radio Network Temporary Identifier.* UE identifier allocated by the SRNC to the UE that has established an RRC connection.
SS7	*Signaling System No. 7.* Architecture for performing out-of-band signaling within a fixed or mobile system.
SSCH	*Secondary Synchronization Channel.* Downlink common physical channel which is part of the SCH that enables the UEs to get frame synchronization with the network. The code sequence composing the SSCH is specific to each cell.
SSDT	*Site Selection Diversity Transmit power control.* Power control technique used during soft handover. This technique is not used nowadays by current UMTS networks.
STTD	*Space Time Transmit Diversity.* Downlink open loop transmit technique where the same signal is transmitted over two uncorrelated antennae after applying a specific encoding.
TBS	*Transport Block Set.* Set of transport blocks in a transport channel which are exchanged between the physical layer and MAC during a TTI.
TDD	*Time Division Duplex.* Technique used in UTRA/TDD and TD-SCDMA where the UE and Node B use the same frequency carrier in downlink and uplink to communicate at different time intervals.
TDMA	*Time Division Multiple Access.* Access method where the signals for different users are transmitted in the same frequency carrier but at different time slots. This is used in GSM, UTRA/TDD and TD-SCDMA radio interfaces.
TD-SCDMA	*Time Division Synchronous Code Division Multiple Access.* Radio access technology in the IMT-2000 frame proposed by China. This is part of 3GPP *Release 4* specifications and is also known as "low chip rate UTRA/TDD" since the chip rate is 1.28 Mcps instead of 3.84 Mcps.
TE	*Terminal Equipment.* Functional part of the UE where the user data is generated in the uplink and processed in the downlink.

TEID   *Tunnel Endpoint Identifier.* Identity used for routing packets between the GGSN and the SGSN and between this and the RNC by using GTP tunneling principles. This is uniquely associated with a PDP context (NSAPI + IMSI) and to a RAB (RAB ID + IMSI).

TF   *Transport Format.* Set of characteristics (dynamic and static) of a specific transport channel (TTI, channel coding type, channel coding rate, etc.).

TFC   *Transport Format Combination.* Combination of all transport formats associated with all active transport channels.

TFCI   *Transport Format Combination Indicator.* Binary representation of current TFC.

TFCS   *Transport Format Combination Set.* Set of TFCs that can be used by the UE.

TGL   *Transmission Gap Length.* Number of consecutive slots where the transmission is halted in a DPCH frame when the compressed mode is activated.

TMSI   *Temporary Mobile Subscriber Identity.* Temporary identity allocated to a UE by the VLR used for security purposes in the radio interface after registration with the CS domain.

TPC   *Transmit Power Control.* Binary representation of the control commands generated by the physical layer to be used by the closed loop power control mechanism.

Transport block   Data unit exchanged between the physical layer and MAC. This is equivalent to "MAC-PDU".

Transport channel Channel offered by layer 1 to MAC for data transport defined by the type of layer 1 operations applied to it (e.g. channel coding type, channel coding rate, bit rate, etc.). Every transport channel is associated with a physical channel.

TSTD   *Time Switched Transmit Diversity.* Open loop transmit technique where the radio signal is alternately transmitted over two uncorrelated antennae. This applies to the SCH exclusively.

TTI   *Transmission Time Interval.* Time interval required to transmit a TBS.

UARFCN   *UTRA Absolute Radio Frequency Channel Number.* Number given to each frequency carrier in the UTRA spectrum.

UE   *User Equipment.* Name given to a mobile terminal in UMTS networks comprising a USIM and an ME.

UICC   *Universal Integrated Circuit Card.* Integrated-Circuit or smart card containing one USIM that can be inserted and removed from the mobile terminal.

UMTS   *Universal Mobile Telecommunications System.* Third generation mobile system defined by a set of 3GPP specifications.

URA	*UTRAN Registration Area.* Area composed of a number of cells that is controlled by the UTRAN and unknown to the core network.
U-RNTI	*UTRAN Radio Network Temporary Identifier.* Temporary identity of a UE allocated by the RNC when an RRC connection has been established. It is used in a DCCH when it is not possible to identify the UE with C-RNTI.
USAT/SAT	*USIM/SIM Application Toolkit.* Set of protocols which enable applications/services residing in the USIM/SIM to control user interface and communication capabilities of the terminal.
USCH	*Uplink Shared Channel.* Uplink transport channel used to send user data and signaling using a time- and code-sharing allocation scheme. This is specific to UTRA/TDD mode.
USIM	*Universal Subscriber Identity Module.* Application residing in the UICC which contains the subscription characteristics of a number of 2G and 3G services. It contains static and dynamic information including authentication, ciphering and integrity parameters.
UTRA	*Universal Terrestrial Radio Access.* Radio access technology in the IMT-2000 framework based on CDMA. UTRA is sometimes referred to as WCDMA and operates in two modes: UTRA/FDD and UTRA/TDD.
UTRAN	*Universal Terrestrial Radio Access Network.* Conceptual term used in 3GPP specifications for identifying the radio access network which consists of one or more RNSs.
Uu	Radio interface between the UE and the UTRAN.
UWC-136	*Universal Wireless Communications 136.* Mobile telecommunication system defined by the Telecommunication Industry Association (TIA) and presented as the 3G evolution of the 2G system IS-136 (US TDMA).
VHE	*Virtual Home Environment.* Concept which depends on several specifications and technologies such as WAP, MexE, USAT, OSA, etc. VHE enables the portability of the user subscribed services across different networks and terminals with the same look and feel.
VLR	*Visitor Location Register.* Database in the CS domain temporarily containing security, authentication and identification data of all UEs that are currently managed by the MSC. One VLR is normally assigned to every MSC.
VPLMN	*Visited PLMN.* PLMN different from the HPLMN where the UE is successfully attached.
WCDMA	*Wideband Code Division Multiple Access.* Term designing radio systems that use CDMA with a frequency bandwidth much larger than that used in narrowband CDMA systems. This is used in UTRA and cdma2000 radio interfaces.

Index